Spring Boot+MVC
实战指南

高洪岩 著

人民邮电出版社

北京

图书在版编目（CIP）数据

Spring Boot+MVC实战指南 ／ 高洪岩著. -- 北京：人民邮电出版社，2022.1
　ISBN 978-7-115-58061-0

　Ⅰ．①S… Ⅱ．①高… Ⅲ．①JAVA语言－程序设计－指南 Ⅳ．①TP312.8-62

中国版本图书馆CIP数据核字(2021)第242227号

内 容 提 要

本书主要讲解如何在Spring Boot框架中开发MVC应用，包括主流的JavaEE框架，如MyBatis、Spring、SpringMVC、FreeMarker和Thymeleaf等。读者可以进行"精要"式学习，正确地进行项目实战，同时汲取JavaEE的思想，并最终将这种思想灵活运用到实际工作中。

本书内容主要涉及MVC框架的原理实现、上传、下载、数据验证、国际化、多模块分组开发、转发/重定向、JSON的解析、将Ajax及JSON和MVC框架进行整合开发，以及MyBatis中映射文件的使用。本书还介绍了Spring中的核心技术（依赖注入与AOP），掌握这两项技术是学习Spring的重中之重。

本书语言简洁，示例丰富，以掌握实用技术为目的，帮助读者迅速掌握使用主流开源JavaEE框架进行开发所需的各种技能。本书适合已具有一定Java编程基础（具有Servlet编程经验）的读者阅读，也可供Java平台下进行各类软件开发的开发人员、测试人员参考。

◆ 著　　高洪岩
　责任编辑　傅道坤
　责任印制　王　郁　焦志炜

◆ 人民邮电出版社出版发行　北京市丰台区成寿寺路11号
　邮编　100164　电子邮件　315@ptpress.com.cn
　网址　https://www.ptpress.com.cn
　北京天宇星印刷厂印刷

◆ 开本：787×1092　1/16
　印张：23.75　　　　　　2022年1月第1版
　字数：577千字　　　　　2022年1月北京第1次印刷

定价：99.90元

读者服务热线：(010)81055410　印装质量热线：(010)81055316
反盗版热线：(010)81055315
广告经营许可证：京东市监广登字20170147号

作者简介

高洪岩，世界 500 强企业高级项目经理，具有 10 余年项目管理与开发经验，在多线程和并发、Android 移动开发、智能报表和分布式处理等企业级架构技术领域深耕多年，深谙 Java 技术开发要点与难点，拥有良好的技术素养和丰富的实践经验，一直在持续关注架构的优化和重构领域，喜欢用技术与理论相结合的方式分享知识，以共同提高。著有《Java 多线程编程核心技术》《Java 并发编程：核心方法与框架》《Java EE 核心框架实战》《NIO 与 Socket 编程技术指南》《虚拟化高性能 NoSQL 存储案例精粹：Redis+Docker》《Java Web 实操》等书籍。

致谢

　　在本书出版的过程中，我要感谢公司领导和同事的大力支持；感谢家人在我写作期间给予的理解与支持；感谢现年3岁多的儿子高晟京，你使我更有动力；最后感谢对本书耗费大量精力的编辑老师们，你们仔细谨慎的工作态度值得我学习。

前言

关于本书

用人单位对 IT 人才的要求越来越趋向于"实战性",也就是要求员工进入软件公司后能立即投入开发的任务中,快速为公司创造巨大的经济价值。本书以此为出发点进行编写。

一本内容精炼而不失实用价值的主流 JavaEE 开源框架图书,应该包含主流框架中常用且重要的内容,这样读者就可以快速上手,根据这些内容探索出一些方向,并在工作和学习中不断拓展和深掘。这就是写作本书的主要目的。

JavaEE 的世界非常庞大,以至于没有任何一本书能把它讲解得非常完整或详细。要想学好 Java 语言或 JavaEE 框架并掌握其中丰富的编码技巧、设计模式以及代码优化,并将它们熟练地综合应用在软件项目中,只有从零开始学习,没有捷径。

本书的章节编排不但涵盖了学习主流 JavaEE 框架所需掌握的核心技术,也涵盖了使用它们进行项目实战的必备知识,旨在希望读者尽快上手,掌握开源 JavaEE 框架的核心内容,正确进行项目实战,汲取 JavaEE 的思想,并最终将这种思想活用到实际工作中。

现在主流的 JavaEE 框架是 SSM 或 Spring Boot,但 Spring Boot 框架仅仅是一个"盒子",一个"封装器",想要实现功能还需要整合其他第三方的框架,如 MyBatis、Spring 或 SpringMVC 等,所以在学习 Spring Boot 之前必须要有 SSM 框架的开发经验。能让一位初入 JavaEE 框架的学习者从零开始到最终掌握这几个框架,一直是我的写作目标。有些 JavaEE 的开源框架的确能非常有效地改善开发效率,但因为使用的人不多,所以导致覆盖面比较窄。而软件公司在招聘时的技术需求大多数情况下是"大众化"的,这就要求应聘者在面试前有主流 JavaEE 框架的学习或使用经验。如果读者找不到合适的教材,导致在学习某一项技术时根本摸不清哪些知识点是常用的,哪些是不常用的,就会极大地降低学习效率。对此,本书为读者提供了一条有效的学习路径。

本书面向的读者

首先,本书适合所有 Java 程序员。JavaEE 作为 Java 开源世界的主流框架,Java 程序员没有理由不学习它们。其次,本书适合希望学习使用这些框架进行编程的在校学生。由于学校的功课很多，一本大部头的框架书需要花费大量的时间研读，而本书可以带领读者快速进入 JavaEE 框架开发的殿堂，同时又不会遗漏应该掌握的核心技能。

本书的结构

第 1 章和第 2 章将介绍 Spring 中的 IoC 和 AOP 技术，包含注入、注入原理、动态代理的实现与 AOP 的原理。

第 3 章将学习最流行的 SpringMVC 框架，体会使用此框架开发一个经典登录功能时使用的技术点，还要掌握分组分模块开发的技术、重定向/转发的使用、JSON+AJAX+SpringMVC 联合开发、上传/下载的实现、数据验证功能的使用、XML 配置文件的处理、业务层 Service 的注入、Model 和 View 对象的使用，以及 HttpSession 在 SpringMVC 中的使用等。

第 4 章将学习基于 SQL 映射的 MyBatis 框架，在本章中可以使用此框架操作主流的数据库，并学习 MyBatis 核心 API 的使用，采用自定义封装法简化 MyBatis 的操作代码，进而加快开发效率。

第 5 章主要讲解 MyBatis 映射相关的知识，包括<sql><resultMap><choose><set>和<foreach>等常用标签、DB 连接信息存储到 Properties 文件的读取、使用 JDBC 数据源、别名 typeAliases 的配置、CLOB 字段的读取以及分页等必备技术点。

第 6 章将介绍主流标签库 FreeMarker 和 Thymeleaf 的使用，以帮助读者实现视图层的静态化处理。

如何使用本书

需要声明的是，本书不是 Java Web 的入门书。学习本书之前，读者要对 Java Web 中的 JSP、Servlet 等 Web 技术有所了解，并尽量能完整地使用 JSP 或 Servlet 开发一个小型项目。在此基础上再阅读本书，读者会发现代码的分层更加明确，结构更加清楚。

对于软件开发，实践才是硬道理，这就需要设计、排错，拥有更多的想法和经验，所以请拿起手中的键盘，练习一下吧！

尽量在自己的计算机中执行本书的所有代码，只读书不动手和自己动手存在天壤之别，两者结合起来才能更深刻地理解框架的各项功能。

如何与作者联系

由于 JavaEE 内容涵盖面广，涉及的知识点非常多，加之作者水平有限，错误之处在所难免，请各位读者赐教和斧正。读者可以通过邮箱 279377921@qq.com 与我联系。

资源与支持

本书由异步社区出品，社区（https://www.epubit.com/）为您提供相关资源和后续服务。

配套资源

本书提供配套资源，要获得相关配套资源，请在异步社区本书页面中点击 配套资源 ，跳转到下载界面，按提示进行操作即可。注意：为保证购书读者的权益，该操作会给出相关提示，要求输入提取码进行验证。

提交勘误

作者和编辑尽最大努力来确保书中内容的准确性，但难免会存在疏漏。欢迎您将发现的问题反馈给我们，帮助我们提升图书的质量。

当您发现错误时，请登录异步社区，按书名搜索，进入本书页面，单击"提交勘误"，输入勘误信息，单击"提交"按钮即可。本书的作者和编辑会对您提交的勘误进行审核，确认并接受后，您将获赠异步社区的 100 积分。积分可用于在异步社区兑换优惠券、样书或奖品。

扫码关注本书

扫描下方二维码,您将会在异步社区微信服务号中看到本书信息及相关的服务提示。

与我们联系

本书责任编辑的电子邮箱是 fundaokun@ptpress.com.cn。

如果您对本书有任何疑问或建议,请您发邮件给我们,并请在邮件标题中注明本书书名,以便我们更高效地做出反馈。

如果您有兴趣出版图书、录制教学视频,或者参与图书技术审校等工作,可以发邮件给本书的责任编辑。

如果您来自学校、培训机构或企业,想批量购买本书或异步社区出版的其他图书,也可以发邮件给我们。

如果您在网上发现有针对异步社区出品图书的各种形式的盗版行为,包括对图书全部或部分内容的非授权传播,请您将怀疑有侵权行为的链接发邮件给我们。您的这一举动是对作者权益的保护,也是我们持续为您提供有价值的内容的动力之源。

关于异步社区和异步图书

"异步社区"是人民邮电出版社旗下 IT 专业图书社区,致力于出版精品 IT 图书和相关学习产品,为作译者提供优质出版服务。异步社区创办于 2015 年 8 月,提供大量精品 IT 图书和电子书,以及高品质技术文章和视频课程。更多详情请访问异步社区官网 https://www.epubit.com。

"异步图书"是由异步社区编辑团队策划出版的精品 IT 专业图书的品牌,依托于人民邮电出版社计算机图书出版的积累和专业编辑团队,相关图书在封面上印有异步图书的 LOGO。异步图书的出版领域包括软件开发、大数据、AI、测试、前端、网络技术等。

异步社区

微信服务号

目录

第 1 章　Spring 5 核心技术之 IoC ········· 1
- 1.1 什么是框架 ········· 1
- 1.2 反射与 XML 操作 ········· 1
  - 1.2.1 基础知识准备——反射 ········· 1
  - 1.2.2 基础知识准备——操作 XML 文件 ········· 13
- 1.3 Spring 框架介绍 ········· 20
- 1.4 Spring 框架的模块组成 ········· 20
- 1.5 控制反转和依赖注入介绍 ········· 21
- 1.6 IoC 容器介绍 ········· 22
- 1.7 AOP 介绍 ········· 22
- 1.8 初步体会 IoC 的优势 ········· 23
  - 1.8.1 传统方式 ········· 23
  - 1.8.2 Spring 方式 ········· 24
  - 1.8.3 依赖注入的原理是反射 ········· 28
- 1.9 在 Spring 中创建 JavaBean ········· 28
  - 1.9.1 使用 \<context:component-scan base-package=""\>创建对象 ········· 29
  - 1.9.2 使用 \<context:component-scan base-package=""\>创建并获取对象 ········· 30
  - 1.9.3 使用"全注解"法创建对象 ········· 30
  - 1.9.4 使用"全注解"法获取对象时出现 NoUniqueBeanDefinitionException 异常的解决办法 ········· 32
  - 1.9.5 使用@ComponentScan (basePackages="")创建并获取对象 ········· 33
  - 1.9.6 使用@ComponentScan (basePackages="")扫描多个包 ········· 34
  - 1.9.7 使用@ComponentScan 的 basePackageClasses 属性进行扫描 ········· 36
  - 1.9.8 使用@ComponentScan 而不使用 basePackages 属性时的效果 ········· 37
  - 1.9.9 解决不同包中有相同类名时出现异常的问题 ········· 38
  - 1.9.10 推荐使用的代码结构 ········· 40
  - 1.9.11 使用@Lazy 注解实现延迟加载 ········· 40
  - 1.9.12 出现 Overriding bean definition 情况时的解决方法 ········· 42
  - 1.9.13 在 IoC 容器中创建单例对象和多例对象 ········· 43
- 1.10 装配 Spring Bean ········· 45
  - 1.10.1 使用注解法注入对象 ········· 45
  - 1.10.2 多实现类的歧义性 ········· 46
  - 1.10.3 使用@Autowired 注解向构造方法的参数进行注入 ········· 49
  - 1.10.4 使用@Autowired 注解向方法的参数进行注入 ········· 50
  - 1.10.5 使用@Autowired 注解向字段进行注入 ········· 50
  - 1.10.6 使用@Inject 注解向字段、方法和构造方法进行注入 ········· 51
  - 1.10.7 使用@Bean 注解向工厂方法的参数进行注入 ········· 53
  - 1.10.8 使用@Autowired (required = false)的写法 ········· 54
  - 1.10.9 使用@Bean 对 JavaBean 的 id 重命名 ········· 56
  - 1.10.10 Spring 上下文的相关知识 ········· 58
  - 1.10.11 BeanFactory 与 ApplicationContext ········· 65
  - 1.10.12 使用注解@Value 进行注入 ········· 65

1.10.13 解决 BeanCurrentlyInCreationException 异常问题 ·················· 67

第 2 章 Spring 5 核心技术之 AOP ·················· 71

2.1 AOP ·················· 71
2.2 AOP 原理之代理设计模式 ·················· 71
2.2.1 静态代理的实现 ·················· 72
2.2.2 使用 JDK 实现动态代理 ·················· 74
2.2.3 使用 Spring 实现动态代理 ·················· 76
2.2.4 使用 cglib 实现动态代理 ·················· 79
2.2.5 使用 javassist 实现动态代理 ·················· 80
2.3 AOP 相关的概念 ·················· 81
2.3.1 横切关注点 ·················· 82
2.3.2 切面 ·················· 82
2.3.3 连接点 ·················· 84
2.3.4 切点 ·················· 84
2.3.5 通知 ·················· 85
2.3.6 织入 ·················· 85
2.4 AOP 核心案例 ·················· 86
2.4.1 实现前置通知、后置通知、返回通知和异常通知 ·················· 86
2.4.2 向前置通知、后置通知、返回通知和异常通知传入 JoinPoint 参数 ·················· 91
2.4.3 实现环绕通知 ·················· 92
2.4.4 使用 bean 表达式 ·················· 92
2.4.5 使用@Pointcut 定义全局切点 ·················· 94
2.4.6 向切面传入参数 ·················· 96
2.4.7 使用@AfterReturning 和 @AfterThrowing 向切面传入参数 ·················· 98
2.4.8 向环绕通知传入参数 ·················· 100
2.4.9 实现多切面的应用 ·················· 102
2.4.10 使用@Order 注解制定切面的运行顺序 ·················· 105

第 3 章 Spring 5 MVC 实战技术 ·················· 106

3.1 简介 ·················· 106
3.2 在 Spring Boot 框架中搭建 Spring MVC 开发环境 ·················· 106
3.2.1 搭建 Spring MVC 开发环境 ·················· 107
3.2.2 搭建 CSS+JavaScript+HTML+JSP 开发环境 ·················· 114
3.3 核心技术 ·················· 118
3.3.1 执行控制层——无传递参数 ·················· 118
3.3.2 执行控制层——有传递参数 ·················· 119
3.3.3 执行控制层——有传递参数简化版 ·················· 120

3.3.4 实现登录功能 ·················· 120
3.3.5 将 URL 参数封装到实体类 ·················· 122
3.3.6 限制提交方式 ·················· 123
3.3.7 控制层方法的参数类型 ·················· 124
3.3.8 控制层方法的返回值类型 ·················· 125
3.3.9 取得 request-response-session 对象 ·················· 126
3.3.10 实现登录失败后的提示信息 ·················· 126
3.3.11 向 Controller 控制层注入 Service 业务逻辑层 ·················· 128
3.3.12 重定向——无传递参数 ·················· 129
3.3.13 重定向——有传递参数 ·················· 130
3.3.14 重定向传递参数——Redirect- Attributes.addAttribute()方法 ·················· 131
3.3.15 重定向传递参数——Redirect- Attributes.addFlashAttribute() 方法 ·················· 132
3.3.16 使用 jackson 库在服务端将 JSON 字符串转换成各种 Java 数据类型 ·················· 133
3.3.17 在控制层返回 JSON 对象 ·················· 137
3.3.18 在控制层返回 JSON 字符串 ·················· 139
3.3.19 使用 HttpServletResponse 对象输出响应字符 ·················· 140
3.3.20 解决日期问题 ·················· 142
3.3.21 单文件上传 1——使用 MultipartHttpServletRequest ·················· 146
3.3.22 单文件上传 2——使用 MultipartFile ·················· 147
3.3.23 单文件上传 3——使用 MultipartFile 并结合实体类 ·················· 148
3.3.24 多文件上传 1——使用 MultipartHttpServletRequest ·················· 149
3.3.25 多文件上传 2——使用 MultipartFile[] ·················· 151
3.3.26 多文件上传 3——使用 MultipartFile[]并结合实体类 ·················· 152
3.3.27 使用 AJAX 实现文件上传 ·················· 154
3.3.28 支持中文文件名的文件下载 ·················· 156
3.3.29 使用@RestController 注解 ·················· 157
3.4 扩展技术 ·················· 158
3.4.1 使用 prefix 和 suffix 简化返回的视图名称 ·················· 158
3.4.2 控制层返回 List 对象及实体 ·················· 159
3.4.3 实现国际化 ·················· 163
3.4.4 处理异常 ·················· 172

3.4.5 方法的参数是 Model 数据类型……177
3.4.6 方法的参数是 ModelMap 数据类型……178
3.4.7 方法的返回值是 ModelMap 数据类型……179
3.4.8 方法的返回值是 ModelAndView 数据类型……180
3.4.9 方法的返回值是 ModelAndView 数据类型（实现重定向）……182
3.4.10 使用@RequestAttribute 和 @SessionAttribute 注解……182
3.4.11 使用@CookieValue 和 @RequestHeader 注解……183
3.4.12 使用@SessionAttributes 注解……184
3.4.13 使用@ModelAttribute 注解实现作用域别名……186
3.4.14 在路径中添加通配符的功能……187
3.4.15 控制层返回 void 数据的情况……188
3.4.16 解决多人开发路径可能重复的问题……189
3.4.17 使用@PathVariable 注解……191
3.4.18 通过 URL 参数访问指定的业务方法……192
3.4.19 使用@GetMapping、@PostMapping、@PutMapping 和 @Delete Mapping 注解……193
3.4.20 使用拦截器……197
3.4.21 Spring 5 MVC 应用 AOP 切面……203

第 4 章 MyBatis 3 核心技术之必备技能……205

4.1 ORM 简介……205
4.2 MyBatis 的优势……206
4.3 使用 JDBC+反射技术实现泛型 DAO……207
4.4 三大核心对象的介绍……213
4.5 三大核心对象的生命周期……213
4.6 使用 MyBatis Generator 插件：单模块……214
　　4.6.1 操作 Oracle 数据库……215
　　4.6.2 操作 MySQL 数据库……222
4.7 使用 MyBatis Generator 插件：多模块……227
　　4.7.1 操作 Oracle 数据库……227
　　4.7.2 操作 MySQL 数据库……232

4.8 自建环境使用 Mapper 接口操作 Oracle-MySQL 数据库……236
　　4.8.1 接口-SQL 映射的对应关系……236
　　4.8.2 针对 Oracle 的 CURD……237
　　4.8.3 针对 MySQL 的 CURD……244
4.9 向 Mapper 接口传入参数类型……250
4.10 从 SQL 映射取得返回值类型……253

第 5 章 MyBatis 3 核心技术之实战技能……255

5.1 实现输出日志……255
5.2 SQL 语句中特殊符号的处理……255
5.3 使用别名……256
　　5.3.1 系统预定义别名……256
　　5.3.2 使用 type-aliases-package 配置设置别名……257
　　5.3.3 别名重复的解决办法……258
5.4 对 yml 文件中的数据库密码进行加密……260
5.5 不同数据库对执行不同 SQL 语句的支持……262
　　5.5.1 使用<databaseIdProvider type="DB_VENDOR">实现执行不同的 SQL 语句……262
　　5.5.2 如果 SQL 映射的 id 值相同，有无 databaseId 的优先级……263
5.6 动态 SQL……264
　　5.6.1 使用<resultMap>标签实现映射……264
　　5.6.2 <resultMap>标签与实体类有参构造方法……265
　　5.6.3 使用${}拼接 SQL 语句……266
　　5.6.4 <sql>标签的使用……266
　　5.6.5 <if>标签的使用……268
　　5.6.6 <where>标签的使用……269
　　5.6.7 针对 Oracle/MySQL 实现 like 模糊查询……270
　　5.6.8 <choose>标签的使用……271
　　5.6.9 <set>标签的使用……272
　　5.6.10 <foreach>标签的使用……273
　　5.6.11 使用<foreach>执行批量插入……274
　　5.6.12 使用<bind>标签对 like 语句进行适配……276
　　5.6.13 使用<trim>标签规范 SQL 语句……278
5.7 读写大文本类型的数据……281

5.7.1 操作 Oracle 数据库……281
5.7.2 操作 MySQL 数据库……283
5.8 实现数据分页……283
5.9 实现一对一级联……284
  5.9.1 数据表结构和内容以及关系……285
  5.9.2 创建实体类……285
  5.9.3 创建 SQL 映射文件……287
  5.9.4 级联解析……288
  5.9.5 根据 ID 查询记录……288
  5.9.6 查询所有记录……289
  5.9.7 对 SQL 语句的执行次数进行优化……289
5.10 实现一对多级联……291
  5.10.1 数据表结构和内容以及关系……291
  5.10.2 创建实体类……291
  5.10.3 创建 SQL 映射文件……293
  5.10.4 级联解析……294
  5.10.5 根据 ID 查询记录……294
  5.10.6 查询所有记录……294
  5.10.7 对 SQL 语句的执行次数进行优化……295
5.11 延迟加载……297
  5.11.1 默认采用立即加载策略……297
  5.11.2 使用全局延迟加载策略与两种加载方式……298
  5.11.3 使用 fetchType 属性设置局部加载策略……304
5.12 缓存的使用……305
  5.12.1 一级缓存……305
  5.12.2 二级缓存……307
  5.12.3 验证 update 语句具有清除二级缓存的特性……308
5.13 Spring 事务传播特性……310
  5.13.1 事务传播特性 REQUIRED……310
  5.13.2 事务传播特性 SUPPORTS……311
  5.13.3 事务传播特性 MANDATORY……311
  5.13.4 事务传播特性 REQUIRES_NEW……312
  5.13.5 事务传播特性 NOT_SUPPORTED……314
  5.13.6 事务传播特性 NEVER……315
  5.13.7 事务传播特性 NESTED……317
  5.13.8 事务传播特性总结……318

第 6 章 模板引擎 FreeMarker 和 Thymeleaf 的使用……320

6.1 使用 FreeMarker 模板引擎……321
  6.1.1 FreeMarker 的优势……321
  6.1.2 FreeMarker 的输出……321
  6.1.3 整合 Spring Boot 与输出常见数据类型……322
  6.1.4 输出布尔值……327
  6.1.5 输出 Date 数据类型……328
  6.1.6 循环集合中的数据……329
  6.1.7 使用 if 命令实现判断……333
  6.1.8 判断 List 的 size 值是否为 0……334
  6.1.9 处理 null 值……335
  6.1.10 实现隔行变色……338
  6.1.11 对象嵌套有 null 值的处理……340
  6.1.12 比较运算符……341
  6.1.13 遗拾增补……342
  6.1.14 填充 select 中的 option……342
  6.1.15 实现自动选中 select 中的 option……343
  6.1.16 实现页面静态化……344
  6.1.17 将 ftlh 文件中的内容输出到内存中……345
6.2 使用 Thymeleaf 模板引擎……346
  6.2.1 整合 Spring Boot 与常见的使用方式……346
  6.2.2 处理复杂数据类型……351
  6.2.3 处理嵌套数据类型……353
  6.2.4 访问 Array……354
  6.2.5 访问 List……355
  6.2.6 访问 Map……356
  6.2.7 访问 request-session-application 作用域……358
  6.2.8 访问 URL 参数值……358
  6.2.9 循环 Array……359
  6.2.10 循环 List……359
  6.2.11 循环 Set……360
  6.2.12 循环 Map……361
  6.2.13 生成 Table……361
  6.2.14 循环生成<input type=text>……362
  6.2.15 获得状态变量……363
  6.2.16 获得状态变量的简化版……364
  6.2.17 实现国际化……365
  6.2.18 处理 URL……366
  6.2.19 处理布尔值……367
  6.2.20 操作属性……367

第 1 章　Spring 5 核心技术之 IoC

本章目标
（1）反射技术的使用
（2）操作 XML 文件
（3）什么是 IoC
（4）什么是 IoC 容器
（5）什么是依赖注入
（6）反射与依赖注入的关系
（7）实现装配 JavaBean

本书针对 Spring、Spring MVC、MyBatis 三大框架的测试环境全部基于 IntelliJ IDEA 和 Spring Boot。

1.1　什么是框架

框架就是软件功能的半成品。框架提供了一个软件项目中通用的功能，将大多数常见的功能进行封装，无须自己重复开发，增加了开发及运行效率。在软件公司中，大多数情况是使用框架开发软件项目。

1.2　反射与 XML 操作

Spring 框架内部大量使用反射与操作 XML 技术，以至于 MyBatis 也高度依赖这两种技术。掌握这两种技术有助于高效理解与学习 Java EE 框架。

本书的全部案例均在 IntelliJ IDEA 开发工具中进行测试，项目类型为 Maven。

1.2.1　基础知识准备——反射

本节介绍反射技术的基本使用，创建 maven-archetype-quickstart 类型的 Maven 项目 reflectTest。创建实体类 Userinfo，代码如下：

```java
package com.ghy.www.entity;

public class Userinfo {

    private long id;
    private String username;
    private String password;

    public Userinfo() {
        System.out.println("public Userinfo()");
    }

    public Userinfo(long id) {
        super();
        this.id = id;
    }

    public Userinfo(long id, String username) {
        super();
        this.id = id;
        this.username = username;
    }

    public Userinfo(long id, String username, String password) {
        super();
        this.id = id;
        this.username = username;
        this.password = password;
        System.out.println("public Userinfo(long id, String username, String password)");
        System.out.println(id + " " + username + " " + password);

    }

    public long getId() {
        return id;
    }

    public void setId(long id) {
        this.id = id;
    }

    public String getUsername() {
        return username;
    }

    public void setUsername(String username) {
        this.username = username;
    }

    public String getPassword() {
        return password;
    }

    public void setPassword(String password) {
        this.password = password;
    }

    public void test() {
```

```java
        System.out.println("public void test1()");
    }

    public void test(String address) {
        System.out.println("public void test2(String address) address=" + address);
    }

    public String test(int age) {
        System.out.println("public String test3(int age) age=" + age);
        return "我是返回值";
    }
}
```

创建实体类 Userinfo2，代码如下：

```java
package com.ghy.www.entity;

public class Userinfo2 {

    private long id;
    private String username;
    private String password;

    private Userinfo2() {
        System.out.println("public Userinfo()");
    }

    public long getId() {
        return id;
    }

    public void setId(long id) {
        this.id = id;
    }

    public String getUsername() {
        return username;
    }

    public void setUsername(String username) {
        this.username = username;
    }

    public String getPassword() {
        return password;
    }

    public void setPassword(String password) {
        this.password = password;
    }
}
```

1. 正常创建对象并调用方法

正常创建对象并调用方法的写法如下：

```java
package com.ghy.www.test1;

import com.ghy.www.entity.Userinfo;
```

```
public class Test1 {
    public static void main(String[] args) {
        Userinfo userinfo = new Userinfo();
        userinfo.setId(100);
        userinfo.setUsername("中国");
        userinfo.setPassword("中国人");

        System.out.println(userinfo.getId());
        System.out.println(userinfo.getUsername());
        System.out.println(userinfo.getPassword());
    }
}
```

但在某些情况下,预创建对象的类名以字符串的形式存储在 XML 文件中,比如 Servlet 技术中的 web.xml 文件部分的示例代码如下:

```
<servlet>
    <servlet-name>InsertUserinfo</servlet-name>
    <servlet-class>controller.InsertUserinfo</servlet-class>
</servlet>
```

Servlet 对象就是由 Tomcat 读取这段 XML 配置代码并解析出"controller.InsertUserinfo"字符串,然后使用反射技术动态创建出 InsertUserinfo 类的对象。可见反射技术无处不在,是学习框架技术非常重要的知识点。

下面我们来模拟一下 Tomcat 处理的过程。

在 resources 文件夹中创建文件 createClassName.txt,内容如下:

```
com.ghy.www.entity.Userinfo
```

创建运行类 Test2,代码如下:

```
package com.ghy.www.test1;

import java.io.IOException;
import java.io.InputStream;

public class Test2 {
    public static void main(String[] args) throws IOException {
        byte[] byteArray = new byte[1000];
        InputStream inputStream = Test2.class.getResourceAsStream("/createClassName.txt");
        int readLength = inputStream.read(byteArray);
        String createClassName = new String(byteArray, 0, readLength);
        System.out.println(createClassName);
        inputStream.close();
    }
}
```

程序运行结果如下:

```
com.ghy.www.entity.Userinfo
```

变量 createClassName 存储的就是类名"com.ghy.www.entity.Userinfo"字符串,这时如果使用图 1-1 所示的代码就是错误的,会出现编译错误,Java 编译器根本不允许使用这种写法创建 com.ghy.www.entity.Userinfo 类的对象,但需求是必须把类名"com.ghy.www.entity.Userinfo"对应的对象创建出来,对于这种情况可以使用反射技术来解决。反射是在运行时获得对象信息的一种技术。

```
public class Test3 {
    public static void main(String[] args) throws IOException {
        byte[] byteArray = new byte[1000];
        InputStream inputStream = Test3.class.getResourceAsStream( name: "/createClassName.txt");
        int readLength = inputStream.read(byteArray);
        String createClassName = new String(byteArray, offset: 0, readLength);
        System.out.println(createClassName);
        new createClassName;
        new createClassName();
        new "com.ghy.www.entity.Userinfo";
        new "com.ghy.www.entity.Userinfo" ();
    }
}
```

图 1-1　错误的代码

2. 获得 Class 类的对象的方式

在使用反射技术之前，必须要获得某一个*.class 类对应 Class 类的对象。

Class 类封装了类的信息（属性、构造方法和方法等），而 Class 类的对象封装了具体的*.class 类中的信息（属性、构造方法和方法等），有了这些信息，就相当于烹饪加工时有了制造食品的原料，就可以创建出类的对象。

有 4 种方式可以获得 Class 类的对象，代码如下：

```
package com.ghy.www.test1;

import com.ghy.www.entity.Userinfo;

public class Test4 {
    public static void main(String[] args) throws ClassNotFoundException {
        Class class1 = Userinfo.class;
        Class class2 = new Userinfo().getClass();
        Class class3 = Userinfo.class.getClassLoader().loadClass("com.ghy.www.entity.Userinfo");
        Class class4 = Class.forName("com.ghy.www.entity.Userinfo");
        System.out.println(class1.hashCode());
        System.out.println(class2.hashCode());
        System.out.println(class3.hashCode());
        System.out.println(class4.hashCode());
    }
}
```

程序运行结果如下：

```
public Userinfo()
460141958
460141958
460141958
460141958
```

从打印结果可以分析出，同一个*.class 类对应的 Class 类的对象是同一个，单例的。

3. 通过 Class 对象获得 Field、Constructor 和 Method 对象

一个*.class 类包含 Field、Constructor 和 Method 信息，可以通过 Class 对象来获取这些信息。示例代码如下：

```java
package com.ghy.www.test1;

import com.ghy.www.entity.Userinfo;

import java.lang.reflect.Constructor;
import java.lang.reflect.Field;
import java.lang.reflect.Method;

public class Test5 {
    public static void main(String[] args) {
        Class userinfoClass = Userinfo.class;
        // 属性列表
        Field[] a = userinfoClass.getDeclaredFields();
        System.out.println(a.length);
        for (int i = 0; i < a.length; i++) {
            System.out.println(a[i].getName());
        }
        System.out.println();
        // 构造方法列表
        Constructor[] b = userinfoClass.getConstructors();
        System.out.println(b.length);
        System.out.println();
        // 方法列表
        Method[] c = userinfoClass.getDeclaredMethods();
        System.out.println(c.length);
        for (int i = 0; i < c.length; i++) {
            System.out.println(c[i].getName());
        }
    }
}
```

程序运行结果如下：

```
3
id
username
password

4

9
getId
test
test
test
getPassword
getUsername
setPassword
setId
setUsername
```

4. 使用 Class.newInstance()方法创建对象

方法 Class.newInstance()调用的是无参构造方法。

示例代码如下：

```
package com.ghy.www.test1;

import com.ghy.www.entity.Userinfo;

public class Test6 {
    public static void main(String[] args)
            throws ClassNotFoundException, InstantiationException, IllegalAccessException {
        // 本实验不用 new 来创建对象，原理就是使用反射技术
        String newObjectName = "com.ghy.www.entity.Userinfo";
        Class class4 = Class.forName(newObjectName);
        // newInstance()开始创建对象
        Userinfo userinfo = (Userinfo) class4.newInstance();
        userinfo.setUsername("美国");
        System.out.println(userinfo.getUsername());
    }
}
```

程序运行结果如下：

```
public Userinfo()
美国
```

5. 对 Field 进行赋值和取值

示例代码如下：

```
package com.ghy.www.test1;

import com.ghy.www.entity.Userinfo;

import java.lang.reflect.Field;

public class Test7 {
    public static void main(String[] args) throws ClassNotFoundException, InstantiationException, IllegalAccessException, NoSuchFieldException, SecurityException {
        String newObjectName = "com.ghy.www.entity.Userinfo";
        Class class4 = Class.forName(newObjectName);
        Userinfo userinfo = (Userinfo) class4.newInstance();

        String fieldName = "username";
        String fieldValue = "法国";
        Field fieldObject = userinfo.getClass().getDeclaredField(fieldName);
        System.out.println(fieldObject.getName());
        fieldObject.setAccessible(true);
        fieldObject.set(userinfo, fieldValue);
        System.out.println(userinfo.getUsername());
        System.out.println(fieldObject.get(userinfo));
    }
}
```

程序运行结果如下：

```
public Userinfo()
username
法国
法国
```

6．获得构造方法对应的 Constructor 对象及调用无参构造方法

示例代码如下：

```java
package com.ghy.www.test1;

import com.ghy.www.entity.Userinfo;

import java.lang.reflect.Constructor;
import java.lang.reflect.InvocationTargetException;

public class Test8 {
    public static void main(String[] args)
            throws ClassNotFoundException, InstantiationException, IllegalAccessException,
NoSuchFieldException, SecurityException, NoSuchMethodException, IllegalArgumentException,
InvocationTargetException {
        String newObjectName = "com.ghy.www.entity.Userinfo";
        Class class4 = Class.forName(newObjectName);
        Constructor constructor1 = class4.getDeclaredConstructor();
        Userinfo userinfo1 = (Userinfo) constructor1.newInstance();
        Userinfo userinfo2 = (Userinfo) constructor1.newInstance();
        Userinfo userinfo3 = (Userinfo) constructor1.newInstance();
        System.out.println(userinfo1.hashCode());
        System.out.println(userinfo2.hashCode());
        System.out.println(userinfo3.hashCode());
    }
}
```

程序运行结果如下：

```
public Userinfo()
public Userinfo()
public Userinfo()
460141958
1163157884
1956725890
```

7．通过 Constructor 对象调用有参构造方法

示例代码如下：

```java
package com.ghy.www.test1;

import java.lang.reflect.Constructor;
import java.lang.reflect.InvocationTargetException;

public class Test9 {
    public static void main(String[] args)
            throws ClassNotFoundException, InstantiationException, IllegalAccessException,
NoSuchFieldException, SecurityException, NoSuchMethodException, IllegalArgumentException,
InvocationTargetException {
```

```
            String newObjectName = "com.ghy.www.entity.Userinfo";
            Class class4 = Class.forName(newObjectName);
            Constructor constructor = class4.getDeclaredConstructor(long.class,
    String.class, String.class);
            System.out.println(constructor.newInstance(11, "a1", "aa1").hashCode());
            System.out.println(constructor.newInstance(12, "a2", "aa2").hashCode());
            System.out.println(constructor.newInstance(13, "a3", "aa3").hashCode());
            System.out.println(constructor.newInstance(14, "a4", "aa4").hashCode());
    }
}
```

程序运行结果如下:

```
public Userinfo(long id, String username, String password)
11 a1 aa1
460141958
public Userinfo(long id, String username, String password)
12 a2 aa2
1163157884
public Userinfo(long id, String username, String password)
13 a3 aa3
1956725890
public Userinfo(long id, String username, String password)
14 a4 aa4
356573597
```

8. 使用反射动态调用无参无返回值方法

示例代码如下:

```
package com.ghy.www.test1;

import com.ghy.www.entity.Userinfo;

import java.lang.reflect.InvocationTargetException;
import java.lang.reflect.Method;

public class Test10 {
    public static void main(String[] args)
            throws ClassNotFoundException, InstantiationException, IllegalAccessException,
NoSuchFieldException, SecurityException, NoSuchMethodException, IllegalArgumentException,
InvocationTargetException {
        String newObjectName = "com.ghy.www.entity.Userinfo";
        Class class4 = Class.forName(newObjectName);
        Userinfo userinfo = (Userinfo) class4.newInstance();
        String methodName = "test";
        Method method = class4.getDeclaredMethod(methodName);
        System.out.println(method.getName());
        method.invoke(userinfo);
    }
}
```

程序运行结果如下:

```
public Userinfo()
test
public void test1()
```

9. 使用反射动态调用有参无返回值方法

示例代码如下:

```java
package com.ghy.www.test1;

import com.ghy.www.entity.Userinfo;

import java.lang.reflect.InvocationTargetException;
import java.lang.reflect.Method;

public class Test11 {
    public static void main(String[] args)
            throws ClassNotFoundException, InstantiationException, IllegalAccessException,
NoSuchFieldException, SecurityException, NoSuchMethodException, IllegalArgumentException,
InvocationTargetException {
        String newObjectName = "com.ghy.www.entity.Userinfo";
        Class class4 = Class.forName(newObjectName);
        Userinfo userinfo = (Userinfo) class4.newInstance();
        String methodName = "test";
        Method method = class4.getDeclaredMethod(methodName, String.class);
        System.out.println(method.getName());
        method.invoke(userinfo, "我的地址在北京!");
    }
}
```

程序运行结果如下:

```
public Userinfo()
test
public void test2(String address) address=我的地址在北京!
```

10. 使用反射动态调用有参有返回值方法

示例代码如下:

```java
package com.ghy.www.test1;

import com.ghy.www.entity.Userinfo;

import java.lang.reflect.InvocationTargetException;
import java.lang.reflect.Method;

public class Test12 {
    public static void main(String[] args)
            throws ClassNotFoundException, InstantiationException, IllegalAccessException,
NoSuchFieldException, SecurityException, NoSuchMethodException, IllegalArgumentException,
InvocationTargetException {
        String newObjectName = "com.ghy.www.entity.Userinfo";
        Class class4 = Class.forName(newObjectName);
```

```
            Userinfo userinfo = (Userinfo) class4.newInstance();
            String methodName = "test";
            Method method = class4.getDeclaredMethod(methodName, int.class);
            System.out.println(method.getName());
            Object returnValue = method.invoke(userinfo, 999999);
            System.out.println("returnValue=" + returnValue);
        }
    }
```

程序运行结果如下:

```
public Userinfo()
test
public String test3(int age) age=999999
returnValue=我是返回值
```

11. 反射破坏了 OOP

创建单例模式，代码如下:

```
package com.ghy.www.objectfactory;

//单例模式
//保证当前进程中只有 1 个 OneInstance 类的对象
public class OneInstance {
    private static OneInstance oneInstance;

    private OneInstance() {
    }

    public static OneInstance getOneInstance() {
        if (oneInstance == null) {
            oneInstance = new OneInstance();
        }
        return oneInstance;
    }
}
```

由于构造方法是 private（私有）的，因此不能 new 实例化对象，只能调用 public static OneInstance getOneInstance()方法获得自身的对象，而且可以保证单例，测试代码如下:

```
package com.ghy.www.test1;

import com.ghy.www.objectfactory.OneInstance;

import java.io.IOException;
import java.lang.reflect.InvocationTargetException;

public class Test13 {
    public static void main(String[] args) throws IOException, ClassNotFoundException,
InstantiationException, IllegalAccessException, NoSuchFieldException, SecurityException,
NoSuchMethodException, IllegalArgumentException, InvocationTargetException {
        OneInstance o1 = OneInstance.getOneInstance();
        OneInstance o2 = OneInstance.getOneInstance();
        OneInstance o3 = OneInstance.getOneInstance();
```

```
        System.out.println(o1);
        System.out.println(o2);
        System.out.println(o3);
    }
}
```

程序运行结果如下：

```
com.ghy.www.objectfactory.OneInstance@1b6d3586
com.ghy.www.objectfactory.OneInstance@1b6d3586
com.ghy.www.objectfactory.OneInstance@1b6d3586
```

但是，使用反射技术会破坏面向对象编程（OOP），导致创建多个 OneInstance 类的对象。
示例代码如下：

```
package com.ghy.www.test1;

import com.ghy.www.objectfactory.OneInstance;

import java.io.IOException;
import java.lang.reflect.Constructor;
import java.lang.reflect.InvocationTargetException;

public class Test14 {
    public static void main(String[] args) throws IOException, ClassNotFoundException,
InstantiationException, IllegalAccessException, NoSuchFieldException, SecurityException,
NoSuchMethodException, IllegalArgumentException, InvocationTargetException {
        Class classRef = Class.forName("com.ghy.www.objectfactory.OneInstance");
        Constructor c = classRef.getDeclaredConstructor();
        c.setAccessible(true);
        OneInstance one1 = (OneInstance) c.newInstance();
        OneInstance one2 = (OneInstance) c.newInstance();
        System.out.println(one1);
        System.out.println(one2);
    }
}
```

程序运行结果如下：

```
com.ghy.www.objectfactory.OneInstance@1b6d3586
com.ghy.www.objectfactory.OneInstance@4554617c
```

这里创建了两个 OneInstance 类的对象，没有保证单例性。虽然构造方法是 private 的，但依然借助反射技术创建了多个对象。

12．方法重载

使用反射可以调用重载方法。
示例代码如下：

```
package com.ghy.www.test1;

import java.lang.reflect.InvocationTargetException;
import java.lang.reflect.Method;

public class Test15 {
```

```
        public void testMethod(String username) {
            System.out.println("public String testMethod(String username)");
            System.out.println(username);
        }

        public void testMethod(String username, String password) {
            System.out.println("public String testMethod(String username, String password)");
            System.out.println(username + " " + password);
        }

        public static void main(String[] args)
                throws ClassNotFoundException, InstantiationException, IllegalAccessException,
NoSuchFieldException, SecurityException, NoSuchMethodException, IllegalArgumentException,
InvocationTargetException {
            Class class15Class = Test15.class;
            Object object = class15Class.newInstance();
            Method method1 = class15Class.getDeclaredMethod("testMethod", String.class);
            Method method2 = class15Class.getDeclaredMethod("testMethod", String.class,
String.class);
            method1.invoke(object, "法国");
            method2.invoke(object, "中国", "中国人");
        }
    }
```

程序运行结果如下：

```
public String testMethod(String username)
法国
public String testMethod(String username, String password)
中国 中国人
```

1.2.2 基础知识准备——操作 XML 文件

提起标记语言，人们首先会想起 HTML，HTML 提供了很多标签来实现 Web 前端界面的设计，但 HTML 中的标签并不允许自定义，如果想定义一些自己独有的标签，HTML 就不再可行了，这时可以使用 XML 来进行实现。

XML（eXtensible Markup Language，可扩展标记语言）可以自定义标记名称与内容，在灵活度上相比 HTML 改善很多，经常用在配置以及数据交互领域。

在开发软件项目时，经常会接触 XML 文件，比如 web.xml 文件中就有 XML 代码，而 Spring 框架也使用 XML 文件存储配置。XML 代码的主要作用就是配置。

创建 maven-archetype-quickstart 类型的 Maven 项目 xmlTest，添加如下依赖：

```
<dependency>
    <groupId>dom4j</groupId>
    <artifactId>dom4j</artifactId>
    <version>1.6.1</version>
</dependency>
```

在 resources 文件夹中创建 struts.xml 配置文件，代码如下：

```
<mymvc>
    <actions>
        <action name="list" class="controller.List">
```

```xml
            <result name="toListJSP">
                /list.jsp
            </result>
            <result name="toShowUserinfoList" type="redirect">
                showUserinfoList.ghy
            </result>
        </action>

        <action name="showUserinfoList" class="controller.ShowUserinfoList">
            <result name="toShowUserinfoListJSP">
                /showUserinfoList.jsp
            </result>
        </action>
    </actions>
</mymvc>
```

1．解析 XML 文件

创建 Reader 类，代码如下：

```java
package com.ghy.www.test1;

import org.dom4j.Attribute;
import org.dom4j.Document;
import org.dom4j.DocumentException;
import org.dom4j.Element;
import org.dom4j.io.SAXReader;

import java.util.List;

public class Reader {
    public static void main(String[] args) {
        try {
            SAXReader reader = new SAXReader();
            Document document = reader.read(reader.getClass()
                    .getResourceAsStream("/struts.xml"));

            Element mymvcElement = document.getRootElement();
            System.out.println(mymvcElement.getName());
            Element actionsElement = mymvcElement.element("actions");
            System.out.println(actionsElement.getName());
            System.out.println("");
            List<Element> actionList = actionsElement.elements("action");
            for (int i = 0; i < actionList.size(); i++) {
                Element actionElement = actionList.get(i);
                System.out.println(actionElement.getName());
                System.out.print("name="
                        + actionElement.attribute("name").getValue());
                System.out.println("action class="
                        + actionElement.attribute("class").getValue());

                List<Element> resultList = actionElement.elements("result");
                for (int j = 0; j < resultList.size(); j++) {
                    Element resultElement = resultList.get(j);
                    System.out.print("   result name="
                            + resultElement.attribute("name").getValue());
                    Attribute typeAttribute = resultElement.attribute("type");
```

```
                    if (typeAttribute != null) {
                        System.out.println(" type=" + typeAttribute.getValue());
                    } else {
                        System.out.println("");
                    }
                    System.out.println("   " + resultElement.getText().trim());
                    System.out.println("");
                }

                System.out.println("");
            }

        } catch (DocumentException e) {
            // 自动生成 catch 块
            e.printStackTrace();
        }
    }
}
```

程序运行结果如下：

```
mymvc
actions

action
name=listaction class=controller.List
  result name=toListJSP
    /list.jsp

  result name=toShowUserinfoList type=redirect
    showUserinfoList.ghy

action
name=showUserinfoListaction class=controller.ShowUserinfoList
  result name=toShowUserinfoListJSP
    /showUserinfoList.jsp
```

2. 创建 XML 文件

创建 Writer 类，代码如下：

```
package com.ghy.www.test1;

import org.dom4j.Document;
import org.dom4j.DocumentHelper;
import org.dom4j.Element;
import org.dom4j.io.OutputFormat;
import org.dom4j.io.XMLWriter;

import java.io.FileWriter;
import java.io.IOException;

public class Writer {

    public static void main(String[] args) {
        try {
            Document document = DocumentHelper.createDocument();
```

```
            Element mymvcElement = document.addElement("mymvc");
            Element actionsElement = mymvcElement.addElement("actions");
            //
            Element listActionElement = actionsElement.addElement("action");
            listActionElement.addAttribute("name", "list");
            listActionElement.addAttribute("class", "controller.List");

            Element toListJSPResultElement = listActionElement
                    .addElement("result");
            toListJSPResultElement.addAttribute("name", "toListJSP");
            toListJSPResultElement.setText("/list.jsp");

            Element toShowUserinfoListResultElement = listActionElement
                    .addElement("result");
            toShowUserinfoListResultElement.addAttribute("name",
                    "toShowUserinfoList");
            toShowUserinfoListResultElement.addAttribute("type", "redirect");
            toShowUserinfoListResultElement.setText("showUserinfoList.ghy");
            //

            Element showUserinfoListActionElement = actionsElement
                    .addElement("action");
            showUserinfoListActionElement.addAttribute("name",
                    "showUserinfoList");
            showUserinfoListActionElement.addAttribute("class",
                    "controller.ShowUserinfoList");

            Element toShowUserinfoListJSPResultElement = showUserinfoListActionElement
                    .addElement("result");
            toShowUserinfoListJSPResultElement.addAttribute("name",
                    "toShowUserinfoListJSP");
            toShowUserinfoListResultElement.setText("/showUserinfoList.jsp");
            //

            OutputFormat format = OutputFormat.createPrettyPrint();
            XMLWriter writer = new XMLWriter(new FileWriter("ghy.xml"), format);
            writer.write(document);
            writer.close();

        } catch (IOException e) {
            e.printStackTrace();
        }

    }

}
```

程序运行后在项目中创建 ghy.xml 文件, 如图 1-2 所示。

3. 修改 XML 文件

创建 Update 类, 代码如下:

```
package com.ghy.www.test1;

import org.dom4j.Attribute;
import org.dom4j.Document;
import org.dom4j.DocumentException;
```

图 1-2　创建的 ghy.xml 文件

```java
import org.dom4j.Element;
import org.dom4j.io.OutputFormat;
import org.dom4j.io.SAXReader;
import org.dom4j.io.XMLWriter;

import java.io.FileWriter;
import java.io.IOException;
import java.util.List;

public class Update {

    public static void main(String[] args) throws IOException {
        try {
            SAXReader reader = new SAXReader();
            Document document = reader.read(reader.getClass().getResourceAsStream("/struts.xml"));
            Element mymvcElement = document.getRootElement();
            Element actionsElement = mymvcElement.element("actions");
            List<Element> actionList = actionsElement.elements("action");
            for (int i = 0; i < actionList.size(); i++) {
                Element actionElement = actionList.get(i);
                List<Element> resultList = actionElement.elements("result");
                for (int j = 0; j < resultList.size(); j++) {
                    Element resultElement = resultList.get(j);
                    String resultName = resultElement.attribute("name").getValue();
                    if (resultName.equals("toShowUserinfoList")) {
                        Attribute typeAttribute = resultElement.attribute("type");
                        if (typeAttribute != null) {
                            typeAttribute.setValue("zzzzzzzzzzzzzzzzzzzzz");
                            resultElement.setText("xxxxxxxxxxxxxxxxxx");
                        }
                    }
                }
            }
            OutputFormat format = OutputFormat.createPrettyPrint();
            XMLWriter writer = new XMLWriter(new FileWriter("src\\ghy.xml"), format);
            writer.write(document);
            writer.close();
        } catch (DocumentException e) {
            e.printStackTrace();
        }
    }
}
```

程序运行后产生的 ghy.xml 文件内容如下：

```xml
<?xml version="1.0" encoding="UTF-8"?>
<mymvc>
    <actions>
        <action name="list" class="controller.List">
            <result name="toListJSP">/list.jsp</result>
            <result name="toShowUserinfoList" type="zzzzzzzzzzzzzzzzzzzzz">xxxxxxxxxxxxxxxxxx</result>
        </action>
        <action name="showUserinfoList" class="controller.ShowUserinfoList">
            <result name="toShowUserinfoListJSP">/showUserinfoList.jsp</result>
```

```
            </action>
        </actions>
</mymvc>
```

成功更改 XML 文件中的属性值与文本内容。

4．删除节点

创建 Delete 类，代码如下：

```java
package com.ghy.www.test1;

import org.dom4j.Document;
import org.dom4j.DocumentException;
import org.dom4j.Element;
import org.dom4j.io.OutputFormat;
import org.dom4j.io.SAXReader;
import org.dom4j.io.XMLWriter;

import java.io.FileWriter;
import java.io.IOException;
import java.util.List;

public class Delete {

    public static void main(String[] args) throws IOException {
        try {
            SAXReader reader = new SAXReader();
            Document document = reader.read(reader.getClass().getResourceAsStream("/struts.xml"));
            Element mymvcElement = document.getRootElement();
            Element actionsElement = mymvcElement.element("actions");
            List<Element> actionList = actionsElement.elements("action");
            for (int i = 0; i < actionList.size(); i++) {
                Element actionElement = actionList.get(i);
                List<Element> resultList = actionElement.elements("result");
                Element resultElement = null;
                boolean isFindNode = false;
                for (int j = 0; j < resultList.size(); j++) {
                    resultElement = resultList.get(j);
                    String resultName = resultElement.attribute("name").getValue();
                    if (resultName.equals("toShowUserinfoList")) {
                        isFindNode = true;
                        break;
                    }
                }
                if (isFindNode == true) {
                    actionElement.remove(resultElement);
                }
            }
            OutputFormat format = OutputFormat.createPrettyPrint();
            XMLWriter writer = new XMLWriter(new FileWriter("src\\ghy.xml"), format);
            writer.write(document);
            writer.close();
        } catch (DocumentException e) {
            e.printStackTrace();
        }
    }

}
```

程序运行后产生的 ghy.xml 文件内容如下：

```xml
<?xml version="1.0" encoding="UTF-8"?>

<mymvc>
    <actions>
        <action name="list" class="controller.List">
            <result name="toListJSP">/list.jsp</result>
        </action>
        <action name="showUserinfoList" class="controller.ShowUserinfoList">
            <result name="toShowUserinfoListJSP">/showUserinfoList.jsp</result>
        </action>
    </actions>
</mymvc>
```

节点被成功删除。

5. 删除属性

创建 DeleteAttr 类，代码如下：

```java
package com.ghy.www.test1;

import org.dom4j.Attribute;
import org.dom4j.Document;
import org.dom4j.DocumentException;
import org.dom4j.Element;
import org.dom4j.io.OutputFormat;
import org.dom4j.io.SAXReader;
import org.dom4j.io.XMLWriter;

import java.io.FileWriter;
import java.io.IOException;
import java.util.List;

public class DeleteAttr {

    public static void main(String[] args) throws IOException {
        try {
            SAXReader reader = new SAXReader();
            Document document = reader.read(reader.getClass().getResourceAsStream("/struts.xml"));
            Element mymvcElement = document.getRootElement();
            Element actionsElement = mymvcElement.element("actions");
            List<Element> actionList = actionsElement.elements("action");
            for (int i = 0; i < actionList.size(); i++) {
                Element actionElement = actionList.get(i);
                List<Element> resultList = actionElement.elements("result");
                for (int j = 0; j < resultList.size(); j++) {
                    Element resultElement = resultList.get(j);
                    String resultName = resultElement.attribute("name").getValue();
                    if (resultName.equals("toShowUserinfoList")) {
                        Attribute typeAttribute = resultElement.attribute("type");
                        if (typeAttribute != null) {
                            resultElement.remove(typeAttribute);
                        }
                    }
                }
            }
```

```
        }
        OutputFormat format = OutputFormat.createPrettyPrint();
        XMLWriter writer = new XMLWriter(new FileWriter("src\\ghy.xml"), format);
        writer.write(document);
        writer.close();
    } catch (DocumentException e) {
        e.printStackTrace();
    }
}
```

程序运行后产生的 ghy.xml 文件内容如下：

```
<?xml version="1.0" encoding="UTF-8"?>

<mymvc>
    <actions>
        <action name="list" class="controller.List">
            <result name="toListJSP">/list.jsp</result>
            <result name="toShowUserinfoList">showUserinfoList.ghy</result>
        </action>
        <action name="showUserinfoList" class="controller.ShowUserinfoList">
            <result name="toShowUserinfoListJSP">/showUserinfoList.jsp</result>
        </action>
    </actions>
</mymvc>
```

XML 文件中的 type 属性成功被删除。

1.3　Spring 框架介绍

Spring 框架简化了 Java EE 开发的流程，它是为了应对企业级应用开发的复杂性而创建的，其强大之处在于对 Java EE 开发进行全方位的简化，对大部分常用的功能进行了封装，比如管理 JavaBean，包含创建及销毁，还提供了基于 Web 的 MVC 分层框架，支持数据库操作、安全验证等功能，但这些功能的实现却要依赖两个技术原理：控制反转（Inversion of Control，IoC）和面向切面编程（Aspect Oriented Programming，AOP）。本书的目的就是使读者学习并掌握 Spring 中的这两个核心技术，以在实际的软件开发中得以运用。

Spring 是一个开放源代码的 Java EE 框架，使用 Spring 简化了 Java EE 开发，提升了 Java EE 软件项目的开发效率，提高开发效率的办法就是使用模块架构，每个模块处理一个功能或者业务。模块架构允许程序员选择使用哪一个模块参与开发，同时为 Java EE 应用程序开发提供集成的容器。在 Spring 框架中提供了 1 个 JavaBean 容器（可以暂时将容器理解为 1 个 List），在该容器中存储不同数据类型的 JavaBean 对象。在容器中，可以将很多种不同功能的 JavaBean 进行整合和集成，以达到多个技术综合应用的目的。

1.4　Spring 框架的模块组成

Spring 框架发展多年，现在已经是一个初具规模的 Java EE 开发平台，在 Spring 5 中的主要模块如下。

（1）Core（核心）模块：依赖注入（dependency injection）、事件处理（events）、资源访问（resources）、国际化（i18n）、验证（validation）、数据绑定（data binding）、数据类型转换（type conversion）、表达式语言（SpEL）、面向切面编程（AOP）。

（2）Testing（测试）模块：模拟对象（mock objects）、TestContext 框架、Spring MVC Test、WebTestClient 测试框架。

（3）Data Access（数据访问）模块：事务处理（transactions）、数据访问对象（DAO）支持、Java 数据库连接（JDBC）、对象关系映射（ORM）、处理 XML（Marshalling XML）。

（4）Spring MVC 和 Spring WebFlux Web 框架模块。

（5）Integration（集成）模块：远程访问（remoting）、Java 消息服务（JMS）、Java 加密体系结构（JCA）、Java 管理扩展（JMX）、邮件处理（email）、任务（task）、执行计划（scheduling）、缓存（cache）。

（6）Languages（语言）模块：支持使用 Kotlin、Groovy、动态语言等进行开发。

Spring 框架的功能可以用在任何 Java EE 服务器中，其核心要点是，保证相同代码在不同 Java EE 容器的可移植性。

1.5　控制反转和依赖注入介绍

在没有 Spring 框架的时候，如果在 A 类中使用 B 类，则必须在 A 类中 new 实例化出 B 类的对象，这就造成了 A 类和 B 类的紧耦合，A 类完全依赖 B 类的功能实现，这样的情况就属于典型的"侵入式开发"。随着软件业务的复杂度提升，当原有的 B 类不能满足 A 类的功能实现时，就需要创建更为高级的 BExt 类，结果就是把所有实例化 B 类的代码替换成 new BExt() 代码，这就产生了源代码的改动，不利于软件运行的稳定性，并不符合商业软件的开发与维护流程。IoC 技术就可以解决这样的问题，其办法就是使用"反射"技术，动态地对一个类中的属性进行反射赋值，对于这样的功能，Spring 形成了一个模块，模块的功能非常强大，并且 Spring 对这种机制进行了命名，叫作控制反转（Inversion of Control，IoC）。

IoC 要达到的目的就是将调用者与被调用者进行分离，让类与类之间的关系进行解耦，这是一种设计思想。解耦的原理如图 1-3 所示。

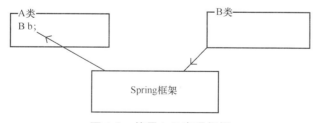

图 1-3　使用 IoC 实现解耦

Spring 框架中的 IoC 技术可以实现 A 类和 B 类的解耦，在 A 类中不再出现 new B() 的情况，实例化 new B() 类的任务由 Spring 框架来进行处理，Spring 框架再使用反射的机制将 B 类的对象赋值给 A 类中的 B b 变量。原来 A 类是主动实例化 B 类的对象，控制方是 A 类，而现在以被动的方式由 Spring 框架来进行实现，控制方现在变成了 Spring 框架，实现了反转，所以此种技术称为"控制反转"。

IoC 是一个理念，也是一种设计思想。A 类中的 B b 对象的值是需要被赋值的，实现的方式是依赖注入（Dependency Injection，DI）。依赖注入是 IoC 思想的实现，侧重于实现，A 类依赖 B 类，B 类的对象由容器进行创建，容器再对 A 类中的 B b 对象进行值的注入。依赖注入在 Java 中的底层技术原理就是"反射"，使用反射技术对某一个类中的属性进行动态赋值，以达到 A 模块和 B 模块之间的解耦。

在 Spring 中，官方将管理 JavaBean 的容器定义为 IoC 容器（IoC container），而在 Spring 中对 IoC 的主要实现方式就是依赖注入，依赖注入的称呼更容易被人所接受。

1.6　IoC 容器介绍

什么是 IoC 容器？前面介绍过，Spring 的依赖注入其实就是对 JavaBean 的属性使用反射技术进行赋值，当有很多的 JavaBean 需要这样的操作时，这些 JavaBean 的管理就成了问题，因为某些 JavaBean 之间需要关联，而某些 JavaBean 之间并不需要关联，而且所有这些 JavaBean 的创建和销毁都要统一的调度，由 Spring 框架管理它们的生命周期，为了方便这种管理，Spring 框架提供了 IoC 容器，对 JavaBean 进行统一组织，便于后期代码的维护。曾经在任意的位置进行 new 实例化任何类对象的情况不复存在了，所有 new 实例化类的对象的任务都要交给 IoC 容器实现，这时，对 JavaBean 的管理就更加规范了。

另外，Spring 的 IoC 容器完全脱离了平台，可以在任何支持 Java 语言的环境中运行，具有极好的可移植性。IoC 容器就是管理 JavaBean 并创建 JavaBean 的一个内存区。

IoC/依赖注入从编程技术上来讲就是将接口和实现相分离，然后使用反射技术对类中的属性进行动态赋值。IoC 容器的职责是管理 JavaBean 的生命周期，处理多个 JavaBean 之间的注入关系。

通过前面的解释，我们已经大概了解 IoC 与 IoC 容器，以及依赖注入的作用与使用场景了。程序员在任何位置创建任何类的对象在 Spring 框架中是不规范的，Spring 框架对 JavaBean 的管理更加具有规划性，比如创建、销毁，还可以动态地对一个属性注入值，通过使用 Spring 的 IoC 容器，使软件项目对 JavaBean 的管理更加统一。

1.7　AOP 介绍

什么是 AOP 呢？AOP 是面向切面编程（Aspect Oriented Programming）。在没有 AOP 技术时，如果我们想对软件项目记录日志，则必须要在关键的业务点写上记录日志的程序代码，日志的信息包含"开始时间""结束时间""执行人"以及"角色"等信息，随着软件项目越来越稳定，曾经的日志代码需要删除，因为输出日志会影响程序运行的效率，这时就要在 Java 源代码中删除记录日志的代码，造成代码的改动。另外，在未来我们有可能还需要记录日志，到时还要重新加入日志的代码，造成源代码的反复更改，不利于软件运行的稳定性。使用 Spring 的 AOP 技术就可以解决这些具有"通用性"的问题，Spring 的 AOP 功能模块具有可插拔性，所以几乎不需要大幅更改代码即可完成前面想要实现的功能。

依赖注入使用的技术原理是"反射"，AOP 使用的技术原理是"动态代理"。代理设计模式是 23 个标准设计模式中的一个。代理设计模式分为静态代理与动态代理。动态代理是在不改变原有代码的基础上，对原有的模块进行功能上的扩展，使原有的模块与扩展后的模块充分解耦，利于软件项目的模块化设计。应用 AOP 的场景可以在不改变 Controller（控制层）代码的基础上加入日志的功能，在不改变 Controller 代码的基础上加入数据库事务的功能等。所以 AOP 主要实现对功能模块进行扩展与模块间的解耦合。

AOP 的详细内容将在第 2 章介绍。

1.8 初步体会 IoC 的优势

前面我们讨论了使用 IoC 可以实现解耦，依赖注入是 IoC 思想的实现，那么如何在 Spring 框架中实现解耦呢？在介绍如何实现解耦之前，我们先看一下使用传统的方法实现一个数据保存功能的弊端。

1.8.1 传统方式

创建测试项目 firstSaveTest。

创建业务类 SaveDBService，代码如下：

```
package com.ghy.www.service;

public class SaveDBService {
    public void saveMethod() {
        System.out.println("将数据保存到数据库");
    }
}
```

创建运行类 Test，代码如下：

```
package com.ghy.www.test1;

import com.ghy.www.service.SaveDBService;

public class Test {
    private SaveDBService service = new SaveDBService();

    public SaveDBService getService() {
        return service;
    }

    public void setService(SaveDBService service) {
        this.service = service;
    }

    public static void main(String[] args) {
        Test test = new Test();
        test.getService().saveMethod();
    }
}
```

程序运行结果如下：

将数据保存到数据库

控制台输出正确的结果，这样的代码结构在大多数的软件项目中被使用，而且有些已经应用到商业项目中，虽然输出的结果是正确的，但从项目的整体设计结构上来看，明显是不合理的，比如以下几点。

（1）源代码反复被修改。本项目将数据保存到数据库，如果换成保存到 XML 文件中，就不得不更改程序，将 SaveDBService 类中的代码改成保存到 XML 文件，或者创建新的 SaveXMLService 类，然后更改代码，变成实例化对象（private SaveXMLService service = new SaveXMLService();），这就造成了程序的改动，不利于项目运行的稳定性。

（2）出现紧耦合。Test 类和 SaveDBService 类产生了紧耦合，Test 类负责创建 SaveDBService 类的对象，不利于软件功能的扩展、测试与复用。

（3）无法保证单例性。SaveDBService 类无法在单实例的情况下被复用，因为它的声明是在 Test 类中，也就是在 Test 类中可以随意地创建出很多 SaveDBService 类的实例，无法保证该类实例的单例性。

（4）无法保证资源被正确释放。如果从 SaveDBService 类中获取一些资源，比如数据库的连接（Connection）、输入输出流（Stream）等，那么 Test 类不得不维护这些资源的开启（open）和关闭（close）。如果出现忘记关闭的情况，则资源不能有效释放，造成资源占用，影响项目运行的稳定性。

以上 4 个缺点造成了这个设计是失败的，那么，根据面向对象三大特性中的"多态"技术，可以把一个 SaveDBService 类改成接口与实现类的模式吗？也就是在 Test 类中声明接口，然后实例化这个接口的实现类（测试代码在 firstSaveTestInterface 项目中）。这样的设计的确比上面的示例要灵活一些，也符合面向接口编程的方式，但这仅仅是将接口和实现进行分离，并没有完全解决业务变更后源代码还要被更改的问题。那么，如果在项目中经常有这样耦合的结构，该如何解决这样的问题？Spring 是如何解决的呢？

在 IoC 容器的帮助下，可以实现松耦合，并且模块之间是分离的，彼此可以互相访问。使用依赖注入技术加上 IoC 容器之后，创建对象的操作是由 IoC 容器来进行控制的，并且也完全基于接口（interface）和实现类（imple）的分离开发。这样将 SaveDBService 类的控制权由原来在 Test 类中转变成在 IoC 容器中，接口的实现类是依赖 IoC 容器进行注入赋值的。下面就使用 Spring 框架来解决这两个类之间的紧耦合问题，把这个问题划上一个句号。

1.8.2　Spring 方式

创建测试项目 spring-first。

添加 pom.xml 配置，代码如下：

```xml
<parent>
    <!-- 从 SpringBoot 继承默认的配置 -->
    <groupId>org.springframework.boot</groupId>
    <artifactId>spring-boot-starter-parent</artifactId>
    <version>2.3.4.RELEASE</version>
</parent>

<dependencies>
    <dependency>
        <groupId>junit</groupId>
        <artifactId>junit</artifactId>
        <version>4.11</version>
        <scope>test</scope>
    </dependency>
```

```xml
    <dependency>
        <groupId>org.springframework.boot</groupId>
        <artifactId>spring-boot-starter-web</artifactId>
    </dependency>
</dependencies>
```

创建一个保存数据的接口，代码如下：

```java
package com.ghy.www.service;

public interface ISaveService {
    public void saveMethod();
}
```

创建一个将数据保存到数据库的实现类，代码如下：

```java
package com.ghy.www.service;

public class SaveDBService implements ISaveService {
    public SaveDBService() {
        System.out.println("Spring 通过反射机制来实例化 SaveDBService 类的对象 " + this);
    }

    @Override
    public void saveMethod() {
        System.out.println("将数据保存到数据库");
    }
}
```

创建一个将数据保存到 XML 文件的实现类，代码如下：

```java
package com.ghy.www.service;

public class SaveXMLService implements ISaveService {
    public SaveXMLService() {
        System.out.println("Spring 通过反射机制来实例化 SaveXMLService 类的对象 " + this);
    }

    @Override
    public void saveMethod() {
        System.out.println("将数据保存到 XML 文件");
    }
}
```

创建运行类 Test，代码如下：

```java
package com.ghy.www.test1;

import com.ghy.www.service.ISaveService;
import org.springframework.context.ApplicationContext;
import org.springframework.context.support.ClassPathXmlApplicationContext;

public class Test {
    private ISaveService service;
    private String username;

    public ISaveService getService() {
        return service;
    }
```

```java
    public void setService(ISaveService service) {
        this.service = service;
        System.out.println("setService(ISaveService service) service=" + service);
    }
    public String getUsername() {
        return username;
    }
    public void setUsername(String username) {
        this.username = username;
    }
    public static void main(String[] args) {
        ApplicationContext context = new ClassPathXmlApplicationContext("applicationContext.xml");
        Test test = context.getBean(Test.class);
        test.getService().saveMethod();
        System.out.println(test.getUsername());
    }
}
```

使用接口 ApplicationContext 可以从 IoC 容器中获得 JavaBean 对象。

创建 IoC 容器可以使用 ClassPathXmlApplicationContext 类加载 applicationContext.xml 配置文件来实现，然后使用 ApplicationContext.getBean()方法从容器中获得 Java 对象。

在 resources 文件夹中创建配置文件 applicationContext.xml，代码如下：

```xml
<?xml version="1.0" encoding="UTF-8"?>
<beans xmlns="http://www.springframework.org/schema/beans"
    xmlns:xsi="http://www.w3.org/2001/XMLSchema-instance"
    xsi:schemaLocation="http://www.springframework.org/schema/beans http://www.springframework.org/schema/beans/spring-beans.xsd">

    <bean id="saveDBService" class="com.ghy.www.service.SaveDBService"></bean>
    <bean id="saveXMLService" class="com.ghy.www.service.SaveXMLService"></bean>
    <bean id="test" class="com.ghy.www.test1.Test">
        <property name="service" ref="saveDBService"></property>
        <property name="username" value="我是 username 的值"></property>
    </bean>

</beans>
```

配置代码如下：

```xml
<bean id="saveDBService" class="com.ghy.www.service.SaveDBService"></bean>
```

它的作用是通过反射机制在 IoC 容器中创建 com.ghy.www.service.SaveDBService 类的对象，对象的别名就是 id 属性值"saveDBService"。

配置代码如下：

```xml
<bean id="test" class="com.ghy.www.test1.Test">
    <property name="service" ref="saveDBService"></property>
    <property name="username" value="我是 username 的值"></property>
</bean>
```

它的作用是创建 com.ghy.www.test1 包中 Test 类的对象，并且对 Test 类对象中的 service 属性进行反射赋值，也就是依赖注入，赋值的方式是通过反射调用 public void setService (ISaveService service)方法。赋予的值来自 ref="saveDBService"配置代码。ref 属性名代表反射赋予的值是复杂对象，是引用数据类型，ref 属性值"saveDBService"是配置代码的 id 值：

```
<bean id="saveDBService" class="com.ghy.www.service.SaveDBService"></bean>
```

配置代码如下：

```
<property name="username" value="我是 username 的值"></property>
```

它的作用是对 username 属性注入值。在 IoC 容器中，如果注入的数据类型是简单数据类型或 String 数据类型，则要使用 value 属性，ref 属性注入的是引用数据类型，这是 value 属性和 ref 属性的区别。

注入关系如图 1-4 所示。

图 1-4　注入关系

图 1-4 要说明的是对 com.ghy.www.test1 包中的 Test 类中的 service 属性注入别名为 saveDBService 的对象，对 username 属性注入字符串"我是 username 的值"。

程序运行结果如下：

```
Spring通过反射机制来实例化SaveDBService类的对象 com.ghy.www.service.SaveDBService@56235b8e
Spring通过反射机制来实例化SaveXMLService类的对象 com.ghy.www.service.SaveXMLService@cd2dae5
setService(ISaveService service) service=com.ghy.www.service.SaveDBService@56235b8e
将数据保存到数据库
我是username的值
```

接口 ISaveService 的实现类不再由 Test 类创建，而是由 Spring 的 IoC 容器进行创建与管理。接口 ISaveService 的实现类与 Test 类完全解耦。

通过使用 Spring 框架，如果业务变更，在需要用 XML 文件保存数据的情况下，只需要创建一个将数据保存到 XML 文件的实现类，并且对 service 属性进行注入，这样就可以非常灵活地将数据保存到数据库或 XML 文件中了，符合"开闭原则"的设计思想，即对扩展开放，对修改关闭。

另外，通过使用 IoC 容器来对创建的 JavaBean 对象进行管理，完全可以非常独立地对各个模块进行依赖注入，这样使用所带来的结果就是软件的 JavaBean 完全复用了，因此使用 Spring 的 IoC 容器技术，可以使复杂的业务逻辑、多变的模块关系、烦琐的软件后期维护及扩展等问题得以解决，这也是 Spring 的"哲学思想"，即用最简单的技术原理（反射）解决复杂的问题。

使用传统方式保存数据时存在前面总结的 4 个问题，但通过使用 Spring 框架，这 4 个问题都一一被解决。

（1）源代码反复被修改。依赖关系不需要更改*.java 文件，而只需要更改 applicationContext.xml

配置文件。

（2）出现紧耦合。A 类和 B 类彻底解耦，B 类由 Spring 框架创建，再向 A 类进行注入。

（3）无法保证单例性。IoC 容器可以保证 JavaBean 的单例性，后面会验证。

（4）无法保证资源被正确释放。结合 Spring 框架中的 AOP 技术，会自动执行 close()方法，资源永远会被自动释放。

1.8.3　依赖注入的原理是反射

Spring 框架的 IoC 容器技术中大量使用了反射。我们可以查看一下堆栈信息，证明使用了反射技术创建对象，如图 1-5 所示。

图 1-5　使用反射技术创建对象

我们继续证明使用反射技术实现依赖注入，如图 1-6 所示。

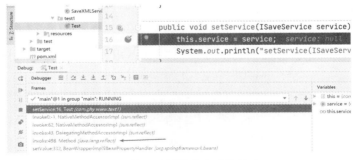

图 1-6　使用反射技术实现依赖注入

下面就对依赖注入的细节进行进一步的介绍，深入理解 IoC 容器控制反转的理念与实现方式。

1.9　在 Spring 中创建 JavaBean

在 Spring 中创建 JavaBean 的实例时并不使用传统的实例化的方式 new Object()，而是使用其他多种途径来创建出类的对象。如果使用 new Object()方式创建对象，那么创建类的对象的目的是达成了,但对对象的管理却非常不方便,以致工程化程度不高,因此,在 Spring 框架中将创建的 JavaBean 对象放入 IoC 容器中,在容器中统一管理这些对象。

在 Spring 框架中使用两种方式创建对象,分别是 XML 声明法以及注解法。但使用 XML 声明法创建对象时容易造成 applicationContext.xml 文件中<bean>声明的配置代码过多而出现"声明爆炸",对代码的后期维护非常不利,使用注解法可以解决这个问题。注解法也是 Spring Boot 默认使用的方式。

1.9.1 使用<context:component-scan base-package="">创建对象

配置代码<context:component-scan base-package="">的作用是在指定的包中扫描符合创建对象的类,如果某些类需要被 Spring 实例化,则类的上方必须使用相关的注解(annotation)。

创建测试项目 annotationCreateBean。

创建实体类 Userinfo,代码如下:

```
package com.ghy.www.entity;

import org.springframework.stereotype.Component;

@Component
public class Userinfo {
    public Userinfo() {
        System.out.println("public Userinfo() " + this.hashCode());
    }
}
```

创建配置文件 applicationContext.xml,代码如下:

```
<?xml version="1.0" encoding="UTF-8"?>
<beans xmlns="http://www.springframework.org/schema/beans"
       xmlns:xsi="http://www.w3.org/2001/XMLSchema-instance"
       xmlns:context="http://www.springframework.org/schema/context"
       xsi:schemaLocation="http://www.springframework.org/schema/beans http://www.springframework.org/schema/beans/spring-beans.xsd http://www.springframework.org/schema/context https://www.springframework.org/schema/context/spring-context.xsd">
    <context:component-scan base-package="com.ghy.www.entity"></context:component-scan>
</beans>
```

注解@Component 的作用就是标识 Userinfo 类是 IoC 容器中的一个组件,能被<context:component-scan base-package="com.ghy.www.entity"></context:component-scan>扫描器所识别并进行自动实例化,最后将 Userinfo 类的对象放入 IoC 容器中。另外,我们也可以不使用注解@Component,转而使用注解@Repository 来进行声明,程序运行结果是没有变化的。两个注解有下列区别。

(1)@Repository 主要用来声明 DAO 层。

(2)@Component 主要用来声明一些通用性的组件。

创建运行类 Test1,代码如下:

```
package com.ghy.www.test;

import org.springframework.context.ApplicationContext;
```

```java
import org.springframework.context.support.ClassPathXmlApplicationContext;

public class Test1 {
    public static void main(String[] args) {
        ApplicationContext context = new ClassPathXmlApplicationContext("applicationContext.xml");
    }
}
```

程序运行结果如下：

```
public Userinfo() 1059063940
```

我们使用注解法成功进行了 Userinfo 类的实例化。

1.9.2 使用<context:component-scan base-package="">创建并获取对象

使用<context:component-scan base-package="">创建对象后，获取对象时使用 getBean()方法，示例代码如下：

```java
package com.ghy.www.test;

import com.ghy.www.entity.Userinfo;
import org.springframework.context.ApplicationContext;
import org.springframework.context.support.ClassPathXmlApplicationContext;

public class Test2 {
    public static void main(String[] args) {
        ApplicationContext context = new ClassPathXmlApplicationContext("applicationContext.xml");
        System.out.println(context.getBean(Userinfo.class).hashCode());
        System.out.println(context.getBean("userinfo").hashCode());
    }
}
```

程序运行结果如下：

```
public Userinfo() 1059063940
1059063940
1059063940
```

1.9.3 使用"全注解"法创建对象

在本节中我们将使用"全注解"的方式创建对象，而不再使用 applicationContext.xml 配置文件。"全注解"法也称为 JavaConfig 法。

配置文件 applicationContext.xml 的根节点是<beans>。<beans>起到全局配置定义的作用，它的子节点<bean>包含了创建 JavaBean 时的细节信息。

在使用"全注解"法创建对象时，与<beans>标记具有相同作用的注解是@Configuration，与<bean>标记具有相同功能的注解是@Bean。

创建测试项目 allAnnotationCreateObject1。

创建实体类 Userinfo，代码如下：

```
package com.ghy.www.entity;

import org.springframework.stereotype.Component;

@Component
public class Userinfo {
    public Userinfo() {
        System.out.println("public Userinfo() " + this.hashCode());
    }
}
```

创建配置类 SpringConfig，代码如下：

```
package com.ghy.www.javaconfig;

import com.ghy.www.entity.Userinfo;
import org.springframework.context.annotation.Bean;
import org.springframework.context.annotation.Configuration;

@Configuration
public class SpringConfig {
    @Bean(name = "u1")
    public Userinfo createUserinfo1() {
        Userinfo userinfo1 = new Userinfo();
        System.out.println("userinfo1 " + userinfo1.hashCode());
        return userinfo1;
    }

    @Bean(name = "u2")
    public Userinfo xxxxxxxxxxxxxxx() {
        Userinfo userinfo2 = new Userinfo();
        System.out.println("userinfo2 " + userinfo2.hashCode());
        return userinfo2;
    }
}
```

使用@Bean 注解声明工厂方法的名称可以是任意的，但必须有返回值，不然会出现没有声明返回值的异常。

创建运行类 Test1，代码如下：

```
package com.ghy.www.test;

import org.springframework.context.ApplicationContext;
import org.springframework.context.annotation.AnnotationConfigApplicationContext;

public class Test1 {
    public static void main(String[] args) {
        ApplicationContext context = new AnnotationConfigApplicationContext("com.ghy.www.javaconfig");
    }
}
```

实例化类（new AnnotationConfigApplicationContext()）时传入参数"com.ghy.www.javaconfig"代表在"com.ghy.www.javaconfig"包中寻找哪个类带有注解@Configuration，再根据该类中的信息创建 JavaBean 对象。

程序运行结果如下：

```
public Userinfo() 1914301543
userinfo1 1914301543
public Userinfo() 1157726741
userinfo2 1157726741
```

我们成功使用"全注解"法创建了两个 Userinfo 类的对象。

> **注意**：使用 new AnnotationConfigApplicationContext("com.ghy.www.javaconfig");扫描配置类时，"com.ghy.www.javaconfig"包中可以有多个配置类。

1.9.4　使用"全注解"法获取对象时出现 NoUniqueBeanDefinitionException 异常的解决办法

在使用"全注解"法获取对象时，如果获取相同类型的对象，就会出现 NoUniqueBeanDefinitionException 异常。

创建运行类 Test2，代码如下：

```java
package com.ghy.www.test;

import com.ghy.www.entity.Userinfo;
import org.springframework.context.ApplicationContext;
import org.springframework.context.annotation.AnnotationConfigApplicationContext;

public class Test2 {
    public static void main(String[] args) {
        ApplicationContext context = new AnnotationConfigApplicationContext("com.ghy.www.javaconfig");
        context.getBean(Userinfo.class);
    }
}
```

程序运行后出现如下异常：

```
Exception in thread "main" org.springframework.beans.factory.NoUniqueBeanDefinitionException: No qualifying bean of type 'com.ghy.www.entity.Userinfo' available: expected single matching bean but found 2: u1,u2
```

这说明有多个对象的数据类型是相同的，解决这个问题的办法是使用 beanId。

创建运行类 Test3，代码如下：

```java
package com.ghy.www.test;

import org.springframework.context.ApplicationContext;
import org.springframework.context.annotation.AnnotationConfigApplicationContext;

public class Test3 {
    public static void main(String[] args) {
        ApplicationContext context = new AnnotationConfigApplicationContext("com.ghy.www.javaconfig");
        System.out.println();
        System.out.println(context.getBean("u2").hashCode());
    }
}
```

对注解@Bean 中的 name 属性赋值，设置每个 Userinfo 类的对象拥有不同的 id 值，并且还要结合 getBean(String)方法。

程序运行结果如下：

```
public Userinfo() 1914301543
userinfo1 1914301543
public Userinfo() 1157726741
userinfo2 1157726741

1157726741
```

1.9.5　使用@ComponentScan(basePackages = "")创建并获取对象

与 XML 配置代码<context:component-scan base-package="packageName"></context:component-scan>作用相同的注解是@ComponentScan(basePackages = "packageName")，该注解也可以进行扫描并实例化。

创建测试项目 ComponentScanTest。

创建实体类 Userinfo，代码如下：

```
package com.ghy.www.entity;

import org.springframework.stereotype.Component;

@Component
public class Userinfo {
    public Userinfo() {
        System.out.println("public Userinfo() " + this.hashCode());
    }
}
```

创建工厂类 SpringConfig，代码如下：

```
package com.ghy.www.javaconfig;

import org.springframework.context.annotation.Bean;
import org.springframework.context.annotation.Configuration;

import java.util.Date;

@Configuration
public class SpringConfig {
    @Bean
    public Date createDate() {
        Date nowDate = new Date();
        System.out.println("createDate " + nowDate.hashCode());
        return nowDate;
    }
}
```

创建运行类 Test1，代码如下：

```
package com.ghy.www.test;

import org.springframework.context.ApplicationContext;
import org.springframework.context.annotation.AnnotationConfigApplicationContext;
```

```
public class Test1 {
    public static void main(String[] args) {
        ApplicationContext context = new AnnotationConfigApplicationContext("com.ghy.
www.javaconfig");
    }
}
```

程序运行结果如下：

```
createDate -1505740341
```

从控制台输出的信息来看，只将 Date 类的对象进行了实例化，并没有创建 Userinfo 类的对象。如果我们想将其他包中的类进行实例化，就需要使用注解：

```
@ComponentScan(basePackages = "com.ghy.www.entity")
```

更改工厂类 SpringConfig，代码如下：

```
package com.ghy.www.javaconfig;

import org.springframework.context.annotation.Bean;
import org.springframework.context.annotation.ComponentScan;
import org.springframework.context.annotation.Configuration;

import java.util.Date;

@Configuration
@ComponentScan(basePackages = "com.ghy.www.entity")
public class SpringConfig {
    @Bean
    public Date createDate() {
        Date nowDate = new Date();
        System.out.println("createDate " + nowDate.hashCode());
        return nowDate;
    }
}
```

程序运行结果如下：

```
public Userinfo() 1625082366
createDate -1505655400
```

注意：使用@ComponentScan 注解可以扫描其他包中的类，那些类包含 Spring Config 配置类。

1.9.6 使用@ComponentScan(basePackages = "")扫描多个包

注解@ComponentScan 支持对多个包进行扫描。

创建测试项目 ComponentScanMorePackage。

创建实体类 Userinfo，代码如下：

```
package com.ghy.www.entity1;

import org.springframework.stereotype.Component;

@Component
public class Userinfo {
    public Userinfo() {
```

```
            System.out.println("public Userinfo() " + this.hashCode());
        }
    }
```

创建实体类 Bookinfo，代码如下：

```
package com.ghy.www.entity2;

import org.springframework.stereotype.Component;

@Component
public class Bookinfo {
    public Bookinfo() {
        System.out.println("public Bookinfo() " + this.hashCode());
    }
}
```

创建工厂类 SpringConfig，代码如下：

```
package com.ghy.www.javaconfig;

import org.springframework.context.annotation.Bean;
import org.springframework.context.annotation.ComponentScan;
import org.springframework.context.annotation.Configuration;

import java.util.Date;

@Configuration
@ComponentScan(basePackages = {"com.ghy.www.entity1", "com.ghy.www.entity2"})
public class SpringConfig {
    @Bean
    public Date createDate() {
        Date nowDate = new Date();
        System.out.println("createDate " + nowDate.hashCode());
        return nowDate;
    }
}
```

注解还可以使用如下写法：

```
@ComponentScan(basePackages = "com.ghy.www.entity1")
@ComponentScan(basePackages = "com.ghy.www.entity2")
```

其作用是相同的。

创建运行类 Test1，代码如下：

```
package com.ghy.www.test;

import org.springframework.context.ApplicationContext;
import org.springframework.context.annotation.AnnotationConfigApplicationContext;

public class Test1 {
    public static void main(String[] args) {
        ApplicationContext context = new AnnotationConfigApplicationContext("com.ghy.www.javaconfig");
    }
}
```

程序运行结果如下：

```
public Userinfo() 384294141
public Bookinfo() 1024597427
createDate -1505250055
```

1.9.7　使用@ComponentScan 的 basePackageClasses 属性进行扫描

注解@ComponentScan 的 basePackageClasses 属性的作用是扫描指定*.class 文件所在的包路径，然后创建该包下以及其子孙包下类的对象。

创建测试项目 basePackageClassesTest。创建的包结构如图 1-7 所示。

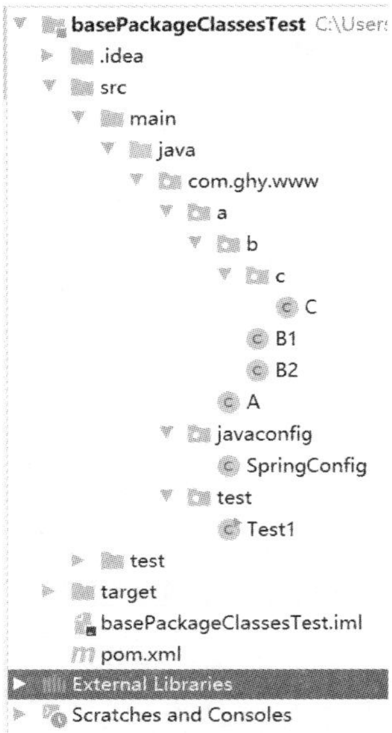

在 com.ghy.www.a 包中有 A 类，在 com.ghy.www.a.b 包中有 B1 类和 B2 类，在 com.ghy.www.a.b.c 包中有 C 类。

下面我们要使用 basePackageClasses 属性扫描 B1 类所在的包及其子孙包中所有的组件。

创建工厂类 SpringConfig，代码如下：

```
package com.ghy.www.javaconfig;

import com.ghy.www.a.b.B1;
import org.springframework.context.annotation.Bean;
import org.springframework.context.annotation.ComponentScan;
import org.springframework.context.annotation.Configuration;

import java.util.Date;

@Configuration
@ComponentScan(basePackageClasses = {B1.class})
public class SpringConfig {
    @Bean
    public Date createDate() {
        Date nowDate = new Date();
        System.out.println("createDate " + nowDate.hashCode());
        return nowDate;
    }
}
```

图 1-7　项目结构

创建运行类 Test1，代码如下：

```
package com.ghy.www.test;

import com.ghy.www.javaconfig.SpringConfig;
import org.springframework.context.ApplicationContext;
import org.springframework.context.annotation.AnnotationConfigApplicationContext;

public class Test1 {
    public static void main(String[] args) {
        ApplicationContext context = new AnnotationConfigApplicationContext(SpringConfig.class);
    }
}
```

程序运行结果如下：

```
public B1()
public B2()
public C()
createDate -1504539077
```

类 A 并没有被实例化，属性 basePackageClasses 只是将类 Date、B1、B2 和 C 进行了实例化，说明该属性实例化的对象为同级及子孙级包。

注解@ComponentScan 还可以指定多个 class 文件所在的包路径并进行扫描，示例代码如下：

```
@ComponentScan(basePackageClasses = {X.class, Y.class})
```

注解@ComponentScan 允许将上面这行代码拆分成两行，下面的写法是有效的：

```
@ComponentScan(basePackageClasses = X.class)
@ComponentScan(basePackageClasses = Y.class)
```

1.9.8 使用@ComponentScan 而不使用 basePackages 属性时的效果

如果在使用注解@ComponentScan 时不使用任何属性，那么注解@ComponentScan 默认扫描的是使用注解@Configuration 的配置类所在的包路径下的所有组件，包含子孙包中的组件。

创建测试项目 ComponentScanTest2，该项目结构如图 1-8 所示。

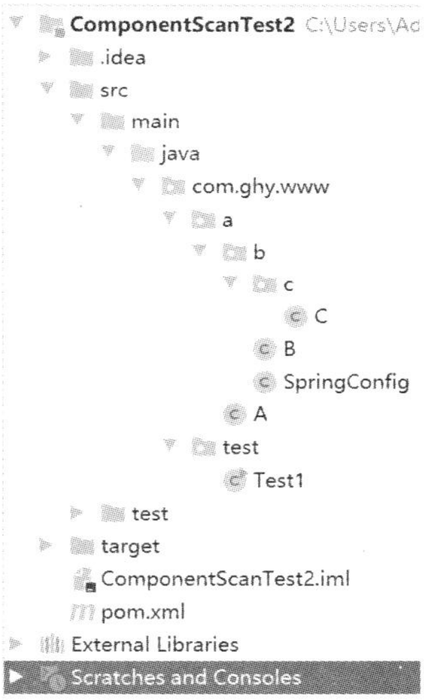

创建工厂类 SpringConfig，代码如下：

```
package com.ghy.www.a.b;

import org.springframework.context.annotation.Bean;
import org.springframework.context.annotation.ComponentScan;
import org.springframework.context.annotation.Configuration;

import java.util.Date;

@Configuration
@ComponentScan
public class SpringConfig {
    @Bean
    public Date createDate() {
        Date nowDate = new Date();
        System.out.println("createDate " + nowDate.hashCode());
        return nowDate;
    }
}
```

图 1-8 项目结构

创建运行类 Test1，代码如下：

```
package com.ghy.www.test;

import com.ghy.www.a.b.SpringConfig;
import org.springframework.context.ApplicationContext;
import org.springframework.context.annotation.AnnotationConfigApplicationContext;

public class Test1 {
    public static void main(String[] args) {
```

```
        ApplicationContext context = new AnnotationConfigApplicationContext(SpringConfig.class);
    }
}
```

程序运行后控制台输出如下:

```
public B()
public C()
createDate -1503759984
```

1.9.9 解决不同包中有相同类名时出现异常的问题

在不同的包中如果有相同的类名,则在扫描时会发生异常:

org.springframework.context.annotation.ConflictingBeanDefinitionException

创建测试项目 packageDiffClassSame。

创建实体类 Userinfo,代码如下:

```
package com.ghy.www.entity1;

import org.springframework.stereotype.Component;

@Component
public class Userinfo {
    public Userinfo() {
        System.out.println("public com.ghy.www.entity1.Userinfo() " + this.hashCode());
    }
}
```

创建实体类 Userinfo,代码如下:

```
package com.ghy.www.entity2;

import org.springframework.stereotype.Component;

@Component
public class Userinfo {
    public Userinfo() {
        System.out.println("public com.ghy.www.entity2.Userinfo() " + this.hashCode());
    }
}
```

创建工厂类 SpringConfig,代码如下:

```
package com.ghy.www.javaconfig;

import org.springframework.context.annotation.ComponentScan;
import org.springframework.context.annotation.Configuration;

@Configuration
@ComponentScan(basePackages = "com.ghy.www.entity1")
@ComponentScan(basePackages = "com.ghy.www.entity2")
public class SpringConfig {
}
```

创建运行类 Test1，代码如下：

```
package com.ghy.www.test;

import org.springframework.context.ApplicationContext;
import org.springframework.context.annotation.AnnotationConfigApplicationContext;

public class Test1 {
    public static void main(String[] args) {
        ApplicationContext context = new AnnotationConfigApplicationContext("com.ghy.www.javaconfig");
    }
}
```

程序运行后控制台输出如下异常信息：

```
Exception in thread "main" org.springframework.beans.factory.BeanDefinitionStoreException:
Failed to parse configuration class [com.ghy.www.javaconfig.SpringConfig]; nested exception
is org.springframework.context.annotation.ConflictingBeanDefinitionException: Annotation-
specified bean name 'userinfo' for bean class [com.ghy.www.entity2.Userinfo] conflicts with
existing, non-compatible bean definition of same name and class [com.ghy.www.entity1.Userinfo]
```

处理异常的方法有以下两种。

（1）将类名设置为不相同。

（2）为组件定义一个别名。

下面我们实现第 2 种方法——更改两个实体类，代码如下：

```
@Component(value = "userinfo1")
public class Userinfo {
```

```
@Component(value = "userinfo2")
public class Userinfo {
```

然后，创建新的运行类，代码如下：

```
package com.ghy.www.test;

import org.springframework.context.ApplicationContext;
import org.springframework.context.annotation.AnnotationConfigApplicationContext;

public class Test2 {
    public static void main(String[] args) {
        ApplicationContext context = new AnnotationConfigApplicationContext("com.ghy.www.javaconfig");
        System.out.println(context.getBean("userinfo1").hashCode());
        System.out.println(context.getBean("userinfo2").hashCode());
    }
}
```

再次运行程序，输出结果如下：

```
public com.ghy.www.entity1.Userinfo() 1344199921
public com.ghy.www.entity2.Userinfo() 2025269734
1344199921
2025269734
```

这样，不同包中有相同类名时出现异常的问题就被解决了。

1.9.10 推荐使用的代码结构

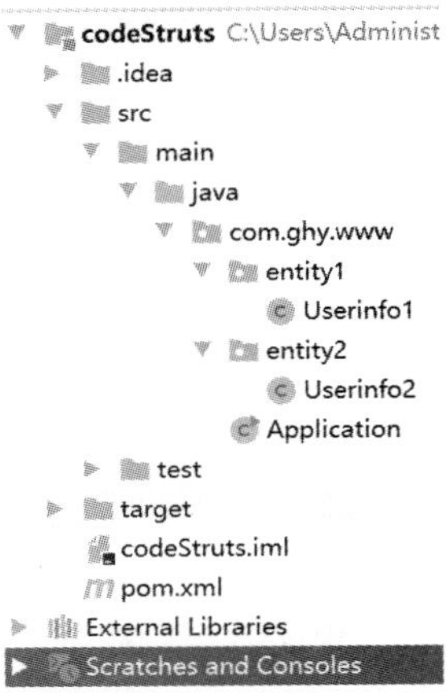

在使用 Spring 注解进行扫描包操作时，好的实践方式是在包的 root 位置启用扫描来对子孙包中的类进行搜索。

创建测试用的项目 codeStruts。项目结构如图 1-9 所示。

类 Application 的主要作用是在 root 位置启用扫描搜索，使用代码 "new AnnotationConfig ApplicationContext("com.ghy.www");" 针对 root 位置和所有子孙包中的组件进行实例化，这样就不需要再使用如下注解对指定的包进行扫描了：

```
@ComponentScan(value="packageName")
```

创建运行类 Application，代码如下：

图 1-9 推荐使用的项目结构

```
package com.ghy.www;

import org.springframework.context.ApplicationContext;
import org.springframework.context.annotation.AnnotationConfigApplicationContext;

public class Application {
    public static void main(String[] args) {
        ApplicationContext context = new AnnotationConfigApplicationContext("com.ghy.www");
    }
}
```

控制台输出结果如下：

```
public Userinfo1()
public Userinfo2()
```

1.9.11 使用@Lazy 注解实现延迟加载

创建测试项目 annotationLazyTest。

创建实体类 Userinfo，代码如下：

```
package com.ghy.www.entity;

import org.springframework.stereotype.Component;

@Component
public class Userinfo {
    public Userinfo() {
        System.out.println("public Userinfo() " + this);
    }
}
```

创建工厂类 SpringConfig，代码如下：

```
package com.ghy.www.javaconfig;

import com.ghy.www.entity.Userinfo;
import org.springframework.context.annotation.Bean;
import org.springframework.context.annotation.Configuration;
```

```
import org.springframework.context.annotation.Lazy;

@Configuration
public class SpringConfig {
    @Bean(name = "userinfo1")
    public Userinfo getUserinfo1() {
        Userinfo userinfo = new Userinfo();
        System.out.println("getUserinfo1 " + userinfo);
        return userinfo;
    }

    @Bean(name = "userinfo2")
    @Lazy(value = true)
    public Userinfo getUserinfo2() {
        Userinfo userinfo = new Userinfo();
        System.out.println("getUserinfo2 " + userinfo);
        return userinfo;
    }
}
```

注解@Lazy 和@Component 可以联合使用，代码如下：

```
@Component
@Lazy(value = true)
```

创建运行类 Test1，代码如下：

```
package com.ghy.www.test;

import org.springframework.context.ApplicationContext;
import org.springframework.context.annotation.AnnotationConfigApplicationContext;

public class Test1 {
    public static void main(String[] args) {
        ApplicationContext context = new AnnotationConfigApplicationContext("com.ghy.www.javaconfig");
    }
}
```

程序运行结果如下：

```
public Userinfo() com.ghy.www.entity.Userinfo@65d6b83b
getUserinfo1 com.ghy.www.entity.Userinfo@65d6b83b
```

上述结果表明程序并没有执行 getUserinfo2()方法来创建 Userinfo 类的对象，说明使用@Lazy(value = true)实现的效果是延迟加载。

创建运行类 Test2，代码如下：

```
package com.ghy.www.test;

import org.springframework.context.ApplicationContext;
import org.springframework.context.annotation.AnnotationConfigApplicationContext;

public class Test2 {
    public static void main(String[] args) {
        ApplicationContext context = new AnnotationConfigApplicationContext("com.ghy.www.javaconfig");
        context.getBean("userinfo2");
    }
}
```

程序运行结果如下：

```
public Userinfo() com.ghy.www.entity.Userinfo@65d6b83b
getUserinfo1 com.ghy.www.entity.Userinfo@65d6b83b
public Userinfo() com.ghy.www.entity.Userinfo@1f3f4916
getUserinfo2 com.ghy.www.entity.Userinfo@1f3f4916
```

上述结果表明程序执行了 getUserinfo2() 方法来创建 Userinfo 类的对象，说明执行代码 context.getBean("userinfo2") 才会创建 Userinfo 类的对象，属于延迟创建。

延迟加载是指当调用 getBean() 方法时 JavaBean 才会创建对象，不调用 getBean() 方法时 JavaBean 不创建对象。

1.9.12　出现 Overriding bean definition 情况时的解决方法

使用"全注解"法会出现 bean 覆盖的情况。

创建测试项目 beanOverriding。

创建实体类 Userinfo，代码如下：

```
package com.ghy.www.entity;

import org.springframework.stereotype.Component;

@Component
public class Userinfo {
    public Userinfo() {
        System.out.println("public Userinfo() " + this);
    }
}
```

创建工厂类 SpringConfig，代码如下：

```
package com.ghy.www.javaconfig;

import com.ghy.www.entity.Userinfo;
import org.springframework.context.annotation.Bean;
import org.springframework.context.annotation.Configuration;

@Configuration
public class SpringConfig {
    @Bean(name = "userinfo")
    public Userinfo getUserinfo() {
        System.out.println("@Bean creator getUser");
        return new Userinfo();
    }
}
```

创建运行类 Test1，代码如下：

```
package com.ghy.www.test;

import org.springframework.context.ApplicationContext;
import org.springframework.context.annotation.AnnotationConfigApplicationContext;

public class Test1 {
    public static void main(String[] args) {
        ApplicationContext context = new AnnotationConfigApplicationContext("com.ghy.
```

```
www");
        }
    }
```

程序运行后，控制台输出结果如下：

```
21:01:22.512 [main] DEBUG org.springframework.beans.factory.support.DefaultListable
BeanFactory - Overriding bean definition for bean 'userinfo' with a different definition:
replacing [Generic bean: class [com.ghy.www.entity.Userinfo]; scope=singleton; abstract=false;
lazyInit=null; autowireMode=0; dependencyCheck=0; autowireCandidate=true; primary=false;
factoryBeanName=null; factoryMethodName=null; initMethodName=null; destroyMethodName=null;
defined in file [C:\Users\Administrator\Desktop\ssm\第1章\beanOverriding\target\classes\
com\ghy\www\entity\Userinfo.class]] with [Root bean: class [null]; scope=; abstract=false;
lazyInit=null; autowireMode=3; dependencyCheck=0; autowireCandidate=true; primary=false;
factoryBeanName=springConfig; factoryMethodName=getUserinfo; initMethodName=null; destroy
MethodName=(inferred); defined in com.ghy.www.javaconfig.SpringConfig]
@Bean creator getUser
public Userinfo() com.ghy.www.entity.Userinfo@7b227d8d
```

控制台输出提示"Overriding bean definition"，而且并不是异常信息，这就会造成出现隐式的问题，本来想创建两个 Userinfo 类的对象，现在却只有一个。

出现"Overriding bean definition"的原因是在工厂方法中使用注解@Bean(name = "userinfo")创建 Userinfo 的别名是 userinfo，而通过扫描创建 Userinfo 的别名也是 userinfo，这就出现了 JavaBean 覆盖的情况，解决的办法就是将@Bean(name = "xxxxxxxxxx")的 id 设置为不同的值。解决此问题的项目是 beanOverridingOK，项目 beanOverridingOK 的输出结果如下：

```
public Userinfo() com.ghy.www.entity.Userinfo@197d671
@Bean creator getUser
public Userinfo() com.ghy.www.entity.Userinfo@7219ec67
```

1.9.13　在 IoC 容器中创建单例对象和多例对象

IoC 容器对 JavaBean 的管理存在作用域，在 Spring 中一共有 7 种作用域：singleton、prototype、request、session、globalSession、application 和 websocket。

Singleton 作用域代表在 IoC 容器中只有一个 JavaBean 的对象，而 prototype 代表当使用 getBean()方法取得一个 JavaBean 时，IoC 容器新建一个指定 JavaBean 的实例并且返回给程序员，每调用一次 getBean()方法相当于执行一次 new 操作，它们的区别仅仅在于 singleton 永远是一个实例，而 prototype 是多实例。

Spring 默认使用 singleton（单例）。

在使用 singleton 时一定要注意 JavaBean 的线程安全问题。非线程安全是指多个线程访问同一个对象的同一个实例变量时，此变量值有可能被覆盖。

我们通过一个实例来具体看一下这两种作用域的区别！

创建测试项目 scopeTest。

单例的示例代码如下：

```
package com.ghy.www.entity;

import org.springframework.context.annotation.Scope;
import org.springframework.stereotype.Component;
```

```java
@Component
@Scope(scopeName = "singleton")
public class Userinfo1 {
    public Userinfo1() {
        System.out.println("public Userinfo1() " + this);
    }
}
```

多例的示例代码如下：

```java
package com.ghy.www.entity;

import org.springframework.context.annotation.Scope;
import org.springframework.stereotype.Component;

@Component
@Scope(scopeName = "prototype")
public class Userinfo2 {
    public Userinfo2() {
        System.out.println("public Userinfo2() " + this);
    }
}
```

创建运行类 Test，代码如下：

```java
package com.ghy.www.test;

import com.ghy.www.entity.Userinfo1;
import com.ghy.www.entity.Userinfo2;
import org.springframework.context.ApplicationContext;
import org.springframework.context.annotation.AnnotationConfigApplicationContext;
import org.springframework.stereotype.Component;

@Component
public class Test {
    public static void main(String[] args) {
        ApplicationContext context = new AnnotationConfigApplicationContext("com.ghy.www");
        System.out.println(context.getBean(Userinfo1.class).hashCode());
        System.out.println(context.getBean(Userinfo1.class).hashCode());
        System.out.println(context.getBean(Userinfo1.class).hashCode());
        System.out.println();
        System.out.println();
        System.out.println(context.getBean(Userinfo2.class).hashCode());
        System.out.println(context.getBean(Userinfo2.class).hashCode());
        System.out.println(context.getBean(Userinfo2.class).hashCode());
    }
}
```

程序运行结果如下：

```
public Userinfo1() com.ghy.www.entity.Userinfo1@5e955596
1586845078
1586845078
1586845078


public Userinfo2() com.ghy.www.entity.Userinfo2@641147d0
1678854096
public Userinfo2() com.ghy.www.entity.Userinfo2@6e38921c
1849201180
```

```
public Userinfo2() com.ghy.www.entity.Userinfo2@64d7f7e0
1691875296
```

当有多个线程对同一个对象的同一个实例变量进行写操作时，要避免出现"非线程安全"问题，所以要使用 prototype（多例），比如 Struts 2 的 Action 必须使用多例，因为 Action 中存在接收前端传入参数的实例变量。而 Spring MVC 中的 Controller 可以使用单例，前端传入的参数是通过方法的参数进行传入的，而不是实例变量。

当没有出现多个线程对同一个对象的同一个实例变量进行写操作时，为了减少内存的使用率，可以使用 singleton（单例）模式。在开发中 Service 类和 DAO 类都使用单例模式，因为这两个类中的实例变量在大多数情况下是只读的。

1.10 装配 Spring Bean

依赖注入会产生类与类之间的关联，产生类之间关联的行为叫作装配（wiring）。本节将使用多种方式来装配 JavaBean 之间的关联。

在 Spring 框架中，一个 JavaBean 不需要创建另一个 JavaBean，JavaBean 之间的关系全部由 IoC 容器进行管理，也就是在 A 类中虽然声明了 B 类的对象，但是不需要实例化 B 类，B 类的对象由容器进行创建，容器还能对 A 类中的 B 类的属性进行赋值，这一切都是由 Spring 框架的 IoC 容器来完成的。

把原来由 A 类实例化 B 类的方式改为由 IoC 容器来管理，这样就使 A 类和 B 类产生了松耦合，有利于代码的可扩展性与后期维护。

Spring 默认按类型（byType）进行匹配，然后实施注入。

1.10.1 使用注解法注入对象

创建测试项目 diTest。

创建实体类 Userinfo，代码如下：

```
package com.ghy.www.entity;

import org.springframework.stereotype.Component;

@Component
public class Userinfo {
    public Userinfo() {
        System.out.println("public Userinfo() " + this.hashCode());
    }
}
```

创建运行类 Test，代码如下：

```
package com.ghy.www.test;

import com.ghy.www.entity.Userinfo;
import org.springframework.beans.factory.annotation.Autowired;
import org.springframework.context.ApplicationContext;
import org.springframework.context.annotation.AnnotationConfigApplicationContext;
import org.springframework.stereotype.Component;
```

```
@Component
public class Test {
    public Test() {
        System.out.println("public Test() " + this.hashCode());
    }

    @Autowired
    private Userinfo userinfo;

    public Userinfo getUserinfo() {
        return userinfo;
    }

    public void setUserinfo(Userinfo userinfo) {
        this.userinfo = userinfo;
        System.out.println("public void setUserinfo(Userinfo userinfo) userinfo=" + userinfo);
    }

    public static void main(String[] args) {
        ApplicationContext context = new AnnotationConfigApplicationContext("com.ghy.www");
        Test test = (Test) context.getBean(Test.class);
        System.out.println("main test " + test.hashCode());
        System.out.println(test.getUserinfo().hashCode());
    }
}
```

程序运行结果如下：

```
public Userinfo() 1295226194
public Test() 1931444790
main test 1931444790
1295226194
```

方法 setUserinfo()并没有被执行，因为使用反射技术直接对 userinfo 变量进行字段赋值。如果将@Autowired 注解放在 public void setUserinfo(Userinfo userinfo)方法之上，则 setUserinfo()方法才会被调用。

1.10.2 多实现类的歧义性

如果对接口类型的变量进行注入，当 IoC 容器发现有多个实现类时，Spring 并不知道应该把哪个实现类对接口进行注入，从而出现 NoUniqueBeanDefinitionException 异常。

创建测试项目 diTestNO。

创建接口 IUserinfoService，代码如下：

```
package com.ghy.www.service;

public interface IUserinfoService {
    public void save();
}
```

创建实现类 UserinfoServiceA，代码如下：

```
package com.ghy.www.service;

import org.springframework.stereotype.Service;
```

```java
@Service
public class UserinfoServiceA implements IUserinfoService {
    public UserinfoServiceA() {
        System.out.println("public Userinfo() " + this.hashCode());
    }

    public void save() {
        System.out.println("将数据保存到A数据库中");
    }
}
```

创建实现类 UserinfoServiceB，代码如下：

```java
package com.ghy.www.service;

import org.springframework.stereotype.Service;

@Service
public class UserinfoServiceB implements IUserinfoService {
    public UserinfoServiceB() {
        System.out.println("public Userinfo() " + this.hashCode());
    }

    public void save() {
        System.out.println("将数据保存到B数据库中");
    }
}
```

创建运行类 Test，代码如下：

```java
package com.ghy.www.test;

import com.ghy.www.service.IUserinfoService;
import org.springframework.beans.factory.annotation.Autowired;
import org.springframework.context.ApplicationContext;
import org.springframework.context.annotation.AnnotationConfigApplicationContext;
import org.springframework.stereotype.Component;

@Component
public class Test {
    public Test() {
        System.out.println("public Test() " + this.hashCode());
    }

    @Autowired
    private IUserinfoService userinfoService;

    public IUserinfoService getUserinfoService() {
        return userinfoService;
    }

    public void setUserinfoService(IUserinfoService userinfoService) {
        this.userinfoService = userinfoService;
    }

    public static void main(String[] args) {
        ApplicationContext context = new AnnotationConfigApplicationContext("com.ghy.www");
        Test test = (Test) context.getBean(Test.class);
```

```
            test.getUserinfoService().save();
    }
}
```

程序运行后出现如下异常信息:

```
org.springframework.beans.factory.NoUniqueBeanDefinitionException: No qualifying bean
of type 'com.ghy.www.service.IUserinfoService' available: expected single matching bean but
found 2: userinfoServiceA,userinfoServiceB
```

上面的结果表明出现了 NoUniqueBeanDefinitionException 异常。因为 Spring 框架不能确定注入哪个实现类对象,所以出现异常,解决的方法有以下 3 种。

(1) 使用@Primary 注解。
(2) 使用@Autowired 注解结合@Qualifier 注解。
(3) 使用@Resource 注解。

1. 使用@Primary 注解

创建测试项目 diTestOK_1。
更改实现类 UserinfoServiceB,代码如下:

```
@Service
@Primary
public class UserinfoServiceB implements IUserinfoService {
```

注解@Primary 表示在遇到相同类型的注入时,当前的组件具有高优先级。

2. 使用@Autowired 注解结合@Qualifier 注解

不使用按类型方式,而是使用 JavaBean 的别名进行注入。
对@Service 组件设置一个别名,使用@Qualifier 注解注入指定别名的组件。
创建测试项目 diTestOK_2。
更改实现类 UserinfoServiceA 和实现类 UserinfoServiceB,代码如下:

```
@Service(value = "userinfoServiceA")
public class UserinfoServiceA implements IUserinfoService {


@Service(value = "userinfoServiceB")
public class UserinfoServiceB implements IUserinfoService {
```

对 JavaBean 设置了别名。
注入的代码如下:

```
@Autowired
@Qualifier(value = "userinfoServiceB")
private IUserinfoService userinfoService;
```

注意:@Qualifier 比@Primary 优先级高。

3. 使用@Resource 注解

javax.annotation.Resource 注解是由 Java EE 规范提供的,而不是 Spring 框架。
创建测试项目 diTestOK_3。

更改实现类 UserinfoServiceA 和实现类 UserinfoServiceB，代码如下：

```
@Service(value = "userinfoServiceA")
public class UserinfoServiceA implements IUserinfoService {

@Service(value = "userinfoServiceB")
public class UserinfoServiceB implements IUserinfoService {
```

运行类代码更改如下：

```
@Resource(name = "userinfoServiceB")
private IUserinfoService userinfoService;
```

注解@Resource 使用 name 属性时，以 byName 的方式进行注入。
注解@Resource 不使用 name 属性时，以 byType 的方式进行注入，示例代码如下：

```
@Resource
private IUserinfoService userinfoService;
```

上面代码的执行结果与下列代码一致，注解@Autowired 就是以 byType 的方式进行注入的：

```
@Autowired
private IUserinfoService userinfoService;
```

1.10.3 使用@Autowired 注解向构造方法的参数进行注入

创建测试项目 autowired_constructor。

创建实体类 Userinfo，代码如下：

```
package com.ghy.www.entity;

import org.springframework.stereotype.Component;

@Component
public class Userinfo {
    public Userinfo() {
        System.out.println("public Userinfo() " + this.hashCode());
    }
}
```

创建实体类 Bookinfo，代码如下：

```
package com.ghy.www.entity;

import org.springframework.beans.factory.annotation.Autowired;
import org.springframework.stereotype.Component;

@Component
public class Bookinfo {
    @Autowired
    public Bookinfo(Userinfo userinfo) {
        System.out.println("public Bookinfo(Userinfo userinfo) userinfo=" + userinfo.hashCode());
    }
}
```

创建配置类 SpringConfig，代码如下：

```java
package com.ghy.www.config;

import org.springframework.context.annotation.ComponentScan;
import org.springframework.context.annotation.Configuration;

@Configuration
@ComponentScan(basePackages = "com.ghy.www.entity")
public class SpringConfig {
}
```

创建运行类 Test，代码如下：

```java
package com.ghy.www.test;

import com.ghy.www.config.SpringConfig;
import org.springframework.context.annotation.AnnotationConfigApplicationContext;

public class Test {
    public static void main(String[] args) {
        new AnnotationConfigApplicationContext(SpringConfig.class);
    }
}
```

程序运行结果如下：

```
public Userinfo() 366590980
public Bookinfo(Userinfo userinfo) userinfo=366590980
```

如果类仅有一个有参构造方法，则可以省略@Autowired 注解，因为 Spring 会自动进行注入。

1.10.4　使用@Autowired 注解向方法的参数进行注入

我们可以使用@Autowired 注解向 setter 方法的参数进行注入。

创建测试项目 autowired_setter。

创建实体类 Bookinfo，代码如下：

```java
package com.ghy.www.entity;

import org.springframework.beans.factory.annotation.Autowired;
import org.springframework.stereotype.Component;

@Component
public class Bookinfo {
    @Autowired
    public void setUserinfo(Userinfo userinfo) {
        System.out.println("public void setUserinfo(Userinfo userinfo) userinfo=" + userinfo.hashCode());
    }
}
```

1.10.5　使用@Autowired 注解向字段进行注入

创建测试项目 autowired_field。

创建实体类 Bookinfo，代码如下：

```
package com.ghy.www.entity;

import org.springframework.beans.factory.annotation.Autowired;
import org.springframework.stereotype.Component;

@Component
public class Bookinfo {
    @Autowired
    private Userinfo userinfo;

    public Userinfo getUserinfo() {
        return userinfo;
    }

    public void setUserinfo(Userinfo userinfo) {
        this.userinfo = userinfo;
    }
}
```

创建运行类 Test，代码如下：

```
package com.ghy.www.test;

import com.ghy.www.config.SpringConfig;
import com.ghy.www.entity.Bookinfo;
import org.springframework.context.ApplicationContext;
import org.springframework.context.annotation.AnnotationConfigApplicationContext;

public class Test {
    public static void main(String[] args) {
        ApplicationContext context = new AnnotationConfigApplicationContext(SpringConfig.class);
        Bookinfo bookinfo = (Bookinfo) context.getBean(Bookinfo.class);
        System.out.println(bookinfo.getUserinfo().hashCode());
    }
}
```

程序运行结果如下：

```
public Userinfo() 6320204
6320204
```

通过前面的测试可知，使用注解@Autowired 可以向字段、方法的参数以及构造方法的参数进行注入。

1.10.6　使用@Inject 注解向字段、方法和构造方法进行注入

@Inject 注解可以实现注入，但需要 javaee-api-8.0.jar 文件，因为@Inject 注解是 Oracle 提供的注解，而@Autowired 是 Spring 官方提供的注解。

使用@Inject 注解向构造方法注入的测试项目为 inject_test1，代码如下：

```
package com.ghy.www.entity;

import org.springframework.beans.factory.annotation.Autowired;
import org.springframework.stereotype.Component;
```

```
import javax.inject.Inject;

@Component
public class Bookinfo {
    @Inject
    public Bookinfo(Userinfo userinfo) {
        System.out.println("public Bookinfo(Userinfo userinfo) userinfo=" + userinfo.hashCode());
    }
}
```

使用@Inject 注解向方法注入的测试项目为 inject_test2，核心代码如下：

```
package com.ghy.www.entity;

import org.springframework.stereotype.Component;

import javax.inject.Inject;

@Component
public class Bookinfo {
    @Inject
    public void setUserinfo(Userinfo userinfo) {
        System.out.println("public void setUserinfo(Userinfo userinfo) userinfo=" + userinfo.hashCode());
    }
}
```

使用@Inject 注解向字段注入的测试项目为 inject_test3，核心代码如下：

```
package com.ghy.www.entity;

import org.springframework.stereotype.Component;

import javax.inject.Inject;

@Component
public class Bookinfo {
    @Inject
    private Userinfo userinfo;

    public Userinfo getUserinfo() {
        return userinfo;
    }

    public void setUserinfo(Userinfo userinfo) {
        this.userinfo = userinfo;
    }
}
```

向接口进行注入：

```
@Inject
private IUserinfoService userinfoService;
```

在有多个实现类时，也会出现 NoUniqueBeanDefinitionException 异常：

```
org.springframework.beans.factory.NoUniqueBeanDefinitionException: No qualifying bean of
type 'com.ghy.www.service.IUserinfoService' available: expected single matching bean but found
2: userinfoServiceA,userinfoServiceB
```

以上的异常结果所涉及的测试代码在项目 inject_test4 中。

为了处理此异常，我们可以使用如下示例代码注入指定别名的 JavaBean 对象：

```
@Inject
@Qualifier(value = "userinfoServiceB")
private IUserinfoService userinfoService;
```

以上测试代码在项目 inject_test5 中。

1.10.7 使用@Bean 注解向工厂方法的参数进行注入

创建测试项目 di-factorymethod。

创建实体类 Userinfo，代码如下：

```
package com.ghy.www.entity;

import org.springframework.stereotype.Component;

@Component
public class Userinfo {
    public Userinfo() {
        System.out.println("public Userinfo() " + this.hashCode());
    }
}
```

创建实体类 Bookinfo，代码如下：

```
package com.ghy.www.entity;

public class Bookinfo {
    public Bookinfo(Userinfo userinfo) {
        System.out.println("public Bookinfo(Userinfo userinfo) userinfo=" + userinfo.hashCode());
    }
}
```

创建配置类 SpringConfig，代码如下：

```
package com.ghy.www.config;

import com.ghy.www.entity.Bookinfo;
import com.ghy.www.entity.Userinfo;
import org.springframework.context.annotation.Bean;
import org.springframework.context.annotation.ComponentScan;
import org.springframework.context.annotation.Configuration;

@Configuration
@ComponentScan(basePackages = "com.ghy.www.entity")
public class SpringConfig {
    @Bean
    public Bookinfo getBookinfo(Userinfo userinfo) {
        System.out.println("public Bookinfo getBookinfo(Userinfo userinfo) userinfo=" +
```

```
userinfo.hashCode());
        return new Bookinfo(userinfo);
    }
}
```

创建运行类 Test，代码如下：

```
package com.ghy.www.test;

import com.ghy.www.config.SpringConfig;
import org.springframework.context.annotation.AnnotationConfigApplicationContext;

public class Test {
    public static void main(String[] args) {
        new AnnotationConfigApplicationContext(SpringConfig.class);
    }
}
```

程序运行结果如下：

```
public Userinfo() 1571967156
public Bookinfo getBookinfo(Userinfo userinfo) userinfo=1571967156
public Bookinfo(Userinfo userinfo) userinfo=1571967156
```

1.10.8　使用@Autowired(required = false)的写法

在注入时，如果找不到符合条件的 JavaBean 对象，控制台会提示异常：

org.springframework.beans.factory.NoSuchBeanDefinitionException: No qualifying bean of type 'com.ghy.www.entity.Userinfo' available: expected at least 1 bean which qualifies as autowire candidate. Dependency annotations: {@org.springframework.beans.factory.annotation.Autowired(required=true)}

创建测试项目 diTest-required1。

创建配置类 SpringConfig，代码如下：

```
package com.ghy.www.config;

import org.springframework.context.annotation.ComponentScan;
import org.springframework.context.annotation.Configuration;

@Configuration
@ComponentScan("com.ghy.www.test")
public class SpringConfig {
}
```

创建实体类 Userinfo，代码如下：

```
package com.ghy.www.entity;

import org.springframework.stereotype.Component;

@Component
public class Userinfo {
    public Userinfo() {
        System.out.println("public Userinfo() " + this.hashCode());
    }
}
```

创建运行类 Test，代码如下：

```
package com.ghy.www.test;

import com.ghy.www.config.SpringConfig;
import com.ghy.www.entity.Userinfo;
import org.springframework.beans.factory.annotation.Autowired;
import org.springframework.context.annotation.AnnotationConfigApplicationContext;
import org.springframework.stereotype.Component;

@Component
public class Test {
    @Autowired
    private Userinfo userinfo;

    public Userinfo getUserinfo() {
        return userinfo;
    }

    public void setUserinfo(Userinfo userinfo) {
        this.userinfo = userinfo;
    }

    public static void main(String[] args) {
        new AnnotationConfigApplicationContext(SpringConfig.class);
    }
}
```

程序运行后出现如下异常信息：

org.springframework.beans.factory.NoSuchBeanDefinitionException: No qualifying bean of type 'com.ghy.www.entity.Userinfo' available: expected at least 1 bean which qualifies as autowire candidate. Dependency annotations: {@org.springframework.beans.factory.annotation.Autowired(required=true)}

该异常信息提示我们没有找到符合条件的 JavaBean 对象。为了避免出现异常，我们可以加入 required = false 属性，代表注入的 userinfo 对象并不是必需的，如果没有对象就不进行注入。创建测试项目 diTest-required2，更改后的代码如下：

```
@Autowired(required = false)
private Userinfo userinfo;
```

程序运行后并没有出现异常。由于没有找到符合条件的记录，因此 userinfo 对象的值为 null。

以下两种写法使用@Autowired(required = false)是无效的，依然会出现 NoSuchBeanDefinitionException 异常，说明注入的对象必须存在。

写法 1 代码如下：

```
package com.ghy.www.entity;

import org.springframework.beans.factory.annotation.Autowired;
import org.springframework.stereotype.Component;

@Component
public class Bookinfo {
    @Autowired(required = false)
    public Bookinfo(Userinfo userinfo) {
        System.out.println("public Bookinfo(Userinfo userinfo) userinfo=" + userinfo.
```

```
        hashCode());
    }
}
```

测试示例在项目 diTest-required3 中。

写法 2 代码如下：

```
package com.ghy.www.config;

import com.ghy.www.entity.Bookinfo;
import com.ghy.www.entity.Userinfo;
import org.springframework.beans.factory.annotation.Autowired;
import org.springframework.context.annotation.Bean;
import org.springframework.context.annotation.ComponentScan;
import org.springframework.context.annotation.Configuration;

@Configuration
@ComponentScan(basePackages = "com.ghy.www.entity")
public class SpringConfig {
    @Bean
    @Autowired(required = false)
    public Bookinfo getBookinfo(Userinfo userinfo) {
        System.out.println("public Bookinfo getBookinfo(Userinfo userinfo) userinfo=" + userinfo.hashCode());
        return new Bookinfo(userinfo);
    }
}
```

测试示例在项目 diTest-required4 中。

1.10.9　使用@Bean 对 JavaBean 的 id 重命名

默认时，使用@Bean 创建的 JavaBean 的 id 就是方法的名称，实际上 id 也可以自定义。

创建测试项目 di-id-is-methodname。

示例代码如下：

```
package com.ghy.www.config;

import org.springframework.context.annotation.Bean;
import org.springframework.context.annotation.ComponentScan;
import org.springframework.context.annotation.Configuration;

import java.util.Date;

@Configuration
@ComponentScan(basePackages = "com.ghy.www.test")
public class SpringConfig {
    @Bean(name = "xxxxxxxxxxxxxxxxxxxxx")
    public Date createDate() {
        Date date = new Date();
        System.out.println("public Date createDate() xxxxxxxxxxxxxxxxxxxxx=" + date.hashCode());
        return date;
    }

    @Bean
```

```
    public Date getDate() {
        Date date = new Date();
        System.out.println("public Date getDate() zzzzzzzzzzzzzzzzzzzzz=" + date.hashCode());
        return date;
    }
}
```

创建运行类 Test，代码如下：

```
package com.ghy.www.test;

import com.ghy.www.config.SpringConfig;
import org.springframework.context.ApplicationContext;
import org.springframework.context.annotation.AnnotationConfigApplicationContext;
import org.springframework.stereotype.Component;

import javax.annotation.Resource;
import java.util.Date;

@Component
public class Test {
    @Resource(name = "xxxxxxxxxxxxxxxxxxxxxxx")
    private Date date1;
    @Resource(name = "getDate")
    private Date date2;

    public Date getDate1() {
        return date1;
    }

    public void setDate1(Date date1) {
        this.date1 = date1;
    }

    public Date getDate2() {
        return date2;
    }

    public void setDate2(Date date2) {
        this.date2 = date2;
    }

    public static void main(String[] args) {
        ApplicationContext context = new AnnotationConfigApplicationContext(SpringConfig.class);
        Test test1 = (Test) context.getBean(Test.class);
        System.out.println("main date1=" + test1.getDate1().hashCode());
        System.out.println("main date2=" + test1.getDate2().hashCode());
    }
}
```

程序运行结果如下：

```
public Date createDate() xxxxxxxxxxxxxxxxxxxxxxx=-885661621
public Date getDate() zzzzzzzzzzzzzzzzzzzzz=-885661642
main date1=-885661621
main date2=-885661642
```

1.10.10 Spring 上下文的相关知识

Spring 的上下文可以理解为 Spring 运行的环境，可以创建多个 Spring 上下文。在默认的情况下，不同上下文中的 JavaBean 对象不可以共享。

1. 创建多个 Spring 上下文

创建多个 Spring 上下文就是创建多个 ApplicationContext 对象。

创建测试项目 context-test1。

创建实体类 Userinfo，代码如下：

```java
package com.ghy.www.entity;

public class Userinfo {
    public Userinfo() {
        System.out.println("public Userinfo() " + this.hashCode());
    }
}
```

创建实体类 Bookinfo，代码如下：

```java
package com.ghy.www.entity;

public class Bookinfo {
    public Bookinfo() {
        System.out.println("public Bookinfo() " + this.hashCode());
    }
}
```

创建配置类 SpringConfig1，代码如下：

```java
package com.ghy.www.config;

import com.ghy.www.entity.Bookinfo;
import org.springframework.context.annotation.Bean;
import org.springframework.context.annotation.Configuration;

@Configuration
public class SpringConfig1 {
    @Bean
    public Bookinfo getBookinfo() {
        Bookinfo bookinfo = new Bookinfo();
        System.out.println("getBookinfo bookinfo=" + bookinfo.hashCode());
        return bookinfo;
    }
}
```

创建配置类 SpringConfig2，代码如下：

```java
package com.ghy.www.config;

import com.ghy.www.entity.Userinfo;
import org.springframework.context.annotation.Bean;
import org.springframework.context.annotation.Configuration;

@Configuration
public class SpringConfig2 {
```

```
    @Bean
    public Userinfo getUserinfo() {
        Userinfo userinfo = new Userinfo();
        System.out.println("getUserinfo userinfo=" + userinfo.hashCode());
        return userinfo;
    }
}
```

创建运行类 Test1，代码如下：

```
package com.ghy.www.test;

import com.ghy.www.config.SpringConfig1;
import com.ghy.www.config.SpringConfig2;
import com.ghy.www.entity.Bookinfo;
import com.ghy.www.entity.Userinfo;
import org.springframework.context.ApplicationContext;
import org.springframework.context.annotation.AnnotationConfigApplicationContext;

public class Test1 {
    public static void main(String[] args) {
        ApplicationContext context1 = new AnnotationConfigApplicationContext(SpringConfig1.class);
        ApplicationContext context2 = new AnnotationConfigApplicationContext(SpringConfig2.class);

        System.out.println("context1=" + context1.getBean(Bookinfo.class).hashCode());
        System.out.println("context2=" + context2.getBean(Userinfo.class).hashCode());
    }
}
```

程序运行结果如下：

```
public Bookinfo() 1131040331
getBookinfo bookinfo=1131040331
public Userinfo() 932285561
getUserinfo userinfo=932285561
context1=1131040331
context2=932285561
```

通过上面的程序，我们成功创建了不同的 Spring 上下文，在自己的上下文中可以获取自己的 JavaBean。

2. 不同 Spring 上下文中的 JavaBean 对象是不共享的

创建运行类 Test2，代码如下：

```
package com.ghy.www.test;

import com.ghy.www.config.SpringConfig1;
import com.ghy.www.config.SpringConfig2;
import com.ghy.www.entity.Bookinfo;
import com.ghy.www.entity.Userinfo;
import org.springframework.context.ApplicationContext;
import org.springframework.context.annotation.AnnotationConfigApplicationContext;

public class Test2 {
    public static void main(String[] args) {
        ApplicationContext context1 = new AnnotationConfigApplicationContext(SpringConfig1.class);
```

```
            ApplicationContext context2 = new AnnotationConfigApplicationContext(Spring
Config2.class);

            System.out.println("context1=" + context1.getBean(Bookinfo.class).hashCode());
            context1.getBean(Userinfo.class);
        }
    }
```

程序运行后出现如下异常信息:

```
public Bookinfo() 1131040331
getBookinfo bookinfo=1131040331
public Userinfo() 932285561
getUserinfo userinfo=932285561
context1=1131040331
Exception in thread "main" org.springframework.beans.factory.NoSuchBeanDefinitionExce
ption: No qualifying bean of type 'com.ghy.www.entity.Userinfo' available
        at org.springframework.beans.factory.support.DefaultListableBeanFactory.getBean
(DefaultListableBeanFactory.java:351)
        at org.springframework.beans.factory.support.DefaultListableBeanFactory.getBean
(DefaultListableBeanFactory.java:342)
        at org.springframework.context.support.AbstractApplicationContext.getBean(Abstract
ApplicationContext.java:1127)
        at com.ghy.www.test.Test2.main(Test2.java:16)
```

该异常信息提示 Userinfo 对象没有被找到。

3. 让多个配置类互相通信

那么,如何使不同上下文中的 JavaBean 可以互相共享呢?

创建测试项目 context-test2。

创建配置类 SpringConfig1,代码如下:

```
package com.ghy.www.config;

import com.ghy.www.entity.Bookinfo;
import org.springframework.context.annotation.Bean;
import org.springframework.context.annotation.Configuration;
import org.springframework.context.annotation.Import;

@Configuration
@Import(SpringConfig2.class)
public class SpringConfig1 {
    @Bean
    public Bookinfo getBookinfo() {
        Bookinfo bookinfo = new Bookinfo();
        System.out.println("getBookinfo bookinfo=" + bookinfo.hashCode());
        return bookinfo;
    }
}
```

我们可以使用如下写法的注解导入其他配置类:

```
@Import(SpringConfig2.class)
```

还可以使用如下写法的注解一次性导入多个配置类,代码如下:

```
@Import({SpringConfig100.class, SpringConfig200.class})
```

创建运行类 Test1，代码如下：

```java
package com.ghy.www.test;

import com.ghy.www.config.SpringConfig1;
import com.ghy.www.entity.Bookinfo;
import com.ghy.www.entity.Userinfo;
import org.springframework.context.ApplicationContext;
import org.springframework.context.annotation.AnnotationConfigApplicationContext;

public class Test1 {
    public static void main(String[] args) {
        ApplicationContext context1 = new AnnotationConfigApplicationContext(SpringConfig1.class);
        System.out.println("context1=" + context1.getBean(Bookinfo.class).hashCode());
        System.out.println("context2=" + context1.getBean(Userinfo.class).hashCode());
    }
}
```

程序运行结果如下：

```
public Userinfo() 1043351526
getUserinfo userinfo=1043351526
public Bookinfo() 937773018
getBookinfo bookinfo=937773018
context1=937773018
context2=1043351526
```

4. JavaBean 的 id 值相同时出现 ConflictingBeanDefinitionException 异常

如果多个上下文中的 JavaBean 的 id 值相同，则会出现 ConflictingBeanDefinitionException 异常。

创建测试项目 context-test3。

创建工厂类 SpringConfig1，代码如下：

```java
package com.ghy.www.config;

import org.springframework.context.annotation.Bean;
import org.springframework.context.annotation.ComponentScan;
import org.springframework.context.annotation.Configuration;
import org.springframework.context.annotation.Import;

@Configuration
@ComponentScan("com.ghy.www.entity1")
@Import(SpringConfig2.class)
public class SpringConfig1 {
}
```

创建工厂类 SpringConfig2，代码如下：

```java
package com.ghy.www.config;

import org.springframework.context.annotation.ComponentScan;
import org.springframework.context.annotation.Configuration;

@Configuration
@ComponentScan("com.ghy.www.entity2")
public class SpringConfig2 {
}
```

创建实体类 Userinfo，代码如下：

```java
package com.ghy.www.entity1;

import org.springframework.stereotype.Component;

@Component("mybean")
public class Userinfo {
    public Userinfo() {
        System.out.println("public Userinfo() " + this.hashCode());
    }
}
```

创建实体类 Bookinfo，代码如下：

```java
package com.ghy.www.entity2;

import org.springframework.stereotype.Component;

@Component("mybean")
public class Bookinfo {
    public Bookinfo() {
        System.out.println("public Bookinfo() " + this.hashCode());
    }
}
```

创建运行类 Test1，代码如下：

```java
package com.ghy.www.test;

import com.ghy.www.config.SpringConfig1;
import org.springframework.context.annotation.AnnotationConfigApplicationContext;

public class Test1 {
    public static void main(String[] args) {
        new AnnotationConfigApplicationContext(SpringConfig1.class);
    }
}
```

程序运行结果如下：

```
Exception in thread "main" org.springframework.beans.factory.BeanDefinitionStoreException: Failed to process import candidates for configuration class [com.ghy.www.config.SpringConfig1]; nested exception is org.springframework.context.annotation.ConflictingBeanDefinitionException: Annotation-specified bean name 'mybean' for bean class [com.ghy.www.entity2.Bookinfo] conflicts with existing, non-compatible bean definition of same name and class [com.ghy.www.entity1.Userinfo]
```

5. 不同上下文中的工厂方法的名称不可以相同

> **注意**：在不同配置类中创建 JavaBean 工厂方法时的名称不能一样，否则会出现"Overriding bean definition"问题。原因是在默认情况下，JavaBean 的 id 和方法名称一样，如果多个工厂方法的名称一样，则不会创建其他的 JavaBean，导致出现找不到对象的诡异问题。

创建测试项目 context-test4。

创建工厂类 SpringConfig1，代码如下：

```java
package com.ghy.www.config;

import com.ghy.www.entity.Bookinfo;
```

```java
import org.springframework.context.annotation.Bean;
import org.springframework.context.annotation.Configuration;
import org.springframework.context.annotation.Import;

@Configuration
@Import(SpringConfig2.class)
public class SpringConfig1 {
    @Bean
    public Bookinfo beanIdSame() {
        Bookinfo bookinfo = new Bookinfo();
        System.out.println("getBookinfo bookinfo=" + bookinfo.hashCode());
        return bookinfo;
    }
}
```

创建工厂类 SpringConfig2，代码如下：

```java
package com.ghy.www.config;

import com.ghy.www.entity.Userinfo;
import org.springframework.context.annotation.Bean;
import org.springframework.context.annotation.Configuration;

@Configuration
public class SpringConfig2 {
    @Bean
    public Userinfo beanIdSame() {
        Userinfo userinfo = new Userinfo();
        System.out.println("getUserinfo userinfo=" + userinfo.hashCode());
        return userinfo;
    }
}
```

创建实体类 Userinfo，代码如下：

```java
package com.ghy.www.entity1;

import org.springframework.stereotype.Component;

public class Userinfo {
    public Userinfo() {
        System.out.println("public Userinfo() " + this.hashCode());
    }
}
```

创建实体类 Bookinfo，代码如下：

```java
package com.ghy.www.entity2;

import org.springframework.stereotype.Component;

public class Bookinfo {
    public Bookinfo() {
        System.out.println("public Bookinfo() " + this.hashCode());
    }
}
```

创建运行类 Test1，代码如下：

```java
package com.ghy.www.test;

import com.ghy.www.config.SpringConfig1;
```

```java
import com.ghy.www.entity.Bookinfo;
import com.ghy.www.entity.Userinfo;
import org.springframework.context.ApplicationContext;
import org.springframework.context.annotation.AnnotationConfigApplicationContext;

public class Test1 {
    public static void main(String[] args) {
        ApplicationContext context1 = new AnnotationConfigApplicationContext(SpringConfig1.class);
        System.out.println("context1=" + context1.getBean(Bookinfo.class).hashCode());
        System.out.println("context2=" + context1.getBean(Userinfo.class).hashCode());
    }
}
```

程序运行结果如下：

```
[main] DEBUG org.springframework.beans.factory.support.DefaultListableBeanFactory - Overriding bean definition for bean 'beanIdSame' with a different definition: replacing [Root bean: class [null]; scope=; abstract=false; lazyInit=null; autowireMode=3; dependencyCheck=0; autowireCandidate=true; primary=false; factoryBeanName=com.ghy.www.config.SpringConfig2; factoryMethodName=beanIdSame; initMethodName=null; destroyMethodName=(inferred); defined in com.ghy.www.config.SpringConfig2] with [Root bean: class [null]; scope=; abstract=false; lazyInit=null; autowireMode=3; dependencyCheck=0; autowireCandidate=true; primary=false; factoryBeanName=springConfig1; factoryMethodName=beanIdSame; initMethodName=null; destroyMethodName=(inferred); defined in com.ghy.www.config.SpringConfig1]
    public Bookinfo() 330084561
    getBookinfo bookinfo=330084561
context1=330084561
Exception in thread "main" org.springframework.beans.factory.NoSuchBeanDefinitionException: No qualifying bean of type 'com.ghy.www.entity.Userinfo' available
        at org.springframework.beans.factory.support.DefaultListableBeanFactory.getBean(DefaultListableBeanFactory.java:351)
        at org.springframework.beans.factory.support.DefaultListableBeanFactory.getBean(DefaultListableBeanFactory.java:342)
        at org.springframework.context.support.AbstractApplicationContext.getBean(AbstractApplicationContext.java:1127)
        at com.ghy.www.test.Test1.main(Test1.java:13)
```

处理这种异常的方法是设置工厂方法的名称时不要重名。

创建测试项目 context-test5。

更改配置类 SpringConfig1，代码如下：

```java
@Configuration
@Import(SpringConfig2.class)
public class SpringConfig1 {
    @Bean
    public Bookinfo createBookinfo() {
```

更改配置类 SpringConfig2，代码如下：

```java
@Configuration
public class SpringConfig2 {
    @Bean
    public Userinfo createUserinfo() {
```

6. 创建 AllConfig 全局配置类

对于多个 Spring 上下文共享 JavaBean 的写法，比较好的代码组织方式是创建一个全局配置类，然后在类的上方使用注解 @Import 导入其他配置类。

创建测试项目 context-test6。

创建全局配置类 AllConfig，代码如下：

```
package com.ghy.www.config;

import org.springframework.context.annotation.Configuration;
import org.springframework.context.annotation.Import;

@Configuration
@Import(value = { SpringConfig1.class, SpringConfig2.class })
public class AllConfig {

}
```

1.10.11　BeanFactory 与 ApplicationContext

包 org.springframework.beans 和 org.springframework.context 是 Spring 框架 IoC 容器的基础。

BeanFactory 接口提供了一种能够管理任何类型对象的高级配置机制，ApplicationContext 是 BeanFactory 的一个子接口，它使得与 Spring 的 AOP 功能更容易集成，支持消息资源处理（用于国际化）、事件发布和应用程序层特定的上下文。

其实 Spring 的 IoC 容器就是一个实现了 BeanFactory 接口的实现类，因为 ApplicationContext 接口继承自 BeanFactory 接口。通过工厂模式取得 JavaBean 对象。

下面我们来看一下 Spring 的 API DOC，如图 1-10 所示。

图 1-10　BeanFactory 接口结构

从图 1-10 中可以发现，ApplicationContext 是 BeanFactory 的子接口，BeanFactory 接口提供了最基本的对象管理功能，而子接口 ApplicationContext 提供了更多附加的功能，如与 Web 整合、支持国际化、事件发布和通知等。

1.10.12　使用注解 @Value 进行注入

如果把其他对象中的属性值作为注入的来源，就使用 #{}。如果把 *.properties 属性文件中 key 对应的 value 作为注入的来源，就使用 ${}。

创建测试项目 valueTest。

创建实体类 Userinfo，代码如下：

```
package com.ghy.www.entity;

public class Userinfo {
    public Userinfo() {
```

```
            System.out.println("public Userinfo() " + this.hashCode());
        }
    }
```

创建属性文件 db.properties，代码如下：

```
dbdriver=oracleDriver
dburl=oracleURL
dbusername=oracleUsername
dbpassword=oraclePassword
```

创建业务类 UserinfoService，代码如下：

```
package com.ghy.www.service;

import com.ghy.www.entity.Userinfo;
import org.springframework.beans.factory.annotation.Value;
import org.springframework.stereotype.Service;

@Service(value = "userinfoService")
public class UserinfoService {

    @Value("UserinfoService 常<量>值&")
    private String username;

    private Userinfo userinfo = new Userinfo();

    public String getUsername() {
        return username;
    }

    public void setUsername(String username) {
        this.username = username;
    }

    public Userinfo getUserinfo() {
        return userinfo;
    }

    public void setUserinfo(Userinfo userinfo) {
        this.userinfo = userinfo;
    }

}
```

创建运行类 Test，代码如下：

```
package com.ghy.www.test;

import com.ghy.www.entity.Userinfo;
import org.springframework.beans.factory.annotation.Value;
import org.springframework.context.ApplicationContext;
import org.springframework.context.annotation.AnnotationConfigApplicationContext;
import org.springframework.context.annotation.ComponentScan;
import org.springframework.context.annotation.Configuration;
import org.springframework.context.annotation.PropertySource;
import org.springframework.stereotype.Component;

@Component
@ComponentScan(basePackages = {"com.ghy.www.test", "com.ghy.www.service"})
```

```
@PropertySource(value = {"classpath:db.properties"})
@Configuration
public class Test {

    @Value("#{userinfoService.userinfo}")
    public Userinfo userinfo;

    @Value("#{userinfoService.username}")
    public String injectStringValue;

    @Value("${dbdriver}")
    public String a;
    @Value("${dburl}")
    public String b;
    @Value("${dbusername}")
    public String c;
    @Value("${dbpassword}")
    public String d;

    public static void main(String[] args) {
        ApplicationContext context = new AnnotationConfigApplicationContext(Test.class);
        Test test = context.getBean(Test.class);
        System.out.println("userinfo=" + test.userinfo.hashCode());
        System.out.println("injectStringValue=" + test.injectStringValue);
        System.out.println("a=" + test.a);
        System.out.println("b=" + test.b);
        System.out.println("c=" + test.c);
        System.out.println("d=" + test.d);
    }
}
```

程序运行结果如下:

```
public Userinfo() 370440646
userinfo=370440646
injectStringValue=UserinfoService 常<量>值&
a=oracleDriver
b=oracleURL
c=oracleUsername
d=oraclePassword
```

1.10.13 解决 BeanCurrentlyInCreationException 异常问题

当 A 类依赖 B 类，而 B 类又依赖 A 类时，就会有在 Spring 中被称为"循环依赖"的情况，循环依赖会产生 BeanCurrentlyInCreationException 异常。

创建测试项目 bean-currently-in-creation-exception。

创建类 A，代码如下:

```
package com.ghy.www.entity;

public class A {

    private B b;

    public A() {
    }
```

```java
    public A(B b) {
        super();
        this.b = b;
    }

    public B getB() {
        return b;
    }

    public void setB(B b) {
        this.b = b;
    }

}
```

创建类 B，代码如下：

```java
package com.ghy.www.entity;

public class B {
    private A a;

    public B() {
    }

    public B(A a) {
        super();
        this.a = a;
    }

    public A getA() {
        return a;
    }

    public void setA(A a) {
        this.a = a;
    }

}
```

创建配置类 SpringConfig，代码如下：

```java
package com.ghy.www.config;

import com.ghy.www.entity.A;
import com.ghy.www.entity.B;
import org.springframework.context.annotation.Bean;
import org.springframework.context.annotation.Configuration;

@Configuration
public class SpringConfig {
    @Bean
    public A getA(B b) {
        return new A(b);
    }

    @Bean
    public B getB(A a) {
```

```
        return new B(a);
    }
}
```

创建运行类 Test，代码如下：

```
package com.ghy.www.test;

import com.ghy.www.config.SpringConfig;
import org.springframework.context.annotation.AnnotationConfigApplicationContext;

public class Test {
    public static void main(String[] args) {
        new AnnotationConfigApplicationContext(SpringConfig.class);
    }
}
```

程序运行后出现异常信息如下：

```
Exception in thread "main" org.springframework.beans.factory.UnsatisfiedDependency
Exception: Error creating bean with name 'getA' defined in com.ghy.www.config.SpringConfig:
Unsatisfied dependency expressed through method 'getA' parameter 0; nested exception is org.
springframework.beans.factory.UnsatisfiedDependencyException: Error creating bean with name
'getB' defined in com.ghy.www.config.SpringConfig: Unsatisfied dependency expressed through
method 'getB' parameter 0; nested exception is org.springframework.beans.factory.BeanCurrently
InCreationException: Error creating bean with name 'getA': Requested bean is currently in
creation: Is there an unresolvable circular reference?
```

循环依赖产生的原因是使用了有参工厂方法：

```
@Bean
public A getA(B b) {
    return new A(b);
}

@Bean
public B getB(A a) {
    return new B(a);
}
```

改用 setter 方法实现注入就可以解决循环依赖的问题。

创建 bean-currently-in-creation-exception-ok 项目。

创建类 A，代码如下：

```
package com.ghy.www.entity;

import org.springframework.beans.factory.annotation.Autowired;
import org.springframework.stereotype.Component;

@Component
public class A {

    private B b;

    public B getB() {
        return b;
    }

    @Autowired
```

```java
    public void setB(B b) {
        this.b = b;
    }

}
```

创建类 B，代码如下：

```java
package com.ghy.www.entity;

import org.springframework.beans.factory.annotation.Autowired;
import org.springframework.stereotype.Component;

@Component
public class B {
    private A a;

    public A getA() {
        return a;
    }

    @Autowired
    public void setA(A a) {
        this.a = a;
    }

}
```

创建运行类 Test，代码如下：

```java
package com.ghy.www.test;

import com.ghy.www.entity.A;
import com.ghy.www.entity.B;
import org.springframework.context.ApplicationContext;
import org.springframework.context.annotation.AnnotationConfigApplicationContext;

public class Test {
    public static void main(String[] args) {
        ApplicationContext context = new AnnotationConfigApplicationContext("com.ghy.www");
        System.out.println(context.getBean(A.class));
        System.out.println(context.getBean(B.class));
    }
}
```

程序运行结果如下：

```
com.ghy.www.entity.A@77e4c80f
com.ghy.www.entity.B@35fc6dc4
```

这样，循环依赖问题就解决了。

第 2 章　Spring 5 核心技术之 AOP

本章目标
（1）什么是 AOP
（2）代理设计模式与 AOP 的关系
（3）实现面向切面编程

2.1　AOP

在以往的程序设计中，当处理记录日志这样的功能时，需要将记录日志相关的代码嵌入业务代码中，这样就造成了功能代码之间严重的紧耦合，不利于代码的后期维护，这种情况可以使用面向切面编程来进行分离。

面向切面编程（Aspect Oriented Programming，AOP）给程序员最直观的感受是它可以在不改动代码的基础上做功能上的增强，比如在原始代码的基础上加入数据库事务处理和记录日志的功能，这种增强的方式是通过代理（proxy）的原理实现的。

那么代理具体是什么呢？我们先来看看生活中的例子，比如你在当当网买书，当当网只需要将书放在邮递包中，并且写上你的地址，快递公司会将书送到你的手中。在这个过程中，快递公司就相当于当当网传书服务的代理。当当网并不负责运送书，运送书的任务由快递公司进行代理，这就是代理模式的一种形象的解释。如果想对某一个类进行功能上的增强而又不改变原始代码，那么只需要将这个类让代理类进行增强，至于增强什么，以及在哪方面增强，由代理类决定。

2.2　AOP 原理之代理设计模式

在学习 AOP 技术之前，我们一定要先了解代理设计模式。Spring 的 AOP 技术的原理就是基于代理设计模式的。下面我们来看一下代理设计模式。

代理设计模式为对象提供一种代理以控制对另一个对象的访问。在某些情况下，一个对象不适合或者不能直接引用另一个对象，而代理可以在两者之间起到中介的作用。

类比到房东—中介—租房者，租房者联系不到房东，必须要通过中介，中介就是代理，帮租房者寻找房子。

代理设计模式可以不改变原始代码的基础上对功能进行增强,使原始对象中的代码与增强代码进行充分解耦。代理模式分为静态代理与动态代理。

在 Java 中要想实现动态代理可以使用 4 种常见方式。

(1) JDK 提供的动态代理。

(2) 使用 cglib 框架。

(3) 使用 Javassist 框架。

(4) 使用 Spring 框架。

本节将实现静态代理,并使用以上 4 种方式实现动态代理。

2.2.1 静态代理的实现

在静态代理中,代理对象与被代理对象必须实现同一个接口,完整保留被代理对象的接口样式,也将一直保持接口不变的原则。

为了学习静态代理的知识,我们来看一个静态代理的代码示例。

创建测试项目 aop-test1。

创建接口,代码如下:

```
package com.ghy.www.service;

public interface ISendBook {
    public void sendBook();
}
```

创建两个接口的实现类,代码如下:

```
package com.ghy.www.service;

public class DangDangBook implements ISendBook {
    @Override
    public void sendBook() {
        System.out.println("当当网图书部门知道你的地址、电话和备注,准备进行配送!");
    }
}

package com.ghy.www.service;

public class JDBook implements ISendBook {
    @Override
    public void sendBook() {
        System.out.println("JD 图书部门知道你的地址、电话和备注,准备进行配送!");
    }
}
```

创建顺风快递送书代理类,代码如下:

```
package com.ghy.www.proxy;

import com.ghy.www.service.ISendBook;

//顺风就是代理,代理就是帮别人做一些事情
//被代理类就是 DangDangBook 或者 JDBook
```

```java
//SFBookProxy就是代理类
public class SFBookProxy implements ISendBook {
    private ISendBook sendBook;// JDBook 或 DangDangBook

    public SFBookProxy(ISendBook sendBook) {
        super();
        this.sendBook = sendBook;
    }

    @Override
    public void sendBook() {
        System.out.println("顺风收件-事务开启-连接开启-增强开始");
        sendBook.sendBook();
        System.out.println("顺风送达-事务提交-连接关闭-增强结束");
    }
}
```

代理类 SFBookProxy 也要实现 ISendBook 接口，以达到行为的一致性。

创建运行类 Test，代码如下：

```java
package com.ghy.www.test;

import com.ghy.www.proxy.SFBookProxy;
import com.ghy.www.service.DangDangBook;
import com.ghy.www.service.ISendBook;
import com.ghy.www.service.JDBook;

public class Test {
    public static void main(String[] args) {
        JDBook jdBook = new JDBook();
        DangDangBook dangdangBook = new DangDangBook();

        ISendBook sf1 = new SFBookProxy(jdBook);
        sf1.sendBook();

        System.out.println();

        ISendBook sf2 = new SFBookProxy(dangdangBook);
        sf2.sendBook();
    }
}
```

程序运行结果如下：

顺风收件-事务开启-连接开启-增强开始
JD图书部门知道你的地址、电话和备注，准备进行配送！
顺风送达-事务提交-连接关闭-增强结束

顺风收件-事务开启-连接开启-增强开始
当当网图书部门知道你的地址、电话和备注，准备进行配送！
顺风送达-事务提交-连接关闭-增强结束

在不改变原有 DangDangBook 类和 JDBook 类的代码的基础上进行功能的增强，在控制台输出信息如下：

JD图书部门知道你的地址、电话和备注，准备进行配送！
当当网图书部门知道你的地址、电话和备注，准备进行配送！

配送之前和之后分别输出如下信息：

顺风收件-事务开启-连接开启-增强开始
顺风送达-事务提交-连接关闭-增强结束

上述示例就是一个典型的日志或事务的功能增强模型，在不改变原始代码的基础上，实现了事务处理，并实现了日志处理。

但从上面这个示例代码中可以发现，静态代理类有自身的主要缺点，即可扩展性不好，因为 SFBookProxy 类绑定了 ISendBook 接口，如果顺风快递想要代理更多的送货任务，则需要创建更多的代理类，如下所示。

（1）配送鼠标代理类：public class SFSendMouseProxy implements ISendMouse。
（2）配送电视代理类：public class SFSendTVProxy implements ISendTV。
（3）配送电话代理类：public class SFSendPhoneProxy implements ISendPhone。
（4）其他代理类。

静态代理的这个最致命的缺点导致其不会应用到实际的软件项目中。那么，有没有一个更好的解决办法不再使代理类绑定接口呢？有，这就是动态代理。

2.2.2 使用 JDK 实现动态代理

静态代理的缺点是代理类绑定了固定的接口，不利于扩展，动态代理则不然。通过动态代理，可以对任何实现某一接口的类进行功能上的增强，而不会出现代理类绑定接口的情况。

在 Java 中动态代理类的对象由 Proxy 类的 newProxyInstance()方法创建，这就说明 Java 实现动态代理的代理类并不像静态代理那样由程序员自己创建，动态代理的代理类的对象是由 JVM 创建的，增强的算法需要由 InvocationHandler 接口实现。

创建测试项目 aop-test2。

创建增强算法类 LogInvocationHandler，代码如下：

```java
package com.ghy.www.hanlder;

import java.lang.reflect.InvocationHandler;
import java.lang.reflect.Method;

public class LogInvocationHandler implements InvocationHandler {
    private Object anyObject;

    public LogInvocationHandler(Object anyObject) {
        super();
        this.anyObject = anyObject;
    }

    @Override
    public Object invoke(Object proxy, Method method, Object[] args) throws Throwable {
        System.out.println("log begin time=" + System.currentTimeMillis());
        method.invoke(anyObject);
        System.out.println("log  end time=" + System.currentTimeMillis());
        return null;
    }
}
```

2.2 AOP 原理之代理设计模式

创建运行类,代码如下:

```
package com.ghy.www.test;

import com.ghy.www.hanlder.LogInvocationHandler;
import com.ghy.www.service.DangDangBook;
import com.ghy.www.service.ISendBook;
import com.ghy.www.service.JDBook;

import java.lang.reflect.Proxy;

public class Test {
    public static void main(String[] args) {
        DangDangBook dangdangBook = new DangDangBook();
        JDBook jdBook = new JDBook();
        {
            LogInvocationHandler handler = new LogInvocationHandler(dangdangBook);
            ISendBook sendBook = (ISendBook) Proxy.newProxyInstance(Test.class.getClassLoader(), dangdangBook.getClass().getInterfaces(), handler);
            sendBook.sendBook();
        }
        System.out.println();
        {
            LogInvocationHandler handler = new LogInvocationHandler(jdBook);
            ISendBook sendBook = (ISendBook) Proxy.newProxyInstance(Test.class.getClassLoader(), dangdangBook.getClass().getInterfaces(), handler);
            sendBook.sendBook();
        }
    }
}
```

程序运行结果如下:

```
log   begin time=1601304808559
当当网图书部门知道你的地址、电话和备注,准备进行配送!
log   end   time=1601304808560

log   begin time=1601304808560
JD 图书部门知道你的地址、电话和备注,准备进行配送!
log   end   time=1601304808560
```

创建代理类的对象 sendBook 的过程如下。

(1) Proxy.newProxyInstance() 方法的参数:原料。

(2) Proxy:工厂类。工厂类 Proxy 使用 3 个原料创建 com.sun.proxy.$Proxy0 代理类的对象。

(3) com.sun.proxy.$Proxy0 类:工厂创建的代理类对象。

(4) sendBook:sendBook 就是 com.sun.proxy.$Proxy0 类的对象,因为是赋值关系。

程序正确运行,总结以下 3 点。

(1) 代理类由 JVM 创建,程序员不需要自己创建代理类,代理类*.java 文件的数量急剧下降。

(2) 代理类不再绑定固定的接口,代理类与接口实现解耦。

(3) 在 InvocationHandler 中,通过反射技术可以更灵活地处理增强算法。

静态代理和动态代理都是针对 public void sendBook() 方法进行增强,在动态代理中虽然使用了反射技术，但 Spring 只支持对方法进行增强，不支持字段级的增强，因为 Spring 官方认为支持字段级的增强违背了面向对象编程（OOP）的思想，所以支持方法的增强是合适的，而且与 Spring 的其他模块进行整合开发时更具标准性。

2.2.3 使用 Spring 实现动态代理

虽然使用 JDK 中原始的动态代理能够实现对目标对象在功能上的增强，但使用时还是不太方便。另外，每个程序员的写法不同，不利于代码风格的统一和代码的后期维护。Spring 框架对动态代理进行了封装，产生了自己的 AOP 框架，称为 Spring AOP。Spring AOP 框架中的常见接口如下。

（1）方法执行前增强要实现 org.springframework.aop.MethodBeforeAdvice 接口。
（2）方法执行后增强要实现 org.springframework.aop.AfterReturningAdvice 接口。
（3）方法执行环绕增强要实现 org.aopalliance.intercept.MethodInterceptor 接口。
（4）异常处理增强要实现 org.springframework.aop.ThrowsAdvice 接口。

创建测试项目 aop-test3。

创建方法执行前的通知类，代码如下：

```java
package com.ghy.www.advice;

import org.springframework.aop.MethodBeforeAdvice;

import java.lang.reflect.Method;

public class MyMethodBeforeAdvice implements MethodBeforeAdvice {
    @Override
    public void before(Method arg0, Object[] arg1, Object arg2) throws Throwable {
        System.out.println("MyMethodBeforeAdvice");
    }
}
```

创建方法执行后的通知类，代码如下：

```java
package com.ghy.www.advice;

import org.springframework.aop.AfterReturningAdvice;

import java.lang.reflect.Method;

public class MyAfterReturningAdvice implements AfterReturningAdvice {
    @Override
    public void afterReturning(Object arg0, Method arg1, Object[] arg2, Object arg3) throws Throwable {
        System.out.println("AfterReturningAdvice");
    }
}
```

创建方法环绕通知类，代码如下：

```java
package com.ghy.www.advice;

import org.aopalliance.intercept.MethodInterceptor;
import org.aopalliance.intercept.MethodInvocation;
```

```java
public class MyMethodInterceptor implements MethodInterceptor {
    @Override
    public Object invoke(MethodInvocation arg0) throws Throwable {
        System.out.println("MethodInterceptor begin");
        Object value = arg0.proceed();
        System.out.println("MethodInterceptor    end");
        return value;
    }
}
```

创建异常处理通知类，代码如下：

```java
package com.ghy.www.advice;

import org.springframework.aop.ThrowsAdvice;

import java.lang.reflect.Method;

public class MyThrowsAdvice implements ThrowsAdvice {
    public void afterThrowing(Method method, Object[] args, Object target, Exception ex) {
        System.out.println("ThrowsAdvice 信息,方法名称=" + method.getName() + "  参数个数:" + args.length + "  原始对象:" + target
                + "  异常信息:" + ex.getMessage());
    }
}
```

创建业务接口 ISendBook，代码如下：

```java
package com.ghy.www.service;

public interface ISendBook {
    public void sendBook();
    public void sendBookError();
}
```

创建业务接口 ISendBook，实现类代码如下：

```java
package com.ghy.www.service;

public class DangDangBook implements ISendBook {
    @Override
    public void sendBook() {
        System.out.println("当当网图书部门知道你的地址、电话和备注，要给你送书了！");
    }

    @Override
    public void sendBookError() {
        System.out.println("当当网图书部门知道你的地址、电话和备注，要给你送书了！");
        Integer.parseInt("a");
    }
}
```

配置文件 applicationContext.xml 的代码如下：

```xml
<bean id="myAfterReturningAdvice"
    class="com.ghy.www.advice.MyAfterReturningAdvice"></bean>
<bean id="myMethodBeforeAdvice"
    class="com.ghy.www.advice.MyMethodBeforeAdvice"></bean>
<bean id="myMethodInterceptor" class="com.ghy.www.advice.MyMethodInterceptor"></bean>
```

```xml
<bean id="myThrowsAdvice" class="com.ghy.www.advice.MyThrowsAdvice"></bean>

<bean id="dangdangbook" class="com.ghy.www.service.DangDangBook"></bean>

<bean id="proxy"
      class="org.springframework.aop.framework.ProxyFactoryBean">
    <property name="interfaces">
        <value>com.ghy.www.service.ISendBook</value>
    </property>
    <property name="target" ref="dangdangbook"></property>
    <property name="interceptorNames">
        <list>
            <value>myMethodBeforeAdvice</value>
            <value>myAfterReturningAdvice</value>
            <value>myMethodInterceptor</value>
            <value>myThrowsAdvice</value>
        </list>
    </property>
</bean>
```

创建运行类 Test，代码如下：

```java
package com.ghy.www.test;

import com.ghy.www.service.ISendBook;
import org.springframework.context.ApplicationContext;
import org.springframework.context.support.ClassPathXmlApplicationContext;

public class Test {
    public static void main(String[] args) {
        ApplicationContext context = new ClassPathXmlApplicationContext("applicationContext.xml");
        ISendBook sendBook = (ISendBook) context.getBean("proxy");
        sendBook.sendBook();
        System.out.println();
        sendBook.sendBookError();
    }
}
```

程序运行结果如下：

```
MyMethodBeforeAdvice
MethodInterceptor begin
当当网图书部门知道你的地址、电话和备注，要给你送书了！
MethodInterceptor   end
AfterReturningAdvice

MyMethodBeforeAdvice
MethodInterceptor begin
当当网图书部门知道你的地址、电话和备注，要给你送书了！
ThrowsAdvice 信息，方法名称=sendBookError 参数个数：0 原始对象：com.ghy.www.service.DangDangBook@4b9e255 异常信息：For input string: "a"
Exception in thread "main" java.lang.NumberFormatException: For input string: "a"
    at java.lang.NumberFormatException.forInputString(NumberFormatException.java:65)
    at java.lang.Integer.parseInt(Integer.java:580)
    at java.lang.Integer.parseInt(Integer.java:615)
    at com.ghy.www.service.DangDangBook.sendBookError(DangDangBook.java:12)
    at sun.reflect.NativeMethodAccessorImpl.invoke0(Native Method)
    at sun.reflect.NativeMethodAccessorImpl.invoke(NativeMethodAccessorImpl.java:62)
```

```
        at sun.reflect.DelegatingMethodAccessorImpl.invoke(DelegatingMethodAccessorImpl.
java:43)
        at java.lang.reflect.Method.invoke(Method.java:498)
        at
        at com.ghy.www.test.Test.main(Test.java:13)
```

在控制台中输出了全部通知类中的功能增强方法,所有通知类都参与了动态代理的算法增强。

2.2.4　使用 cglib 实现动态代理

创建测试项目 aop-test4。

在 pom.xml 文件中添加依赖配置:

```xml
<dependency>
    <groupId>cglib</groupId>
    <artifactId>cglib</artifactId>
    <version>2.2.2</version>
</dependency>

<dependency>
    <groupId>org.ow2.asm</groupId>
    <artifactId>asm</artifactId>
    <version>8.0</version>
</dependency>
```

创建业务类,代码如下:

```java
package com.ghy.www.service;

public class UserinfoService {
    public void save() {
        System.out.println("将数据保存到数据库!");
    }
}
```

创建 cglib 强化类,代码如下:

```java
package com.ghy.www.interceptor;

import net.sf.cglib.proxy.MethodInterceptor;
import net.sf.cglib.proxy.MethodProxy;

import java.lang.reflect.Method;

public class MyMethodInterceptor implements MethodInterceptor {
    private Object object;

    public MyMethodInterceptor(Object object) {
        super();
        this.object = object;
    }

    @Override
    public Object intercept(Object arg0, Method arg1, Object[] arg2, MethodProxy arg3)
throws Throwable {
        System.out.println("begin");
        Object returnValue = arg1.invoke(object, arg2);
        System.out.println("  end");
```

```
            return returnValue;
        }
}
```

创建运行类,代码如下:

```
package com.ghy.www.test;

import com.ghy.www.interceptor.MyMethodInterceptor;
import com.ghy.www.service.UserinfoService;
import net.sf.cglib.proxy.Enhancer;

public class Test {
    public static void main(String[] args) {
        // Enhancer 类是 cglib 中的增强工具类
        Enhancer enhancer = new Enhancer();
        // cglib 使用继承原理来实现代理,所以要设置父类
        enhancer.setSuperclass(UserinfoService.class);
        // 设置回调对象
        enhancer.setCallback(new MyMethodInterceptor(new UserinfoService()));
        // 创建代理对象
        UserinfoService service = (UserinfoService) enhancer.create();
        // 调用代理对象中的业务方法
        service.save();
    }
}
```

程序运行结果如下:

```
begin
将数据保存到数据库!
  end
```

当要增强的目标对象没有接口时,就要使用 cglib 并根据继承原理来实现代理增强。

2.2.5 使用 javassist 实现动态代理

创建测试项目 aop-test5。

在 pom.xml 文件中添加依赖配置:

```xml
<dependency>
    <groupId>org.javassist</groupId>
    <artifactId>javassist</artifactId>
    <version>3.21.0-GA</version>
</dependency>
```

创建业务类,代码如下:

```
package com.ghy.www.service;

public class UserinfoService {
    public void save() {
        System.out.println("将数据保存到数据库!");
    }
}
```

创建 javassist 强化类，代码如下：

```java
package com.ghy.www.handler;
import java.lang.reflect.Method;

import javassist.util.proxy.MethodHandler;

public class MyMethodHandler implements MethodHandler {
    private Object object;

    public MyMethodHandler(Object object) {
        super();
        this.object = object;
    }

    @Override
    public Object invoke(Object arg0, Method arg1, Method arg2, Object[] arg3) throws Throwable {
        System.out.println("begin");
        Object returnValue = arg1.invoke(object, arg3);
        System.out.println("  end");
        return returnValue;
    }

}
```

创建运行类，代码如下：

```java
package com.ghy.www.test;

import com.ghy.www.handler.MyMethodHandler;
import com.ghy.www.service.UserinfoService;
import javassist.util.proxy.Proxy;
import javassist.util.proxy.ProxyFactory;

public class Test {
    public static void main(String[] args) throws InstantiationException, IllegalAccessException {
        ProxyFactory proxyFactory = new ProxyFactory();
        proxyFactory.setSuperclass(UserinfoService.class);

        Class classRef = proxyFactory.createClass();
        UserinfoService service = (UserinfoService) classRef.newInstance();
        ((Proxy) service).setHandler(new MyMethodHandler(new UserinfoService()));
        service.save();
    }
}
```

程序运行结果如下：

```
begin
将数据保存到数据库！
  end
```

AOP 与代理的关系如图 2-1 所示。代理模式分为静态和动态代理，实现动态代理可以有四种方式。AOP 的内部原理就是动态代理。

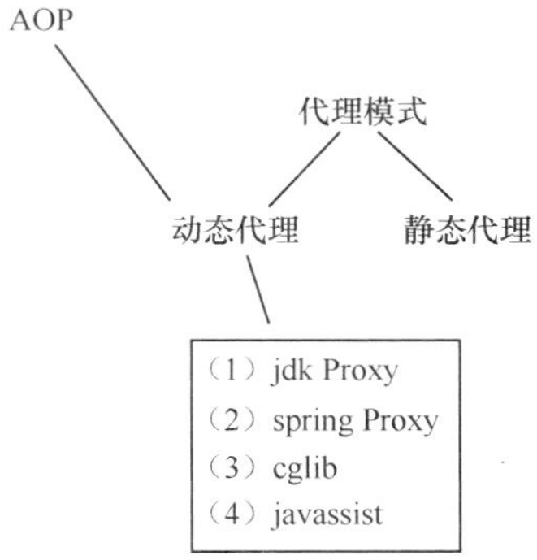

2.3　AOP 相关的概念

Spring 框架的 AOP 是基于动态代理的技术，掌握动态代理

图 2-1　AOP 与代理的关系

就是掌握 AOP 技术的前提。AOP 现在已经成为一个独立的技术,掌握并理解 AOP 相关的概念有助于我们学习 AOP 的知识,这些概念不只存在于 Spring 框架,在其他的 AOP 框架中也存在。然而,AOP 相关的概念并不是特别直观,如果我们不了解它们,将会对后续的学习造成影响。

2.3.1 横切关注点

在开发软件项目时,我们需要关注一些"通用"功能,比如需要计算方法的执行时间,有没有访问资源的权限,记录一些常规的日志,这些日志的内容包括哪位用户正在登录,在什么时间执行了哪些操作,操作的结果是正常还是出现异常等信息,这些"通用"功能的代码大多数交织在 Service 业务对象中,代码结构如图 2-2 所示。

在大多数 Service 业务类的代码中会重复出现这些"通用"功能的代码,这些代码与业务代码混合,两者之间产生了密不可分的紧耦合,不利于软件的后期维护与扩展。为了解决这个问题,程序员就要关注这些"通用"功能,它们在 Spring AOP 中被称为"横切关注点"(cross-cutting concern)。把"通用"功能的功能代码从 Service 业务代码中分离出来正是 AOP 所要达到的核心目的:解耦。

依赖注入中的"解耦"主要是对象与对象之间进行解耦,而 AOP 中的"解耦"主要是将"横切关注点"相关的代码与"业务代码"进行分离,从而实现解耦。虽然两者都是解耦,但性质是不一样的。

常用的横切关注点如图 2-3 所示。

```
public class UserinfoService {
    public void save() {
        (1)是否有权执行此方法
        (2)常规日志记录
        (3)方法执行开始时间
            业务代码
        (4)方法执行结束时间
    }
}
```

图 2-2 交织在一起的"面条式"代码

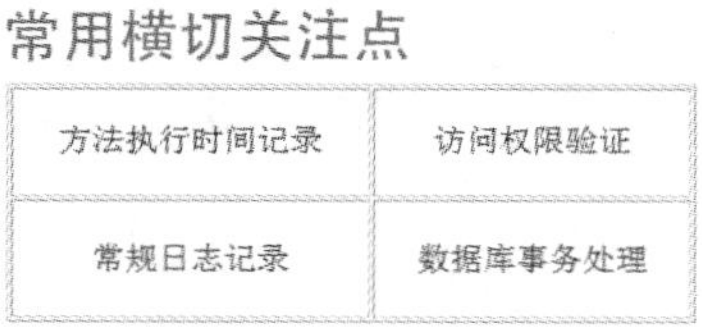

常用横切关注点

方法执行时间记录	访问权限验证
常规日志记录	数据库事务处理

图 2-3 常用的横切关注点

横切关注点就是"通用"功能。

2.3.2 切面

在介绍什么是切面(aspect)之前,我们先了解一下汉字中的"切"字的含义,如图 2-4 所示。

图 2-4 关于"切"字的解释

汉字中的"切"是指将物品分成若干部分,这个特性和 AOP 是一致的,效果如图 2-5 所示。

2.3 AOP 相关的概念

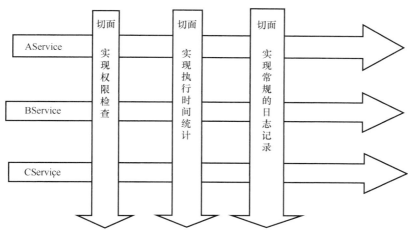

图 2-5　使用切的方法将流程分成若干部分

在 Aservice、BService 和 CService 中都需要 3 个切面中的功能，要把这些功能从 Service 业务类中提取出来。当 Service 业务类中的代码执行时动态地对切面中的功能代码进行调用，也就实现了对"横切关注点"通用功能的代码的复用，达到了解耦的目的。AOP 可以让一组类的对象拥有相同的行为。

虽然使用"切"的手段能将流程分成若干部分，但 Service 业务类的执行流程是不会被中断的，比如 AService 的流程代码不会因切面的存在而被中断，而是会一直运行到最后。在这个过程中，切面就类似于过滤器（filter）：拦截下来，进行处理，然后放行。不过，过滤器拦截的是请求（request），而切面拦截的是方法的调用。虽然两者运行时的特性基本一致，但目的不一样。

汉字中的"切"以及在 Spring 中如何应用"切"对执行的方法进行拦截已经解释完毕，那么，什么是"切面"呢？"切面"就是对"横切关注点"的模块化，也就是将"横切关注点"的功能代码提取出来并放入一个单独的类中进行统一处理。这个类就是"切面类"，也可称为切面。在 AOP 中，主要就是针对"切面"进行代码设计。

"切面"对"横切关注点"模块化的示例代码如图 2-6 所示。

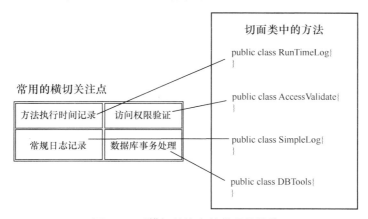

图 2-6　对横切关注点的代码模块化

将横切关注点中的功能代码放入切面类的方法中，可以实现横切关注点的模块化，切面类中的方法可以被很多 Java 类以共享的方式进行访问。

2.3.3 连接点

前面介绍了切面，那么，什么是连接点呢？连接点（join point）是指在软件执行过程中能够插入切面的一个点。连接点可以出现在调用方法前、调用方法后、方法抛出异常时和方法返回值之后等。在这些连接点中可以插入在切面中定义的通用功能，从而添加新的软件行为。

连接点示意图如图 2-7 所示。

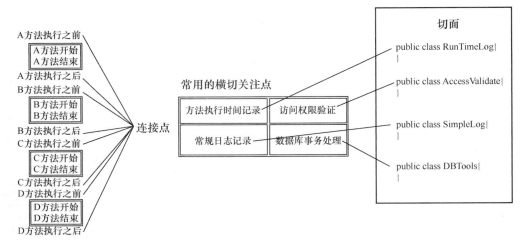

图 2-7 连接点

在软件系统中存在多个连接点，在这些连接点处可以插入切面，比如在执行 B 方法之前插入日志的功能（SimpleLog()），在执行 C 方法之前和之后插入记录执行时间的功能（RunTimeLog()）等。

2.3.4 切点

考虑到软件运行的效率，对所有的连接点应用切面是不现实的。在大多数情况下，我们只想针对"部分连接点"应用切面，这些"部分连接点"称为"切点"（pointcut），在切点上应用切面。

切点示意图如图 2-8 所示。

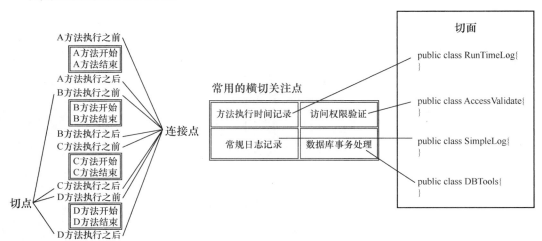

图 2-8 切点

切点是缩小连接点数量的范围，只针对某几个切点进行切面的参与，切点就是精确的定义在什么位置放置切面。

2.3.5 通知

切点指出了应用切面的精准位置，什么时候应用切面是由通知（advice）决定的，比如可以在执行方法之前、执行方法之后、出现异常时应用切面，这个应用切面的时机可称为通知。

通知示意图如图 2-9 所示。

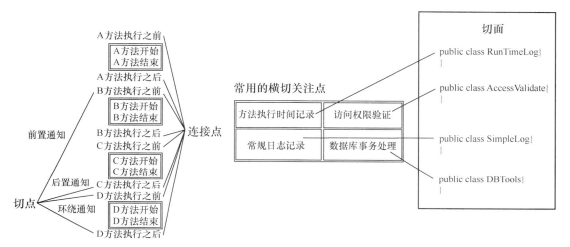

图 2-9 通知

在 Spring 的 AOP 中，通知分为以下 5 种。
（1）前置通知（before）：方法被调用之前。
（2）后置通知（after）：方法被调用之后。
（3）环绕通知（around）：方法被调用之前与之后。
（4）返回通知（after-returning）：方法返回了值。
（5）异常通知（after-throwing）：方法出现了异常。

在这 5 种通知类型中，都可以应用切面中的功能代码，关于这 5 种通知的使用，在下面会专门介绍。

切面包含通知和切点，是两者的结合。通知和切点是切面的基本元素。
（1）通知：定义了在什么时机应用切面。
（2）切点：定义了可以在哪些连接点上放置切面。

2.3.6 织入

织入（weaving）就是把切面、切点和通知整合起来，应用到目标对象中。

在 Spring 的 AOP 中，织入的原理是由 JVM 创建代理对象，在代理对象中调用原始对象中的方法，再结合增强算法实现。由于 Spring 的 AOP 技术基于动态代理，因此 Spring 中的 AOP 只支持方法的连接点，不支持字段级的连接点。

织入示意图如图 2-10 所示。织入使切面应用到目标对象中。

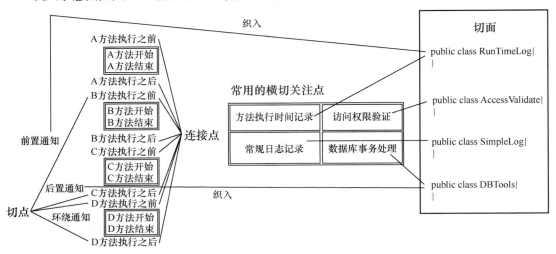

图 2-10　织入

2.4　AOP 核心案例

如果一个类实现了接口，那么 Spring 使用 JDK 的动态代理完成 AOP（推荐）。
如果一个类没有实现接口，那么 Spring 使用 cglib 完成 AOP。
AOP 技术不是 Spring 框架特有的，而 Spring 框架却高度依赖 AOP 技术实现功能。
在介绍完 AOP 相关的概念后，下面我们利用若干示例代码来介绍一下 AOP 的程序设计。

2.4.1　实现前置通知、后置通知、返回通知和异常通知

本节使用注解的方式来实现前置通知、后置通知、返回通知和异常通知，从运行的结果来看，与代理设计模式是一致的。
创建测试项目 aop-test6。
创建切面类，代码如下：

```
package com.ghy.www.aspect;

import org.aspectj.lang.annotation.*;
import org.springframework.stereotype.Component;

@Component
@Aspect
public class AspectObject {
    @Before(value = "execution(* com.ghy.www.service.UserinfoService.*(..))")
    public void before() {
        System.out.println("public void before()");
    }

    @After(value = "execution(* com.ghy.www.service.UserinfoService.*(..))")
    public void after() {
        System.out.println("public void after()");
```

2.4 AOP 核心案例

```
    }

    @AfterReturning(value = "execution(* com.ghy.www.service.UserinfoService.*(..))")
    public void afterReturning() {
        System.out.println("public void afterReturning()");
    }

    @AfterThrowing(value = "execution(* com.ghy.www.service.UserinfoService.*(..))")
    public void afterThrowing() {
        System.out.println("public void afterThrowing()");
    }
}
```

对切面表达式的解释如图 2-11 所示。

图 2-11　切面表达式 execution 解释

注意：包名后面可以写类名或接口名。

切面表达式 execution 的语法形式如下：

```
execution(modifiers-pattern?
        ret-type-pattern
        declaring-type-pattern?name-pattern(param-pattern)
        throws-pattern?)
```

（1）modifiers-pattern 代表修饰符，如 public、private。
（2）ret-type-pattern 代表返回值类型。
（3）declaring-type-pattern 代表包路径。
（4）name-pattern 代表方法。
（5）param-pattern 参数。
（6）throws-pattern 代表异常。

问号（?）表示可以忽略，也就是除 ret-type-pattern 之外，其他是可选的。

常用切点表达式的示例如下。

（1）execution(public * *(..))：匹配任何 public 方法。
（2）execution(* set*(..))：匹配任何以 set 开头的方法。
（3）execution(* com.xyz.service.AccountService.*(..))：匹配 com.xyz.service.AccountService 类或接口中的所有方法。
（4）execution(* com.xyz.service.*.*(..))：匹配 com.xyz.service 包中的任何类或接口中的任何方法。
（5）execution(* com.xyz.service..*.*(..))：匹配 com.xyz.service 和子孙包中的任何类或接口中的任何方法。
（6）execution(public void addUser(com.ghy.www.entity.User))：严格匹配。
（7）execution(public void *(com.entity.User))：表示任意方法名，但参数类型是 User。

（8）execution(public void save(..))：表示任意参数。
（9）execution(public * save())：表示任意返回值类型。
（10）execution(* com.ghy.UserBiz.*(..))：UserBiz 类或接口中的任意方法。
（11）execution(* set(..))：任意 set 方法。

在 Spring 中，增强方式通过通知进行实现。通知的应用场景如图 2-12 所示。

增强类型	应用场景
前置增强（before advice）	权限控制、记录调用日志
后置增强（after returning advice）	统计分析结果数据
异常增强（after throwing advice）	通过日志记录方法异常信息
最终增强（after finally advice）	释放资源
环绕增强（around advice）	缓存、性能日志、权限、事务管理

图 2-12　通知的应用场景

通知的运行结果总结如图 2-13 所示。

```
5种不同的增强：
aop:before（前置增强）：在方法执行之前执行增强。
aop:after-returning（后置增强）：在方法正常执行完成之后执行增强。
aop:throwing（异常增强）：在方法抛出异常退出时执行增强。
aop:after（最终增强）：在方法执行之后执行，相当于在finally里面执行；可以通过配置throwing来
获得拦截到的异常信息。
aop:around（环绕增强）：最强大的一种增强类型。环绕增强可以在方法调用前后完成自定义的行
为，环绕通知有两个要求。
    1. 方法必须要返回一个Object（返回的结果）。
    2. 方法的第一个参数必须是ProceedingJoinPoint（可以继续向下连接的连接点）。
```

图 2-13　通知运行结果总结

切面类中的功能代码就是对"横切关注点"的模块化。我们把这些通用功能的代码统一管理，便于复用。

注意：下面的实现类没有实现任何接口。

创建业务类，代码如下：

```java
package com.ghy.www.service;

import org.springframework.stereotype.Service;

@Service
public class UserinfoService {
    public void method1() {
        System.out.println("method1 run !");
    }

    public String method2() {
        System.out.println("method2 run !");
        return "我是返回值A";
    }

    public String method3() {
```

```
            System.out.println("method3 run !");
            Integer.parseInt("a");
            return "我是返回值B";
    }
}
```

Service 业务类中的 3 种方法分别实现了无返回值、有返回值和出现异常情况。

创建配置类，代码如下：

```
package com.ghy.www.config;

import org.springframework.context.annotation.ComponentScan;
import org.springframework.context.annotation.Configuration;
import org.springframework.context.annotation.EnableAspectJAutoProxy;

@Configuration
@EnableAspectJAutoProxy
@ComponentScan(basePackages = {"com.ghy.www.aspect", "com.ghy.www.service"})
public class SpringConfig {
}
```

创建运行类，代码如下：

```
package com.ghy.www.test;

import com.ghy.www.config.SpringConfig;
import com.ghy.www.service.UserinfoService;
import org.springframework.context.ApplicationContext;
import org.springframework.context.annotation.AnnotationConfigApplicationContext;

public class Test {
    public static void main(String[] args) {
        ApplicationContext context = new AnnotationConfigApplicationContext(SpringConfig.class);
        UserinfoService service = (UserinfoService) context.getBean(UserinfoService.class);
        service.method1();
        System.out.println();
        System.out.println();
        System.out.println("main get method2 returnValue=" + service.method2());
        System.out.println();
        System.out.println();
        System.out.println("main get method3 returnValue=" + service.method3());
    }
}
```

程序运行结果如下：

```
public void before()
method1 run !
public void afterReturning()
public void after()


public void before()
method2 run !
public void afterReturning()
public void after()
main get method2 returnValue=我是返回值A
```

```
public void before()
method3 run !
public void afterThrowing()
public void after()
Exception in thread "main" java.lang.NumberFormatException: For input string: "a"
```

在无异常的情况下,程序运行顺序如下。

(1)前置通知@Before。

(2)业务类中的方法。

(3)返回通知@AfterReturning。

(4)后置通知@After。

(5)返回值 A。

在有异常的情况下,程序运行顺序如下。

(1)前置通知@Before。

(2)业务类中的方法。

(3)异常通知@AfterThrowing。

(4)后置通知@After。

在有异常出现时,返回通知@AfterReturning 的信息没有在控制台中输出,说明方法内部出现了程序运行流程的中断,程序不再继续执行,于是没有"返回通知",也就是如果方法返回而不引发异常,通知@AfterReturning 就会被执行。

后置通知@After 和返回通知@AfterReturning 在输出效果上非常类似,这两者在本质上的区别如下。

(1)@After 着重点在于方法执行完毕,也就是方法执行完成的方式无论是正常或者是异常返回,都将执行@After 通知,相当于 finally 语句块。

(2)@AfterReturning 着重点在于方法无异常地执行完毕后返回而继续执行后面的方法,在返回的同时可以使用@AfterReturning 取得返回值。这个功能对于@After 是不具有的,使用@AfterReturning 取得返回值的示例代码如下:

```
@Aspect
public class AfterReturningExample {
    @AfterReturning(
        value = "execution(* service.UserinfoService.*(..))", returning="retVal")
    public void doAccessCheck(Object retVal) {
        // ...
    }
}
```

对于此测试,在后面介绍。

注意:在本节中的业务类并没有实现任何接口。如果业务类实现了接口,在获得业务类时,必须使用 context.getBean("beanId")的写法,而不能使用 context.getBean(Bean.class)的写法,否则会出现找不到 JavaBean 的异常。出现异常的原因是根据 context.getBean(Bean.class)写法获得的对象的真实数据类型是 Bean 的代理类 BeanProxy,Bean.class 和 BeanProxy.class 不是同一个类型,所以出现异常。此实验在项目 aop-test7 中进行测试。

2.4.2　向前置通知、后置通知、返回通知和异常通知传入 JoinPoint 参数

JoinPoint 参数可以实现对连接点信息的获取。

创建测试项目 aop-test8。

创建切面类，代码如下：

```
package com.ghy.www.aspect;

import java.util.Arrays;

import org.aspectj.lang.JoinPoint;
import org.aspectj.lang.Signature;
import org.aspectj.lang.annotation.After;
import org.aspectj.lang.annotation.AfterReturning;
import org.aspectj.lang.annotation.AfterThrowing;
import org.aspectj.lang.annotation.Aspect;
import org.aspectj.lang.annotation.Before;
import org.springframework.stereotype.Component;

@Component
@Aspect
public class AspectObject {

    @Before(value = "execution(* com.ghy.www.service.UserinfoService.*(..))")
    public void before(JoinPoint point) {
        Signature signature = point.getSignature();
        System.out.println("public void before()" + " 对象：" + point.getThis() + " 方法：" + signature.getName() + " 参数："
                + Arrays.toString(point.getArgs()));
    }

    @After(value = "execution(* com.ghy.www.service.UserinfoService.*(..))")
    public void after(JoinPoint point) {
        Signature signature = point.getSignature();
        System.out.println("public void after()" + " 对象：" + point.getThis() + " 方法：" + signature.getName() + " 参数："
                + Arrays.toString(point.getArgs()));
    }

    @AfterReturning(value = "execution(* com.ghy.www.service.UserinfoService.*(..))")
    public void afterReturning(JoinPoint point) {
        Signature signature = point.getSignature();
        System.out.println("public void afterReturning()" + " 对象：" + point.getThis() +
                " 方法：" + signature.getName()
                + " 参数：" + Arrays.toString(point.getArgs()));
    }

    @AfterThrowing(value = "execution(* com.ghy.www.service.UserinfoService.*(..))")
    public void afterThrowing(JoinPoint point) {
        Signature signature = point.getSignature();
        System.out.println("public void afterThrowing()" + " 对象：" + point.getThis() +
                " 方法：" + signature.getName()
                + " 参数：" + Arrays.toString(point.getArgs()));
    }

}
```

程序运行后，通过 JoinPoint 对象可以输出对象和方法的名称以及参数信息，如下：

```
public void before() 对象：com.ghy.www.service.UserinfoService@306f16f3 方法：method1 参数：[]
method1 run !
public void afterReturning() 对象：com.ghy.www.service.UserinfoService@306f16f3 方法：method1 参数：[]
public void after() 对象：com.ghy.www.service.UserinfoService@306f16f3 方法：method1 参数：[]


public void before() 对象：com.ghy.www.service.UserinfoService@306f16f3 方法：method2 参数：[]
method2 run !
public void afterReturning() 对象：com.ghy.www.service.UserinfoService@306f16f3 方法：method2 参数：[]
public void after() 对象：com.ghy.www.service.UserinfoService@306f16f3 方法：method2 参数：[]
main get method2 returnValue=我是返回值 A


public void before() 对象：com.ghy.www.service.UserinfoService@306f16f3 方法：method3 参数：[]
method3 run !
public void afterThrowing() 对象：com.ghy.www.service.UserinfoService@306f16f3 方法：method3 参数：[]
public void after() 对象：com.ghy.www.service.UserinfoService@306f16f3 方法：method3 参数：[]
Exception in thread "main" java.lang.NumberFormatException: For input string: "a"
```

2.4.3 实现环绕通知

环绕通知是指在执行方法之前和之后都可以应用切面。

创建测试项目 aop-test9。

创建切面类，代码如下：

```java
package com.ghy.www.aspect;

import org.aspectj.lang.ProceedingJoinPoint;
import org.aspectj.lang.annotation.Around;
import org.aspectj.lang.annotation.Aspect;
import org.springframework.stereotype.Component;

@Component
@Aspect
public class AspectObject {
    @Around(value = "execution(* com.ghy.www.service.UserinfoService.*(..))")
    public Object around(ProceedingJoinPoint point) {
        Object returnObject = null;
        try {
            System.out.println("开始");
            returnObject = point.proceed();
            System.out.println("结束");
        } catch (Throwable e) {
            e.printStackTrace();
        }
        return returnObject;
    }
}
```

2.4.4 使用 bean 表达式

在 Spring 中，提供了 bean 表达式，可以用它限制切面应用的目标对象。

创建测试项目 aop-test10。
创建切面类，代码如下：

```
package com.ghy.www.aspect;

import org.aspectj.lang.ProceedingJoinPoint;
import org.aspectj.lang.annotation.Around;
import org.aspectj.lang.annotation.Aspect;
import org.springframework.stereotype.Component;

@Component
@Aspect
public class AspectObject {
    @Around(value = "execution(* com.ghy.www.service..*(..)) and bean(service1)")
    public Object around(ProceedingJoinPoint point) {
        Object returnObject = null;
        try {
            System.out.println("开始");
            returnObject = point.proceed();
            System.out.println("结束");
        } catch (Throwable e) {
            e.printStackTrace();
        }
        return returnObject;
    }
}
```

切点表达式"service.."中的".."代表任意子孙级的包。
表达式"and bean(service1)"的作用是切面必须要应用于 bean 的 id 值为 service1 的对象。
创建业务类 UserinfoSeviceA，代码如下：

```
package com.ghy.www.service;

import org.springframework.stereotype.Service;

@Service(value = "service1")
public class UserinfoServiceA {
    public void method1() {
        System.out.println("methodA run !");
    }
}
```

创建业务类 UserinfoSeviceB，代码如下：

```
package com.ghy.www.service;

import org.springframework.stereotype.Service;

@Service(value = "service2")
public class UserinfoServiceB {
    public void method1() {
        System.out.println("methodB run !");
    }
}
```

创建运行类，代码如下：

```
package com.ghy.www.test;

import com.ghy.www.config.SpringConfig;
```

```java
import com.ghy.www.service.UserinfoServiceA;
import com.ghy.www.service.UserinfoServiceB;
import org.springframework.context.ApplicationContext;
import org.springframework.context.annotation.AnnotationConfigApplicationContext;

public class Test {
    public static void main(String[] args) {
        ApplicationContext context = new AnnotationConfigApplicationContext(SpringConfig.class);
        UserinfoServiceA serviceA = (UserinfoServiceA) context.getBean(UserinfoServiceA.class);
        serviceA.method1();
        System.out.println();
        UserinfoServiceB serviceB = (UserinfoServiceB) context.getBean(UserinfoServiceB.class);
        serviceB.method1();
    }
}
```

程序运行结果如下：

```
开始
methodA run !
结束

methodB run !
```

可见，只对 bean 的 id 值为 service1 的对象应用了切面。

2.4.5　使用@Pointcut 定义全局切点

在上文的多处使用了同样的 execution 表达式，可以将 execution 表达式进行全局化，以减少冗余的配置。

创建测试项目 aop-test11。

创建切面类，代码如下：

```java
package com.ghy.www.aspect;

import org.aspectj.lang.annotation.*;
import org.springframework.stereotype.Component;

@Component
@Aspect
public class AspectObject {
    @Pointcut(value = "execution(* com.ghy.www.service.UserinfoService.*(..))")
    public void publicPointcut() {

    }

    @Before(value = "publicPointcut()")
    public void before() {
        System.out.println("public void before()");
    }

    @After(value = "publicPointcut()")
    public void after() {
```

```java
        System.out.println("public void after()");
    }

    @AfterReturning(value = "publicPointcut()")
    public void afterReturning() {
        System.out.println("public void afterReturning()");
    }

    @AfterThrowing(value = "publicPointcut()")
    public void afterThrowing() {
        System.out.println("public void afterThrowing()");
    }
}
```

全局化配置如下：

```java
@Pointcut(value = "execution(* com.ghy.www.service.UserinfoService.*(..))")
```

由于依附于 public void publicPointcut() 方法，在引用这个全局切点（Pointcut）时引用方法名称即可。

创建业务类，代码如下：

```java
package com.ghy.www.service;

import org.springframework.stereotype.Service;

@Service
public class UserinfoService {
    public void method1() {
        System.out.println("method1 run !");
    }

    public String method2() {
        System.out.println("method2 run !");
        return "我是返回值A";
    }

    public String method3() {
        System.out.println("method3 run !");
        Integer.parseInt("a");
        return "我是返回值B";
    }
}
```

创建运行类，代码如下：

```java
package com.ghy.www.test;

import com.ghy.www.config.SpringConfig;
import com.ghy.www.service.UserinfoService;
import org.springframework.context.ApplicationContext;
import org.springframework.context.annotation.AnnotationConfigApplicationContext;

public class Test {
    public static void main(String[] args) {
        ApplicationContext context = new AnnotationConfigApplicationContext(SpringConfig.class);
        UserinfoService service = (UserinfoService) context.getBean(UserinfoService.class);
        service.method1();
        System.out.println();
```

```
            System.out.println();
            System.out.println("main get method2 returnValue=" + service.method2());
            System.out.println();
            System.out.println();
            System.out.println("main get method3 returnValue=" + service.method3());
        }
    }
```

程序运行结果如下：

```
public void before()
method1 run !
public void afterReturning()
public void after()


public void before()
method2 run !
public void afterReturning()
public void after()
main get method2 returnValue=我是返回值A


public void before()
method3 run !
public void afterThrowing()
public void after()
Exception in thread "main" java.lang.NumberFormatException: For input string: "a"
```

2.4.6　向切面传入参数

如果 Service 业务类的方法有参数，则切面类能获得参数值，可以进行日志处理。

创建测试项目 aop-test12。

创建切面类，代码如下：

```
package com.ghy.www.aspect;

import org.aspectj.lang.annotation.*;
import org.springframework.stereotype.Component;

import java.util.Date;

@Component
@Aspect
public class AspectObject {
    // @Pointcut 注解的解释如下
    // (1)@Pointcut 注解的功能是声明1个切点表达式
    // (2)由于切面要取得调用方法时传入的参数，因此要使用 args 表达式：args(xxxxxx)来进行获取
    // (3)与@Pointcut 关联的方法的参数名称必须和 args() 表达式中的 xxxxxx 一样
    @Pointcut(value = "execution(* com.ghy.www.service.UserinfoService.method1(int)) && args(xxxxxx)")
    public void methodAspect1(int xxxxxx) {
    }

    // @Before 注解的解释如下
```

```java
// (1)属性value = "methodAspect(ageabc)"是引用方法：public void methodAspect(int xxxxxx)
// 引用时方法的参数名称可以不一样，一个是xxxxxx，另一个是ageabc
// (2)与@Before关联的方法：public void method1Before(int ageabc)
// 中的参数名称必须和@Before(value = "methodAspect(ageabc)")
// 配置中的方法的参数名称一样
// (3)@Pointcut 和@Before交接的关联点在方法的名称methodAspect，
// 不包含参数的命名统一性
@Before(value = "methodAspect1(ageabc)")
public void method1Before(int ageabc) {
    System.out.println("切面：public void method1Before(int ageabc) ageabc=" + ageabc);
}


@Pointcut(value = "execution(* com.ghy.www.service.UserinfoService.method2(String,String,int,java.util.Date)) && args(u,p,a,i)")
public void methodAspect2(String u, String p, int a, Date i) {
}

@Before(value = "methodAspect2(uu,pp,aa,ii)")
public void method2Before(String uu, String pp, int aa, Date ii) {
    System.out.println("切面：public void method2Before(String uu, String pp, int aa, Date ii) uu=" + uu + " pp=" + pp
            + " aa=" + aa + " ii=" + ii);
}
}
```

创建业务类，代码如下：

```java
package com.ghy.www.service;

import org.springframework.stereotype.Service;

import java.util.Date;

@Service
public class UserinfoService {
    public void method1(int ageage) {
        System.out.println("method1 age=" + ageage);
    }

    public String method2(String username, String password, int age, Date insertdate) {
        System.out.println(
                "method2 username=" + username + " password=" + password + " age=" + age +
" insertdate=" + insertdate);
        return "我是返回值method2";
    }
}
```

创建运行类，代码如下：

```java
package com.ghy.www.test;

import com.ghy.www.config.SpringConfig;
import com.ghy.www.service.UserinfoService;
import org.springframework.context.ApplicationContext;
import org.springframework.context.annotation.AnnotationConfigApplicationContext;
```

```
    import java.util.Date;

    public class Test {
        public static void main(String[] args) {
            ApplicationContext context = new AnnotationConfigApplicationContext(SpringConfig.class);
            UserinfoService service = (UserinfoService) context.getBean(UserinfoService.class);
            service.method1(100);
            System.out.println();
            System.out.println();
            System.out.println("main get method2 returnValue=" + service.method2("中国", "中国人", 123, new Date()));
        }
    }
```

程序运行结果如下：

```
切面：public void method1Before(int ageabc) ageabc=100
method1 age=100


切面：public void method2Before(String uu, String pp, int aa, Date ii) uu=中国 pp=中国人 aa=123 ii=Mon Sep 28 23:52:27 CST 2020
method2 username=中国 password=中国人 age=123 insertdate=Mon Sep 28 23:52:27 CST 2020
main get method2 returnValue=我是返回值method2
```

2.4.7　使用@AfterReturning 和@AfterThrowing 向切面传入参数

创建测试项目 aop-test13。

创建切面类，代码如下：

```
package com.ghy.www.aspect;

import org.aspectj.lang.annotation.*;
import org.springframework.stereotype.Component;

import java.util.Date;

@Component
@Aspect
public class AspectObject {
    @Pointcut(value = "execution(* com.ghy.www.service.UserinfoService.method1(int)) && args(xxxxxx)")
    public void methodAspect1(int xxxxxx) {
    }

    @Pointcut(value = "execution(* com.ghy.www.service.UserinfoService.method2(String,String,int,java.util.Date)) && args(u,p,a,i)")
    public void methodAspect2(String u, String p, int a, Date i) {
    }

    @Pointcut(value = "execution(* com.ghy.www.service.UserinfoService.*(..))")
    public void methodAspect3() {
    }

    @Before(value = "methodAspect1(xxxxxx)")
```

```java
        public void method1Before(int xxxxxx) {
            System.out.println("切面: public void method1Before(int xxxxxx) xxxxxx=" + xxxxxx);
        }

        @Before(value = "methodAspect2(u, p, a, i)")
        public void method2Before(String u, String p, int a, Date i) {
            System.out.println("切面: public void method2Before(String u, String p, int a, Date i) u=" + u + " p=" + p + " a="
                    + a + " i=" + i);
        }

        @AfterReturning(value = "methodAspect3()", returning = "returnParam")
        public void method3AfterReturning(Object returnParam) {
            System.out.println("public void method3AfterReturning(Object returnParam) returnParam=" + returnParam);
        }

        @AfterThrowing(value = "methodAspect3()", throwing = "t")
        public void method4AfterThrowing(Throwable t) {
            System.out.println("public void method4AfterThrowing(Throwable t) t=" + t);
        }
    }
```

创建业务类，代码如下：

```java
package com.ghy.www.service;

import org.springframework.stereotype.Service;

import java.util.Date;

@Service
public class UserinfoService {
    public void method1(int ageage) {
        System.out.println("method1 age=" + ageage);
    }

    public String method2(String username, String password, int age, Date insertdate) {
        System.out.println(
                "method2 username=" + username + " password=" + password + " age=" +
                        age + " insertdate=" + insertdate);
        Integer.parseInt("a");
        return "我是返回值method2";
    }
}
```

创建运行类，代码如下：

```java
package com.ghy.www.test;

import com.ghy.www.config.SpringConfig;
import com.ghy.www.service.UserinfoService;
import org.springframework.context.ApplicationContext;
import org.springframework.context.annotation.AnnotationConfigApplicationContext;

import java.util.Date;

public class Test {
    public static void main(String[] args) {
```

```
                ApplicationContext context = new AnnotationConfigApplicationContext(SpringConfig.
class);
                UserinfoService service = (UserinfoService) context.getBean(UserinfoService.class);
                service.method1(100);
                System.out.println();
                System.out.println();
                System.out.println("main get method2 returnValue=" + service.method2("中国",
"中国人", 123, new Date()));
        }
}
```

程序运行结果如下：

```
切面: public void method1Before(int xxxxxx) xxxxxx=100
method1 age=100
public void method3AfterReturning(Object returnParam) returnParam=null



切面: public void method2Before(String u, String p, int a, Date i) u=中国 p=中国人 a=123
i=Mon Sep 28 23:56:09 CST 2020
    method2 username=中国 password=中国人 age=123 insertdate=Mon Sep 28 23:56:09 CST 2020
    public void method4AfterThrowing(Throwable t) t=java.lang.NumberFormatException: For
input string: "a"
    Exception in thread "main" java.lang.NumberFormatException: For input string: "a"
```

2.4.8　向环绕通知传入参数

下面我们将实现向环绕通知传入参数。

创建测试项目 aop-test14。

创建切面类，代码如下：

```
package com.ghy.www.aspect;

import org.aspectj.lang.ProceedingJoinPoint;
import org.aspectj.lang.annotation.*;
import org.springframework.stereotype.Component;

import java.util.Date;

@Component
@Aspect
public class AspectObject {
    @Pointcut(value = "execution(* com.ghy.www.service.UserinfoService.method1(int))
&& args(xxxxxx)")
    public void methodAspect1(int xxxxxx) {
    }

    @Pointcut(value = "execution(* com.ghy.www.service.UserinfoService.method2(String,
String,int,java.util.Date)) && args(u,p,a,i)")
    public void methodAspect2(String u, String p, int a, Date i) {
    }

    @Pointcut(value = "execution(* com.ghy.www.service.UserinfoService.*(..))")
    public void methodAspect3() {
    }
```

```java
    @Around(value = "methodAspect1(xxxxxx)")
    public void method1Around(ProceedingJoinPoint point, int xxxxxx) throws Throwable {
        System.out.println("切面开始: public void method1Before(ProceedingJoinPoint point, int xxxxxx) xxxxxx=" + xxxxxx);
        point.proceed();
        System.out.println("切面结束: public void method1Before(ProceedingJoinPoint point, int xxxxxx) xxxxxx=" + xxxxxx);
    }

    @Around(value = "methodAspect2(u, p, a, i)")
    public Object method2Around(ProceedingJoinPoint point, String u, String p, int a, Date i) throws Throwable {
        Object returnValue = null;
        System.out.println(
                "切面开始: public void method2Before(ProceedingJoinPoint point, String u, String p, int a, Date i) u=" + u
                        + " p=" + p + " a=" + a + " i=" + i);
        returnValue = point.proceed();
        System.out.println(
                "切面开始: public void method2Before(ProceedingJoinPoint point, String u, String p, int a, Date i) u=" + u
                        + " p=" + p + " a=" + a + " i=" + i);
        return returnValue;
    }

    @AfterReturning(value = "methodAspect3()", returning = "returnParam")
    public void method3AfterReturning(Object returnParam) {
        System.out.println("public void method3AfterReturning(Object returnParam) returnParam=" + returnParam);
    }

    @AfterThrowing(value = "methodAspect3()", throwing = "t")
    public void method4AfterThrowing(Throwable t) {
        System.out.println("public void method4AfterThrowing(Throwable t) t=" + t);
    }
}
```

注意: 在切面中不要对异常进行捕获（catch），要将异常抛给（throws）Spring 框架进行后续处理，这样会使@AfterThrowing 通知得到执行。

创建业务类，代码如下：

```java
package com.ghy.www.service;

import org.springframework.stereotype.Service;

import java.util.Date;

@Service
public class UserinfoService {
    public void method1(int ageage) {
        System.out.println("method1 age=" + ageage);
    }

    public String method2(String username, String password, int age, Date insertdate) {
        System.out.println(
                "method2 username=" + username + " password=" + password + " age=" + age + " insertdate=" + insertdate);
```

```java
        Integer.parseInt("a");
        return "我是返回值method2";
    }
}
```

创建运行类,代码如下:

```java
package com.ghy.www.test;

import com.ghy.www.config.SpringConfig;
import com.ghy.www.service.UserinfoService;
import org.springframework.context.ApplicationContext;
import org.springframework.context.annotation.AnnotationConfigApplicationContext;

import java.util.Date;

public class Test {
    public static void main(String[] args) {
        ApplicationContext context = new AnnotationConfigApplicationContext(SpringConfig.class);
        UserinfoService service = (UserinfoService) context.getBean(UserinfoService.class);
        service.method1(100);
        System.out.println();
        System.out.println();
        System.out.println("main get method2 returnValue=" + service.method2("中国", "中国人", 123, new Date()));
    }
}
```

程序运行结果如下:

```
切面开始: public void method1Before(ProceedingJoinPoint point, int xxxxxx) xxxxxx=100
method1 age=100
public void method3AfterReturning(Object returnParam) returnParam=null
切面结束: public void method1Before(ProceedingJoinPoint point, int xxxxxx) xxxxxx=100


切面开始: public void method2Before(ProceedingJoinPoint point, String u, String p, int a, Date i) u=中国 p=中国人 a=123 i=Mon Sep 28 23:59:42 CST 2020
    method2 username=中国 password=中国人 age=123 insertdate=Mon Sep 28 23:59:42 CST 2020
    public void method4AfterThrowing(Throwable t) t=java.lang.NumberFormatException: For input string: "a"
Exception in thread "main" java.lang.NumberFormatException: For input string: "a"
```

2.4.9 实现多切面的应用

在系统中多个切面可以一起运行。

创建测试项目 aop-test15。

创建切面类 AspectObject1,代码如下:

```java
package com.ghy.www.aspect;

import org.aspectj.lang.ProceedingJoinPoint;
import org.aspectj.lang.annotation.Around;
import org.aspectj.lang.annotation.Aspect;
import org.springframework.stereotype.Component;
```

```java
@Component
@Aspect
public class AspectObject1 {
    @Around(value = "execution(* com.ghy.www.service.UserinfoService.*(..))")
    public Object around(ProceedingJoinPoint point) {
        Object returnObject = null;
        try {
            System.out.println("开始1");
            returnObject = point.proceed();
            System.out.println("结束1");
        } catch (Throwable e) {
            e.printStackTrace();
        }
        return returnObject;
    }
}
```

创建切面类 AspectObject2，代码如下：

```java
package com.ghy.www.aspect;

import org.aspectj.lang.ProceedingJoinPoint;
import org.aspectj.lang.annotation.Around;
import org.aspectj.lang.annotation.Aspect;
import org.springframework.stereotype.Component;

@Component
@Aspect
public class AspectObject2 {
    @Around(value = "execution(* com.ghy.www.service.UserinfoService.*(..))")
    public Object around(ProceedingJoinPoint point) {
        Object returnObject = null;
        try {
            System.out.println("开始2");
            returnObject = point.proceed();
            System.out.println("结束2");
        } catch (Throwable e) {
            e.printStackTrace();
        }
        return returnObject;
    }
}
```

创建切面类 AspectObject3，代码如下：

```java
package com.ghy.www.aspect;

import org.aspectj.lang.ProceedingJoinPoint;
import org.aspectj.lang.annotation.Around;
import org.aspectj.lang.annotation.Aspect;
import org.springframework.stereotype.Component;

@Component
@Aspect
public class AspectObject3 {
    @Around(value = "execution(* com.ghy.www.service.UserinfoService.*(..))")
    public Object around(ProceedingJoinPoint point) {
        Object returnObject = null;
        try {
```

```java
            System.out.println("开始 3");
            returnObject = point.proceed();
            System.out.println("结束 3");
        } catch (Throwable e) {
            e.printStackTrace();
        }
        return returnObject;
    }
}
```

创建业务类，代码如下：

```java
package com.ghy.www.service;

import org.springframework.stereotype.Service;

@Service
public class UserinfoService {
    public void method1() {
        System.out.println("method1 run !");
    }

    public String method2() {
        System.out.println("method2 run !");
        return "我是返回值 A";
    }

    public String method3() {
        System.out.println("method3 run !");
        Integer.parseInt("a");
        return "我是返回值 B";
    }
}
```

创建运行类，代码如下：

```java
package com.ghy.www.test;

import com.ghy.www.config.SpringConfig;
import com.ghy.www.service.UserinfoService;
import org.springframework.context.ApplicationContext;
import org.springframework.context.annotation.AnnotationConfigApplicationContext;

public class Test {
    public static void main(String[] args) {
        ApplicationContext context = new AnnotationConfigApplicationContext(SpringConfig.class);
        UserinfoService service = (UserinfoService) context.getBean(UserinfoService.class);
        service.method1();
    }
}
```

程序运行结果如下：

```
开始 1
开始 2
开始 3
method1 run !
```

结束 3
结束 2
结束 1

如果我们想制定切面运行的顺序，可以使用@Order 注解。

2.4.10 使用@Order 注解制定切面的运行顺序

在系统中多个切面可以一起运行，并且可以使用@Order 注解制定切面运行的顺序。

创建测试项目 aop-test16。

创建切面类，代码如下：

```
@Component
@Aspect
@Order(value = 3)
public class AspectObject1 {

@Component
@Aspect
@Order(value = 2)
public class AspectObject2 {

@Component
@Aspect
@Order(value = 1)
public class AspectObject3 {
```

程序运行结果如下：

```
开始 3
开始 2
开始 1
method1 run !
结束 1
结束 2
结束 3
```

第 3 章 Spring 5 MVC 实战技术

本章目标
（1）掌握常用注解
（2）结合 AJAX 和 JSON
（3）转发与重定向
（4）上传与下载文件
（5）实现国际化
（6）结合 AOP 切面

3.1 简介

什么是 MVC 模式？MVC 模式是一种开发方式，它的主要用途是将组件进行隔离分层。其中，M 表示模型，模型中包含传递的数据。在软件项目中，M 常常被定义为业务模型，也就是业务/服务层，V 表示视图层，也就是用什么组件显示数据，常用的是 HTML 和 JSP 等文件，C 为控制层，表示软件在大方向上的执行流程以及用哪个视图对象将数据展示给客户。因此 MVC 模式是指将不同功能的组件进行隔离和分层，这样有利于代码的后期维护。

什么是软件框架呢？软件框架就是软件功能的半成品，提供了针对某一个领域所写代码的基本模型，对大多数通用功能进行封装，程序员只需要在这些半成品上继续开发，这样可以提高程序员的开发效率，缩短软件设计的整体周期，统一并规范软件的整体架构。所有程序员使用一种方式进行开发，有利于新成员快速加入开发进程。

Spring 5 MVC 框架是现在主流的 Java Web 服务端 MVC 分层框架，它在功能及代码执行效率上进行了优化和增强。现阶段越来越多的软件公司在使用 Spring 5 MVC 框架开发软件项目，Spring 5 MVC 提供了很多功能性的注解，这样可以方便程序员代码的设计与后期维护，实现组件的松耦合，有利于软件模块间的设计，最重要的是减少了 XML 配置文件中的代码量。

3.2 在 Spring Boot 框架中搭建 Spring MVC 开发环境

Spring Boot 是由 Pivotal 团队开发的框架。Spring Boot 框架基本上没有提供任何应用层技术，其设计的目的是简化 Spring 应用的环境搭建和开发过程，以及其他框架的集成，比如使用 Spring Boot 框架可以简化 SSM 和 SSH 框架整合的步骤，甚至整合的 XML 配置文件都可以省略，因为大部分工作已经由 Spring Boot 完成。我们可以将 Spring Boot 理解成一个盒子，在盒子中放入其他框架来完成软件的设计。这说明想要开发软件项目，只会 Spring Boot 是不够的，我们还要掌握 Spring、Spring MVC 和 MyBatis 等主流架框，然后使用 Spring Boot 简化开发环境的搭建。这个观念是大多数 Java 初学者不知道的。很多初学者认为，现在 Spring Boot 这么流行，可以在不掌握 Spring、Spring MVC 和 MyBatis 等主流架框的前提下直接学习 Sping Boot，结果导致处处踩"坑"，不得不回过头来学习 Spring、Spring MVC 和 MyBatis 等主流架框，增加了学习的时间成本。因此学习技术是有顺序的，学技术前要仔细调研。

本节将使用 Spring Boot 开发基于 MVC 模式的 Web 项目。

3.2.1 搭建 Spring MVC 开发环境

本节实现在 Spring Boot 中搭建 Spring MVC 开发环境。

1．创建 Maven Web Project

创建 Maven Web Project，如图 3-1 所示。

图 3-1　创建 Maven Web Project

选择 "maven-archetype-webapp" 类型的 Web 项目，单击 Next 按钮继续，出现图 3-2 所示的界面。

项目名称为 springmvc_1，单击 Next 按钮继续，出现图 3-3 所示的界面，进行相关配置。

单击 Finish 按钮完成 Maven Web Project 的创建。

项目结构如图 3-4 所示。

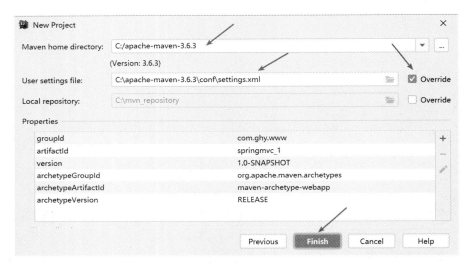

图 3-2　配置项目

图 3-3　配置 Maven

手动创建 java 和 resources 文件夹，并设置文件夹类型，然后创建包"com.ghy.www"，创建后的项目结构如图 3-5 所示。

图 3-4　项目结构

图 3-5　创建文件夹和包

2. 在 pom.xml 文件中添加 Spring Boot 相关配置

在 pom.xml 文件中添加核心配置如下：

```xml
<parent>
    <!-- 从 Spring Boot 继承默认的配置 -->
    <groupId>org.springframework.boot</groupId>
    <artifactId>spring-boot-starter-parent</artifactId>
    <version>2.3.4.RELEASE</version>
</parent>

<properties>
    <project.build.sourceEncoding>UTF-8</project.build.sourceEncoding>
    <maven.compiler.source>1.8</maven.compiler.source>
    <maven.compiler.target>1.8</maven.compiler.target>
</properties>

<dependencies>
    <dependency>
        <groupId>junit</groupId>
        <artifactId>junit</artifactId>
        <version>4.11</version>
        <scope>test</scope>
    </dependency>

    <!-- 引入 Spring Boot 对 Web 开发的依赖 -->
    <dependency>
        <groupId>org.springframework.boot</groupId>
        <artifactId>spring-boot-starter-web</artifactId>
    </dependency>
</dependencies>
```

3. Spring Boot 中常用的 Starter

Spring Boot 提供了许多 Starter 来将依赖的 jar 包文件添加到类路径中。Starter 就是将某一个功能依赖的所有组件的 JAR 文件进行打包。只要引用这个 Starter，该 Starter 依赖的所有组件的 JAR 文件就会被自动下载到项目中。由于当前正在开发一个 Web 应用程序，因此添加了一个 spring-boot-starter-web 依赖。

在 pom.xml 文件中，必须指定 spring-boot-starter-parent 版本号，其他 Starter 版本号最终由 spring-boot-starter-parent 来决定，而不需要显式地指定。

有关 Spring Boot 支持的完整 Starter 列表，请参看官方文档，如图 3-6 所示。

4. 官方推荐的项目结构

Spring Boot 官方并没有强制规定 Spring Boot 的项目结构，但推荐使用的项目结构如图 3-7 所示。

在 com.example.myapplication 包下创建 Application.java（启动类），执行 Application.java（启动类）后会自动扫描子孙包中的资源，并不需要显式地指定扫描某个包。

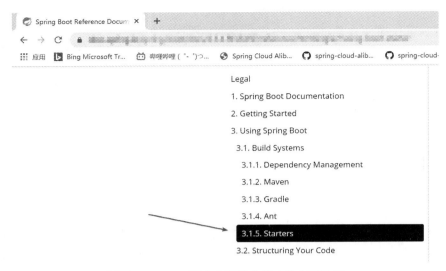

图 3-6　Starter 列表在帮助文档中的索引位置

```
com
 +- example
     +- myapplication
         +- Application.java
         |
         +- customer
         |   +- Customer.java
         |   +- CustomerController.java
         |   +- CustomerService.java
         |   +- CustomerRepository.java
         |
         +- order
             +- Order.java
             +- OrderController.java
             +- OrderService.java
             +- OrderRepository.java
```

图 3-7　官方推荐使用的项目结构

5．创建 Spring MVC 控制层

创建控制层，代码如下：

```java
package com.ghy.www.controller;

import org.springframework.stereotype.Controller;
import org.springframework.web.bind.annotation.RequestMapping;
import org.springframework.web.bind.annotation.ResponseBody;

@Controller
public class TestController {
    @RequestMapping("test")
    @ResponseBody
    public String test() {
        System.out.println("执行了public String test()");
        return "我是public String test()的返回值";
    }
}
```

6. 创建运行类

创建运行类，代码如下：

```
package com.ghy.www;

import org.springframework.boot.SpringApplication;
import org.springframework.boot.autoconfigure.SpringBootApplication;

@SpringBootApplication
public class Application {
    public static void main(String[] args) {
        SpringApplication.run(Application.class, args);
    }
}
```

注解@SpringBootApplication 表示创建的类是 Spring Boot 应用，是 Spring Boot 项目的启动类。

7. 运行项目

运行 Application 类其实就是启动 Tomcat，如图 3-8 所示。

```
: Tomcat initialized with port(s): 8080 (http)
: Starting service [Tomcat]
: Starting Servlet engine: [Apache Tomcat/9.0.38]
: Initializing Spring embedded WebApplicationContext
: Root WebApplicationContext: initialization completed in 730 ms
: Initializing ExecutorService 'applicationTaskExecutor'
: Tomcat started on port(s): 8080 (http) with context path ''
: Started Application in 1.35 seconds (JVM running for 2.436)
```

图 3-8 启动 Tomcat

Spring Boot 内置 Tomcat，使用的默认端口是 8080。

执行控制层地址 test，程序运行结果如图 3-9 所示。

图 3-9 成功运行控制层

通过以上步骤，我们可以发现，Spring Boot 简化了 Spring MVC 开发过程。

（1）省略了配置 web.xml。

（2）省略了配置 springMVC-servlet.xml。

（3）不需要显式地指定扫描某个包。

在 Spring Boot 项目中直接开发 Application.java 和×××Controller.java 就可以了，因为所有与环境配置相关的工作都由 Spring Boot 进行了封装。程序员只需要把精力放在业务的实现上，而不必面对烦琐的 XML 配置工作。Spring Boot 简化了程序员搭建项目的步骤，提高了开发效率。

8. 在 CMD 中启动项目

在生产环境中，我们不可能在服务器里安装 IntelliJ IDEA 后再启动项目，而通常是在 CMD 中启动项目。

在 CMD 中进入项目的根目录，输入并执行命令 mvn spring-boot:run，如图 3-10 所示。

```
C:\Users\Administrator\Desktop\ssm\第3章\springmvc_1>mvn spring-boot:run
[INFO] Scanning for projects...
Downloading from mynexus: http://localhost:8081/repository/maven-group/org/codehaus/mojo/bu
Downloaded from mynexus: http://localhost:8081/repository/maven-group/org/codehaus/mojo/bui
Downloading from mynexus: http://localhost:8081/repository/maven-group/org/codehaus/mojo/mo
```

图 3-10　执行命令

启动项目后可以正常访问控制层。

注意：可以使用<Ctrl+C>组合键结束进程。

9. 创建可执行 WAR/JAR 文件

我们建议将 Spring Boot Web 项目打包成可以自动运行的 WAR/JAR 文件，直接运行 WAR/JAR 文件就可以启动 Web 项目。WAR/JAR 中已经内嵌了 Tomcat。

我们可以通过创建一个完全独立的可执行 WAR/JAR 文件来运行程序。可执行 WAR/JAR 文件也称为"fat WAR/JAR"文件，其内部包含已编译的类，以及运行项目所需要的所有依赖项（包含 JAR、XML、properties 和 JSP/HTML 文件等资源）。

在 pom.xml 文件中添加配置如下：

```xml
<build>
    <plugins>
        <plugin>
            <groupId>org.springframework.boot</groupId>
            <artifactId>spring-boot-maven-plugin</artifactId>
        </plugin>
    </plugins>
</build>
```

在 CMD 中进入项目的根目录，通过执行命令 mvn package 进行打包。

控制台出现的信息如图 3-11 所示。

```
--- maven-war-plugin:3.2.2:war (default-war) @ springmvc_1 ---
Packaging webapp
Assembling webapp [springmvc_1] in [C:\Users\Administrator\Desktop\ssm\第3章\springmvc_1\target\springmvc_1]
Processing war project
Copying webapp resources [C:\Users\Administrator\Desktop\ssm\第3章\springmvc_1\src\main\webapp]
Webapp assembled in [213 msecs]
Building war: C:\Users\Administrator\Desktop\ssm\第3章\springmvc_1\target\springmvc_1.war

--- spring-boot-maven-plugin:2.3.4.RELEASE:repackage (repackage) @ springmvc_1 ---
Replacing main artifact with repackaged archive
------------------------------------------------------------------------
BUILD SUCCESS
------------------------------------------------------------------------
Total time:  2.714 s
```

图 3-11　成功创建 WAR 文件

3.2 在 Spring Boot 框架中搭建 Spring MVC 开发环境

如果我们想创建 JAR 文件，则需要更改 pom.xml 文件中的如下配置。

```
<packaging>jar</packaging>
```

然后，在 CMD 中进入项目的根目录，通过执行 mvn package 命令进行打包。

```
mvn package
```

生成的 JAR 文件如图 3-12 所示。
生成的 WAR 文件和 JAR 文件都可以自动运行。
WAR 文件和 JAR 文件所在的位置如图 3-13 所示。

图 3-12　成功创建 JAR 文件　　　　　图 3-13　成功创建 JAR 和 WAR 文件

在 CMD 中使用命令

```
java -jar springmvc_1.war
```

或者

```
java -jar springmvc_1.jar
```

启动项目并运行控制层，就会发现启动后的项目可以正常运行控制层。

注意：可以使用<Ctrl+C>组合键结束进程。

如果不在 pom.xml 文件中添加图 3-14 所示的配置，则在运行项目时会出现图 3-15 所示的异常。

```
<plugins>
    <plugin>
        <groupId>org.springframework.boot</groupId>
        <artifactId>spring-boot-maven-plugin</artifactId>
    </plugin>
</plugins>
```

图 3-14　必备配置

```
C:\Users\Administrator\Desktop\ssm\第3章\springmvc_1\target>java -jar springmvc_1.jar
springmvc_1.jar中没有主清单属性
```

```
C:\Users\Administrator\Desktop\ssm\第3章\springmvc_1\target>java -jar springmvc_1.war
springmvc_1.war中没有主清单属性
```

图 3-15　出现异常

3.2.2　搭建 CSS+JavaScript+HTML+JSP 开发环境

我们按照创建 springmvc_1 项目的步骤创建 springmvc_2 项目，并把 springmvc_1 项目中的相关资源复制到 springmvc_2 项目中。

1．初步配置 pom.xml 文件

在 pom.xml 文件中添加如下依赖配置。

```xml
<!-- 添加 JSTL 依赖 -->
<dependency>
    <groupId>javax.servlet</groupId>
    <artifactId>jstl</artifactId>
</dependency>
```

2．创建控制层

创建控制层，代码如下：

```java
package com.ghy.www.controller;

import org.springframework.stereotype.Controller;
import org.springframework.web.bind.annotation.RequestMapping;

import javax.servlet.http.HttpServletRequest;
import java.util.ArrayList;
import java.util.List;

@Controller
public class TestController {
    @RequestMapping("htmlTest")
    public String htmlTest() {
        System.out.println("执行了public String htmlTest()");
        return "test.html";
    }

    @RequestMapping("jspTest")
    public String jspTest(HttpServletRequest request) {
        System.out.println("执行了public String jspTest()");
        List listString = new ArrayList();
        listString.add("中国人1");
        listString.add("中国人2");
        listString.add("中国人3");

        request.setAttribute("listString", listString);

        return "test.jsp";
    }
}
```

3.2 在 Spring Boot 框架中搭建 Spring MVC 开发环境

3. 创建 CSS 文件、JavaScript 文件、HTML 文件和 JSP 文件

创建的 CSS 文件 mycss.css，代码如下：

```css
.myStyle {
    color: red;
    font-size: 100px;
}
```

创建的 JavaScript 文件 myjs.js，代码如下：

```javascript
setTimeout(function() {
    alert("我是js脚本，我是自动运行的！");
}, 3000);
```

创建的 HTML 文件 test.html，代码如下：

```html
<!DOCTYPE html>
<html lang="en">
    <head>
        <meta charset="UTF-8">
        <title>Title</title>
        <script type="text/javascript" src="js/myjs.js"></script>
        <link rel="stylesheet" type="text/css" href="css/mycss.css"/>
    </head>
    <body>
        我是html文件！
        <br/>
        <h1 class="myStyle">我是美化的文字</h1>
    </body>
</html>
```

创建的 JSP 文件 test.jsp 文件，代码如下：

```jsp
<%@ page language="java" contentType="text/html; charset=utf-8"
    pageEncoding="utf-8" %>
<%@ taglib uri="http://java.sun.com/jsp/jstl/core" prefix="c" %>
<!DOCTYPE html>
<html>
    <head>
        <meta charset="utf-8">
        <title>Insert title here</title>
        <script type="text/javascript" src="js/myjs.js"></script>
        <link rel="stylesheet" type="text/css" href="css/mycss.css"/>
    </head>
    <body>
        我是jsp文件！
        <br/>
        <h1 class="myStyle">我是美化的文字</h1>
        <br/>
        循环输出List中的数据：
        <br/>
        <c:forEach var="eachString" items="${listString}">
            ${eachString}<br/>
        </c:forEach>
    </body>
</html>
```

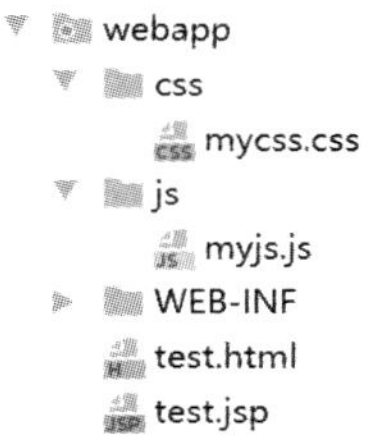

以上创建的文件的位置如图 3-16 所示。

图 3-16 项目结构

4. 创建运行类

创建运行类，代码如下：

```
package com.ghy.www;

import org.springframework.boot.SpringApplication;
import org.springframework.boot.autoconfigure.SpringBootApplication;

@SpringBootApplication
public class Application {
    public static void main(String[] args) {
        SpringApplication.run(Application.class, args);
    }
}
```

5. 运行项目

启动项目，执行控制层地址 htmlTest，结果如图 3-17 所示。

图 3-17　成功执行 HTML 文件

执行控制层地址 jspTest，结果如图 3-18 所示。

图 3-18　没有执行 JSP 文件而是进行下载

我们发现,输入地址后并没有执行 JSP 文件,而是进行了下载。我们可以通过在 pom.xml 文件中添加如下依赖解决上述问题。

```xml
<!-- 解决下载JSP文件而不是执行的问题 -->
<dependency>
    <groupId>org.apache.tomcat.embed</groupId>
    <artifactId>tomcat-embed-jasper</artifactId>
    <scope>provided</scope>
</dependency>
```

重启 Tomcat 服务,再次执行控制层地址 jspTest,结果如图 3-19 所示。

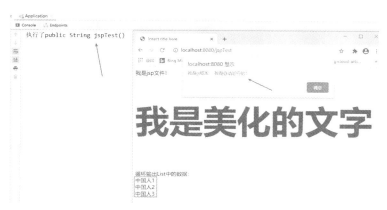

图 3-19　成功执行 JSP 文件

6. 实现项目首页

我们可以对 Web 项目设置首页。首页的起始点不是 index.jsp,而是默认执行一个控制层,取得数据后转发到 main.jsp,最后将数据通过 main.jsp 中显示。

创建 main.jsp 文件,代码如下:

```jsp
<%@ page language="java" contentType="text/html; charset=utf-8"
    pageEncoding="utf-8" %>
<!DOCTYPE html>
<html>
    <head>
        <meta charset="utf-8">
        <title>Insert title here</title>
    </head>
    <body>
        我是main.jsp文件!
    </body>
</html>
```

创建控制层,代码如下:

```java
package com.ghy.www.controller;

import org.springframework.stereotype.Controller;
import org.springframework.web.bind.annotation.RequestMapping;

@Controller
public class MainPageController {
    @RequestMapping("/")
```

```
    public String htmlTest() {
        System.out.println("执行了/");
        return "main.jsp";
    }
}
```

在浏览器中输入并打开网址：http://localhost:8080/。

控制台和浏览器中显示的内容如图 3-20 所示。

图 3-20　出现首页

3.3　核心技术

对于本节中介绍的技术，读者必须要掌握，因为经常会在软件项目中使用它们。

3.3.1　执行控制层——无传递参数

本示例将实现在浏览器上访问控制层的 URL 后执行对应控制层中的代码，然后转发到 JSP 文件的效果，"请求—响应"模型的全部过程都在此示例中进行了体现。

创建项目 noparam。

创建控制层，代码如下：

```
package com.ghy.www.controller;

import org.springframework.stereotype.Controller;
import org.springframework.web.bind.annotation.RequestMapping;

//@Controller 注解表示该 Java 类是控制层
@Controller
public class TestController {
    // 通过@RequestMapping 注解可以用指定的 URL
    // 访问该控制层中与 URL 关联的业务方法
    @RequestMapping(value = "helloWorld")
    public String helloWorldMethod() {
        System.out.println("run helloWorld Method!~");
        return "hello.jsp";
    }
}
```

创建 hello.jsp 文件，该文件的核心代码如下：

```jsp
<%@ page language="java" contentType="text/html; charset=utf-8"
    pageEncoding="utf-8" %>
<%@ taglib uri="http://java.sun.com/jsp/jstl/core" prefix="c" %>
<!DOCTYPE html>
<html>
    <head>
        <meta charset="utf-8">
        <title>Insert title here</title>
    </head>
    <body>
        this is hello.jsp
    </body>
</html>
```

启动项目并执行控制层地址 helloWorld，结果如图 3-21 所示。

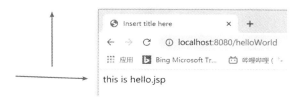

图 3-21 成功转发到 JSP 文件

3.3.2 执行控制层——有传递参数

本示例将实现在浏览器上访问控制层的 URL 后执行对应控制层中的代码，并且向控制层传递参数，然后转发到 JSP 文件的效果。

创建项目 hasparam。

创建控制层，代码如下：

```java
package com.ghy.www.controller;

import org.springframework.stereotype.Controller;
import org.springframework.web.bind.annotation.RequestMapping;
import org.springframework.web.bind.annotation.RequestParam;

@Controller
public class TestController {
    @RequestMapping(value = "helloWorld")
    public String helloWorldMethod(@RequestParam("username") String u) {
        System.out.println("hello " + u);
        return "hello.jsp";
    }
}
```

注解代码

```
@RequestParam("username") String u
```

的作用是取得 URL 中参数名是 username 的值，再将参数值传递给参数 u。

启动项目并执行控制层地址 helloWorld?username=i_like_spring,控制台输出的结果如下:

```
hello i_like_spring
```

控制层成功地从 URL 中获得参数值。

3.3.3 执行控制层——有传递参数简化版

从请求(request)中获得参数值可以进行简化,即注解@RequestParam 可以省略。

创建项目 simplehasparam。

(1)创建控制层,代码如下:

```java
package com.ghy.www.controller;

import org.springframework.stereotype.Controller;
import org.springframework.web.bind.annotation.RequestMapping;

@Controller
public class TestController {
    @RequestMapping(value = "helloWorld")
    public String helloWorldMethod(String username, String password) {
        System.out.println("hello " + username + " " + password);
        return "hello.jsp";
    }
}
```

URL 中的同名参数将要自动传给控制层方法中的同名参数,并且不需要@RequestParam 注解。

(2)部署项目,执行控制层地址 helloWorld?username=like&password=java,控制台输出的结果如下:

```
hello like java
```

控制层成功地从 URL 中获得参数值。

3.3.4 实现登录功能

基于前面的介绍,可以使用 Spring MVC 实现登录功能,以整体的角度强化 Spring 5 MVC 的使用。

创建项目 logintest。

(1)本示例要创建 3 个 JSP 文件,其中登录界面对应的 login.jsp 文件中的核心代码如下:

```jsp
<%@ page language="java" contentType="text/html; charset=utf-8"
    pageEncoding="utf-8" %>
<%@ taglib uri="http://java.sun.com/jsp/jstl/core" prefix="c" %>
<!DOCTYPE html>
<html>
    <head>
        <meta charset="utf-8">
        <title>Insert title here</title>
    </head>
    <body>
        post:
```

```
            <br/>
            <form action="login" method="post">
                username:<input type="text" name="username">
                <br/>
                password:<input type="text" name="password">
                <br/>
                <input type="submit" value="submit">
            </form>
            <br/>
            get:
            <br/>
            <form action="login" method="get">
                username:<input type="text" name="username">
                <br/>
                password:<input type="text" name="password">
                <br/>
                <input type="submit" value="submit">
            </form>
        </body>
</html>
```

（2）登录成功界面对应的 ok.jsp 文件中的核心代码如下：

```
<body>
    welcome:${param.username}
</body>
```

（3）登录失败界面对应的 no.jsp 文件中的核心代码如下：

```
<body>
    登录失败！
</body>
```

（4）创建控制层，核心代码如下：

```
package com.ghy.www.controller;

import org.springframework.stereotype.Controller;
import org.springframework.web.bind.annotation.RequestMapping;

@Controller
public class UserinfoController {
    @RequestMapping(value = "login")
    public String loginMethod(String username, String password) {
        if (username.equals("a") && password.equals("aa")) {
            return "ok.jsp";
        } else {
            return "no.jsp";
        }
    }
}
```

（5）启动项目，输入并打开网址：http://localhost:8080/login.jsp，会显示登录界面，如图 3-22 所示。

可以分别在 post 或 get 的 "username" 中输入 a，在 "password" 中输入 aa，以便使用 post 或 get 提交请求实现成功登录的效果。单击登录界面中的两个 "submit" 按钮，将会显示登录成功界面，如图 3-23 所示。

图 3-22　登录界面

如果在"username"和"password"中不输入任何内容，直接单击"submit"按钮，就会出现登录失败界面，如图 3-24 所示。

图 3-23 登录成功界面　　　　图 3-24 登录失败界面

在默认的情况下控制层可以处理 get 请求和 post 请求。

3.3.5 将 URL 参数封装到实体类

我们可以将 URL 中的参数值封装到实体类。

本示例通过项目 paramtoentity 实现。

（1）创建封装 URL 参数的实体类，代码如下：

```java
package com.ghy.www.entity;

public class Userinfo {
    private String username;
    private String password;

    public Userinfo() {
    }

    public Userinfo(String username, String password) {
        super();
        this.username = username;
        this.password = password;
    }

    public String getUsername() {
        return username;
    }

    public void setUsername(String username) {
        this.username = username;
    }

    public String getPassword() {
        return password;
    }

    public void setPassword(String password) {
        this.password = password;
    }
}
```

（2）创建控制层，代码如下：

```
package com.ghy.www.controller;

import com.ghy.www.entity.Userinfo;
import org.springframework.stereotype.Controller;
import org.springframework.web.bind.annotation.RequestMapping;

@Controller
public class UserinfoController {
    @RequestMapping(value = "login")
    public String loginMethod(Userinfo userinfo) {
        System.out.println("username=" + userinfo.getUsername());
        System.out.println("password=" + userinfo.getPassword());
        return "index.jsp";
    }
}
```

（3）执行控制层地址 login?username=123&password=456，控制台输出的内容如下：

```
username=123
password=456
```

URL 中的参数值被成功地封装到 Userinfo 实体类。

3.3.6 限制提交方式

在上文中，我们使用 Spring 5 MVC 实现了登录功能，在默认情况下，控制层允许以 post 和 get 方式进行提交，但标准的登录功能使用的是 post，绝大多数不允许以 get 方式进行提交，这就需要在控制层中对提交方式进行限制。这个需求在项目 requestmethodtype 中实现。

（1）创建控制层，代码如下：

```
package com.ghy.www.controller;

import org.springframework.stereotype.Controller;
import org.springframework.web.bind.annotation.RequestMapping;
import org.springframework.web.bind.annotation.RequestMethod;

@Controller
public class UserinfoController {
    @RequestMapping(value = "login", method = RequestMethod.POST)
    public String loginMethod(String username, String password) {
        if (username.equals("a") && password.equals("aa")) {
            return "ok.jsp";
        } else {
            return "no.jsp";
        }
    }
}
```

我们可以使用属性 method = RequestMethod.POST 限制提交方式必须是 post。

（2）部署项目，输入并打开网址：http://localhost:8080/login.jsp。

以 post 方式提交的表单能实现登录成功或失败的效果。但以 get 方式提交表单时却出现异常，如图 3-25 所示。

图 3-25 以 get 方式提交时不被支持

使用注解@RequestMapping(value = "login", method = RequestMethod.POST)可以限制提交方式，这样有利于规范代码。

3.3.7 控制层方法的参数类型

前面的控制层方法的声明如下：

```
public String loginMethod(String username, String password)
```

或

```
public String loginMethod(Userinfo userinfo)
```

控制层方法的参数类型是 String 或实体类，其实 Spring MVC 控制层方法的参数还可以是如下常见的数据类型，如表 3-1 所示。

表 3-1　　　　　　　　　　控制层方法的参数类型

控制层方法的参数类型	解释
WebRequest NativeWebRequest	可以访问 request 的 parameters，request 和 session 的 attributes，而不需要使用 Servlet API
javax.servlet.ServletRequest javax.servlet.ServletResponse MultipartRequest MultipartHttpServletRequest	使用指定的 request 或 response 对象
javax.servlet.http.HttpSession	使用指定的 HttpSession 对象 注意：访问 HttpSession 不是线程安全的。如果有多个请求同时访问 HttpSession 对象，则需要将 RequestMappingHandlerAdapter 类的 synchronizeOnSession 属性设置为 true
HttpMethod	request 请求的 method 方式
java.util.Locale	当前请求的区域
java.util.TimeZone + java.time.ZoneId	当前请求关联的 Zone
java.io.InputStream java.io.Reader	访问 request body 原始的数据
java.io.OutputStream java.io.Writer	访问 response body 原始的数据
@PathVariable	访问 URI 模板变量
@MatrixVariable	访问以 name-value 形式存在于 URI 路径中的片段
@RequestParam	访问 request 中的 parameters。该注解是可选的
@RequestHeader	访问 request headers 中的数据

控制层方法的参数类型	解释
@CookieValue	访问 Cookie
@RequestBody	访问 request body
HttpEntity<B>	访问 request 中的 headers 和 body
@RequestPart	处理"multipart/form-data"请求中的 part
java.util.Map org.springframework.ui.Model org.springframework.ui.ModelMap	用于与 View 层的交互
RedirectAttributes	在重定向时添加 attributes 处理,有两种用法: (1)可以将数据放在 query string 中 (2)结合 flash attributes 将数据存储到临时的空间,当重定向结束后删除临时空间中的数据
@ModelAttribute	访问已存在的 attribute
Errors BindingResult	访问的 Errors 来自数据验证和绑定,或者来自对@RequestBody 或@RequestPart 的验证。一个 Errors 或 BindingResult 参数要声明在验证方法参数之后
类级别的@SessionAttributes	定义 HttpSession 中的 attributes,在处理完成后触发清理
@SessionAttribute	访问 session 中的 attribute。与作为类级的@SessionAttributes 声明的结果而存储在会话中的 Model 属性形成对比
@RequestAttribute	访问 request 中的 attributes

3.3.8 控制层方法的返回值类型

控制层方法可以返回如下常见的数据类型,而不仅仅是 String,如表 3-2 所示。

表 3-2　　　　　　　　　　控制层方法的返回值类型

控制层方法的返回值类型	解释
@ResponseBody	通过 HttpMessageConverters 转换返回值,并且写入 response
HttpEntity<B> ResponseEntity<B>	Response 包括完整的 headers 和 body,通过 HttpMessageConverters 转换返回值,并且写入 response
HttpHeaders	返回 response headers,但不包括 body
String	使用 ViewResolver 解析的视图名称
View	返回 View 实例
java.util.Map org.springframework.ui.Model	要添加到隐式 Model 中的属性,并通过 RequestToViewNameTranslator 确定视图名称
@ModelAttribute	要添加到隐式 Model 中的属性,并通过 RequestToViewNameTranslator 确定视图名称
ModelAndView	用于确定 View 和 attributes
void	如果具有 void 返回值或返回 null 值的方法具有 ServletResponse、OutputStream 参数或@ResponseStatus 注解,则视为已完全处理响应

3.3.9 取得 request-response-session 对象

有时，我们需要在控制层取得 HttpServletRequest、HttpServletResponse 和 HttpSession 对象，从而调用这 3 个对象的方法。

创建项目 request-response-session。

（1）创建控制层，代码如下：

```java
package com.ghy.www.controller;

import org.springframework.stereotype.Controller;
import org.springframework.web.bind.annotation.RequestMapping;

import javax.servlet.http.HttpServletRequest;
import javax.servlet.http.HttpServletResponse;
import javax.servlet.http.HttpSession;

@Controller
public class TestController {
    @RequestMapping(value = "test")
    public String loginMethod(HttpServletRequest request, HttpServletResponse response, HttpSession session) {
        System.out.println(request);
        System.out.println(response);
        System.out.println(session);
        request.setAttribute("requestKey", "request 大中国");
        session.setAttribute("sessionKey", "session 大中国");
        System.out.println(request.getSession().getServletContext().getRealPath("/"));
        return "index.jsp";
    }
}
```

（2）JSP 代码如下：

```
<body>
    ${requestKey}
    <br/>
    ${sessionKey}
</body>
```

（3）执行控制层路径 test，控制台和浏览器会显示相关的数据信息，如图 3-26 所示。

```
org.apache.catalina.connector.RequestFacade@591dbf60
org.apache.catalina.connector.ResponseFacade@789367a8
org.apache.catalina.session.StandardSessionFacade@7d01c676
C:\Users\Administrator\Desktop\ssm\第 3 章\request-response-session\src\main\webapp\
```

图 3-26 输出结果

3.3.10 实现登录失败后的提示信息

虽然 Spring 5 MVC 框架提供的验证框架可以作为前端参数的数据有效性验证，但是，如果存在业务型的验证，仍然需要程序员以手写代码的方式进行处理。本示例演示如何用手动的方式来验证前端传递过来的登录数据，当登录失败后，在前端显示提示信息。

创建测试项目 loginerrormessage。

（1）创建登录界面，代码如下：

```jsp
<%@ page language="java" contentType="text/html; charset=utf-8"
    pageEncoding="utf-8" %>
<%@ taglib uri="http://java.sun.com/jsp/jstl/core" prefix="c" %>
<!DOCTYPE html>
<html>
    <head>
        <meta charset="utf-8">
        <title>Insert title here</title>
    </head>
    <body>
        <form action="login" method="post">
            username:<input type="text" name="username">${message.usernameisnull}
            <br/>
            password:<input type="text" name="password">${message.passwordisnull}
            <br/>
            <input type="submit" value="submit">
        </form>
    </body>
</html>
```

（2）创建登录成功界面，代码如下：

```jsp
<body>
    welcome:${param.username}
</body>
```

（3）创建控制层，其核心代码如下：

```java
package com.ghy.www.controller;

import org.springframework.stereotype.Controller;
import org.springframework.web.bind.annotation.RequestMapping;

import javax.servlet.http.HttpServletRequest;
import java.util.HashMap;
import java.util.Map;

@Controller
public class UserinfoController {
    public Map loginValidateMethod(String username, String password) {
        Map map = new HashMap();
        if (username == null || "".equals(username)) {
            map.put("usernameisnull", "账号为空！");
        }
        if (password == null || "".equals(password)) {
            map.put("passwordisnull", "密码为空！");
        }
        return map;
    }

    @RequestMapping(value = "login")
    public String loginMethod(String username, String password, HttpServletRequest request) {
        Map map = loginValidateMethod(username, password);
```

```
            if (map.size() > 0) {
                request.setAttribute("message", map);
                return "login.jsp";
            } else {
                return "ok.jsp";
            }
        }
    }
```

上面的代码只是验证了表单为空，其实上述代码还可以进行相应的修改，用于验证身份证号格式、邮箱格式、电话号码格式等信息。

（4）如果不输入账号和密码，那么提交表单后返回 login.jsp 页面时会显示出错信息，如图 3-27 所示。

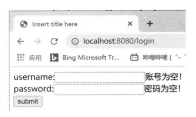

图 3-27　显示出错信息

3.3.11　向 Controller 控制层注入 Service 业务逻辑层

前面的示例都是在 Controller 控制层中进行业务的处理，下面向 Controller 控制层注入 Service 业务逻辑层来实现严格的 MVC 分层。示例代码在项目 injectservice 中。

（1）创建业务接口，代码如下：

```
package com.ghy.www.service;

public interface IUserinfoService {
    public String getUsername();
}
```

（2）创建业务类 A，代码如下：

```
package com.ghy.www.service;

import org.springframework.stereotype.Service;

@Service(value = "serviceA")
public class UserinfoServiceA implements IUserinfoService {
    @Override
    public String getUsername() {
        return "业务逻辑层的大中国 A";
    }
}
```

（3）创建业务类 B，代码如下：

```
package com.ghy.www.service;

import org.springframework.stereotype.Service;

@Service(value = "serviceB")
public class UserinfoServiceB implements IUserinfoService {
    @Override
    public String getUsername() {
        return "业务逻辑层的大中国 B";
    }
}
```

使用注解@Service，表示该类是一个业务对象。

(4)创建控制层,代码如下:

```
package com.ghy.www.controller;

import com.ghy.www.service.IUserinfoService;
import org.springframework.beans.factory.annotation.Autowired;
import org.springframework.beans.factory.annotation.Qualifier;
import org.springframework.stereotype.Controller;
import org.springframework.web.bind.annotation.RequestMapping;

@Controller
public class TestController {
    @Autowired
    @Qualifier(value = "serviceB")
    private IUserinfoService userinfoService;

    @RequestMapping(value = "test")
    public String test() {
        System.out.println(userinfoService.getUsername());
        return "index.jsp";
    }
}
```

(5)运行项目,控制台输出的结果如下:

业务逻辑层的大中国B

3.3.12 重定向——无传递参数

本节要实现在两个控制层中进行重定向操作,并且在重定向时不向目的控制层传递参数。创建测试项目 redirect1。

(1)创建控制层,代码如下:

```
package com.ghy.www.controller;

import org.springframework.stereotype.Controller;
import org.springframework.web.bind.annotation.RequestMapping;

import javax.servlet.http.HttpServletRequest;
import java.util.ArrayList;
import java.util.List;

@Controller
public class TestController {
    @RequestMapping(value = "login")
    public String loginMethod(String username) {
        System.out.println("loginMethod username=" + username);
        return "redirect:/listString.spring";// 重定向无传参
    }

    @RequestMapping(value = "listString")
    public String listStringMethod(HttpServletRequest request) {
        System.out.println("listStringMethod");
        List list = new ArrayList();
        list.add("中国1");
        list.add("中国2");
```

```
        list.add("中国3");
        list.add("中国4");
        request.setAttribute("list", list);
        return "listString.jsp";
    }
}
```

对于重定向的关键代码，就是在返回字符串中加入前缀"redirect:/"（表示这个操作是重定向）。

（2）文件 listString.jsp 的核心代码如下：

```
<%@ page language="java" contentType="text/html; charset=utf-8"
    pageEncoding="utf-8" %>
<%@ taglib uri="http://java.sun.com/jsp/jstl/core" prefix="c" %>
<!DOCTYPE html>
<html>
    <head>
        <title>Title</title>
    </head>
    <body>
        <c:forEach var="eachString" items="${list}">
            ${eachString}
            <br/>
        </c:forEach>
    </body>
</html>
```

（3）部署项目，在浏览器中执行控制层地址 login?username=123，随后，浏览器的地址栏会发生变化，即重定向到网址 listString，证明重定向到其他控制层成功，如图 3-28 所示。

图 3-28　控制层重定向到其他控制层（无传参）

3.3.13　重定向——有传递参数

本节要实现在两个控制层中进行重定向操作，并且在重定向时向目的控制层传递参数。创建测试项目 redirect2。

（1）创建控制层，代码如下：

```
@Controller
public class UserinfoController {
    @RequestMapping(value = "login")
    public String loginMethod() throws UnsupportedEncodingException {
        System.out.println("loginMethod run !");
        String username = java.net.URLEncoder.encode("我是中文我是参数", "utf-8");
        return "redirect:/listString.spring?xxxxxxxxxx=" + username;
    }
```

```
        @RequestMapping(value = "listString")
        public String listStringMethod(String xxxxxxxxxx, HttpServletRequest request) throws
UnsupportedEncodingException {
            xxxxxxxxxx = java.net.URLDecoder.decode(xxxxxxxxxx, "utf-8");
            System.out.println("listStringMethod xxxxxxxxxx=" + xxxxxxxxxx);

            List list = new ArrayList();
            list.add("中国1");
            list.add("中国2");
            list.add("中国3");
            list.add("中国4");

            request.setAttribute("list", list);

            return "listString.jsp";
        }
    }
```

（2）部署项目，在浏览器中，执行控制层地址 login，随后，浏览器的地址栏会发生变化，重定向到网址：listString?xxxxxxxxxx=%E6%88%91%E6%98%AF%E4%B8%AD%E6%96%87%E6%88%91%E6%98%AF%E5%8F%82%E6%95%B0，证明重定向到其他控制层成功，如图 3-29 所示。

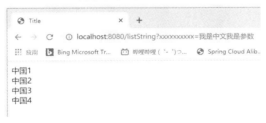

图 3-29　控制层重定向到其他控制层（有传参）

3.3.14　重定向传递参数——RedirectAttributes.addAttribute()方法

在重定向传递参数时，使用拼接 URL 的形式

```
return "redirect:/listString?xxxxxxxxxx=" + username;
```

不太标准与规范，可以使用 RedirectAttributes 类来替代，RedirectAttributes 类是由 Spring MVC 框架提供的。

创建测试项目 redirect3。

（1）创建控制层，代码如下：

```
package com.ghy.www.controller;

import org.springframework.stereotype.Controller;
import org.springframework.web.bind.annotation.RequestMapping;
import org.springframework.web.servlet.mvc.support.RedirectAttributes;

@Controller
public class TestController {
```

```
@RequestMapping(value = "a")
public String a(RedirectAttributes attr) {
    System.out.println("into a method");
    attr.addAttribute("username", "abc");
    attr.addAttribute("age", 123);
    return "redirect:/b";
}

@RequestMapping(value = "b")
public String b(String username, String age) {
    System.out.println("into b method");
    System.out.println("username:" + username);
    System.out.println("age:" + age);
    return "index.jsp";
}
```

（2）部署项目，在浏览器中执行控制层地址 a，随后，浏览器的地址栏发生变化，重定向到网址 b?username=abc&age=123，证明重定向到其他控制层成功，如图 3-30 所示。

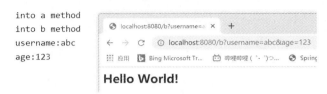

图 3-30　控制层重定向到其他控制层（用方法 1 传参）

3.3.15　重定向传递参数——RedirectAttributes.addFlashAttribute()方法

在正常情况下，重定向的参数存在于 URL，但 Spring 5 MVC 框架还提供了一种将重定向的参数放入 HttpSession 的技术，URL 中不再出现参数值。这样就可以以透明的方式实现重定向传递参数，参数值暂存在 HttpSession，重定向结束后从 HttpSession 中自动清除数据，提高了数据的安全性。

创建测试项目 redirect4。

（1）创建控制层，代码如下：

```
package com.ghy.www.controller;

import org.springframework.stereotype.Controller;
import org.springframework.web.bind.annotation.ModelAttribute;
import org.springframework.web.bind.annotation.RequestMapping;
import org.springframework.web.servlet.mvc.support.RedirectAttributes;

@Controller
public class TestController {
    @RequestMapping(value = "a")
    public String a(RedirectAttributes attr) {
        System.out.println("into a method");
        attr.addFlashAttribute("address", "地址");
        attr.addFlashAttribute("note", "备注");
        return "redirect:/b";
    }
```

```
        @RequestMapping(value = "b")
        public String b(@ModelAttribute("address") String address, @ModelAttribute("note") String note) {
            System.out.println("into b method");
            System.out.println("address:" + address);
            System.out.println("note:" + note);
            return "index.jsp";
        }
    }
```

（2）部署项目，在浏览器中执行控制层地址 a，随后，浏览器的地址栏会发生变化，即重定向到网址 b，证明重定向到其他控制层成功，如图 3-31 所示。

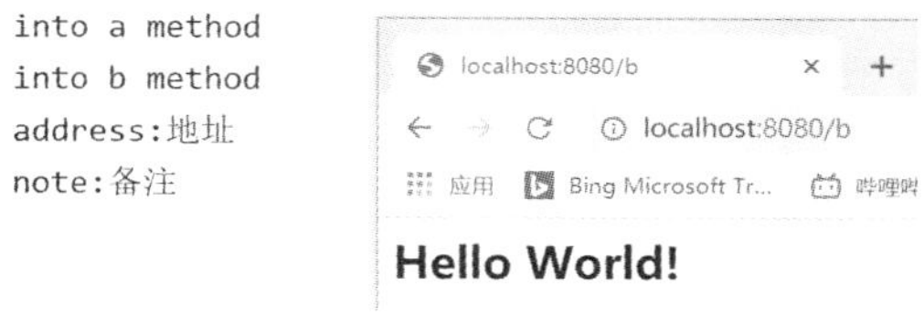

图 3-31 控制层重定向到其他控制层（用方法 2 传参）

3.3.16 使用 jackson 库在服务端将 JSON 字符串转换成各种 Java 数据类型

本节实现将 JSON 字符串转换成各种 Java 数据类型，内部依赖 jackson 库，因为 Spring 5 MVC 默认使用的 JSON 解析类库就是 jackson 库。

创建测试项目 jsontoobject。

（1）创建实体类，代码如下：

```
package com.ghy.www.entity;

public class Userinfo {
    private String username;
    private String password;

    public Userinfo() {
    }

    public Userinfo(String username, String password) {
        super();
        this.username = username;
        this.password = password;
    }

    public String getUsername() {
        return username;
    }

    public void setUsername(String username) {
        this.username = username;
    }

    public String getPassword() {
        return password;
    }

    public void setPassword(String password) {
        this.password = password;
    }
}
```

（2）创建控制层，代码如下：

```java
package com.ghy.www.controller;

import com.ghy.www.entity.Userinfo;
import org.springframework.stereotype.Controller;
import org.springframework.web.bind.annotation.RequestBody;
import org.springframework.web.bind.annotation.RequestMapping;

import java.util.LinkedHashMap;
import java.util.List;
import java.util.Map;

@Controller
public class TestController {
    @RequestMapping(value = "test1")
    public String test1(@RequestBody Userinfo userinfo) {
        System.out.println(userinfo.getUsername());
        System.out.println(userinfo.getPassword());
        return "index.jsp";
    }

    @RequestMapping(value = "test2")
    public String test2(@RequestBody List<String> listData) {
        for (int i = 0; i < listData.size(); i++) {
            System.out.println(listData.get(i));
        }
        return "index.jsp";
    }

    @RequestMapping(value = "test3")
    public String test3(@RequestBody List<LinkedHashMap> listData) {
        for (int i = 0; i < listData.size(); i++) {
            Map map = listData.get(i);
            System.out.println(map.get("username") + " " + map.get("password"));
        }
        return "index.jsp";
    }

    @RequestMapping(value = "test4")
    public String test4(@RequestBody Map map) {
        System.out.println(map.get("username"));
        List<Map> workList = (List) map.get("work");
        for (int i = 0; i < workList.size(); i++) {
            Map eachWorkMap = workList.get(i);
            System.out.println(eachWorkMap.get("address"));

        }
        Map schoolMap = (Map) map.get("school");
        System.out.println(schoolMap.get("name"));
        System.out.println(schoolMap.get("address"));
        return "index.jsp";
    }

    @RequestMapping(value = "test5")
    public String test5(@RequestBody Map map) {
        List list1 = (List) map.get("myArray");
        System.out.println(((Map) list1.get(0)).get("username1"));
```

```java
            System.out.println(((Map) list1.get(1)).get("username2"));
            List list2 = (List) list1.get(2);
            System.out.println(list2.get(0));
            System.out.println(list2.get(1));
            System.out.println(list2.get(2));

            List list3 = (List) list2.get(3);
            for (int i = 0; i < list3.size(); i++) {
                System.out.println(list3.get(i));
            }

            System.out.println(((Map) map.get("myObject")).get("username"));

            List<Map> list4 = (List) ((Map) map.get("myObject1")).get("address");
            for (int i = 0; i < list4.size(); i++) {
                Map eachMap = list4.get(i);
                System.out.println(eachMap.get("name"));

            }
            return "index.jsp";
        }
    }
```

使用@RequestBody 注解后，前端只要向 Controller 提交一个符合 JSON 格式的 request body，Spring 5 MVC 就会自动将其转换成 Java 的各种数据类型。

（3）创建 JSP 文件，代码如下：

```jsp
<%@ page language="java" contentType="text/html; charset=utf-8"
    pageEncoding="utf-8" %>
<%@ taglib uri="http://java.sun.com/jsp/jstl/core" prefix="c" %>
<!DOCTYPE html>
<html>
    <head>
        <script src="jquery3.5.1.js">
        </script>
        <script>
            function Userinfo(username, password) {
                this.username = username;
                this.password = password;
            }

            function test1() {
                var userinfo = new Userinfo("中国", "中国人");
                var jsonString = JSON.stringify(userinfo);
                $.ajax({
                    "type": "post",
                    "url": "test1?t=" + new Date().getTime(),
                    "data": jsonString,
                    "contentType": "application/json"
                });
            }

            function test2() {
                var myArray = new Array();
                myArray[0] = "中国1";
                myArray[1] = "中国2";
                myArray[2] = "中国3";
```

```javascript
        myArray[3] = "中国 4";

        var jsonString = JSON.stringify(myArray);
        $.ajax({
            "type": "post",
            "url": "test2?t=" + new Date().getTime(),
            "data": jsonString,
            "contentType": "application/json"
        });
    }

    function test3() {
        var myArray = new Array();
        myArray[0] = new Userinfo("中国 1", "中国人 1");
        myArray[1] = new Userinfo("中国 2", "中国人 2");
        myArray[2] = new Userinfo("中国 3", "中国人 3");
        myArray[3] = new Userinfo("中国 4", "中国人 4");

        var jsonString = JSON.stringify(myArray);
        $.ajax({
            "type": "post",
            "url": "test3?t=" + new Date().getTime(),
            "data": jsonString,
            "contentType": "application/json"
        });
    }

    function test4() {
        var jsonObject = {
            "username": "accp",
            "work": [{
                "address": "address1"
            }, {
                "address": "address2"
            }],
            "school": {
                "name": "tc",
                "address": "pjy"
            }
        }

        var jsonString = JSON.stringify(jsonObject);
        $.ajax({
            "type": "post",
            "url": "test4?t=" + new Date().getTime(),
            "data": jsonString,
            "contentType": "application/json"
        });
    }

    function test5() {
        var userinfo = {
            "myArray": [{
                "username1": "usernameValue11"
            }, {
                "username2": "usernameValue22"
            }, ["abc", 123, true, [123, 456]]],
```

```
            "myObject": {
                "username": "大中国"
            },
            "myObject1": {
                "address": [{
                    "name": "name1"
                }, {
                    "name": "name2"
                }]
            },
        };

        var jsonString = JSON.stringify(userinfo);
        $.ajax({
            "type": "post",
            "url": "test5?t=" + new Date().getTime(),
            "data": jsonString,
            "contentType": "application/json"
        });
    }
    </script>
</head>
<body>
    <input type="button" value="sendAjax1" onclick="javascript:test1()">
    <br/>
    <input type="button" value="sendAjax2" onclick="javascript:test2()">
    <br/>
    <input type="button" value="sendAjax3" onclick="javascript:test3()">
    <br/>
    <input type="button" value="sendAjax4" onclick="javascript:test4()">
    <br/>
    <input type="button" value="sendAjax5" onclick="javascript:test5()">
</body>
</html>
```

（4）运行项目，单击5个按钮，将分别在控制台输出相应的信息，说明已经将JSON字符串成功转换成Java的不同数据类型。

3.3.17 在控制层返回JSON对象

有时，我们需要在控制层以response的方式返回JSON对象，如返回学生列表等信息。

创建测试项目returnjsonobject。

（1）创建控制层，代码如下：

```
package com.ghy.www.controller;

import com.ghy.www.entity.Userinfo;
import org.springframework.stereotype.Controller;
import org.springframework.web.bind.annotation.RequestBody;
import org.springframework.web.bind.annotation.RequestMapping;
import org.springframework.web.bind.annotation.ResponseBody;

@Controller
public class TestController {
    @RequestMapping(value = "test1", produces = "application/json")
    @ResponseBody
    public Userinfo test1(@RequestBody Userinfo userinfo) {
```

```
        System.out.println(userinfo.getUsername());
        System.out.println(userinfo.getPassword());
        Userinfo returnUserinfo = new Userinfo();
        returnUserinfo.setUsername("返回的账号");
        returnUserinfo.setPassword("返回的密码");
        return returnUserinfo;
    }
}
```

注解@ResponseBody 可以将 Userinfo 类的对象转换成 JSON 字符串并放在 ResponseBody 中，再传给浏览器，浏览器根据属性 produces = "application/json"中的配置来决定接收的数据类型。

（2）创建 JSP 文件，代码如下：

```
<%@ page language="java" contentType="text/html; charset=utf-8"
    pageEncoding="utf-8" %>
<%@ taglib uri="http://java.sun.com/jsp/jstl/core" prefix="c" %>
<!DOCTYPE html>
<html>
    <head>
        <script src="jquery3.5.1.js">
        </script>
        <script>
            function Userinfo(username, password) {
                this.username = username;
                this.password = password;
            }

            function test1() {
                var userinfo = new Userinfo("中国", "中国人");
                var jsonString = JSON.stringify(userinfo);
                $.ajax({
                    "type": "post",
                    "url": "test1?t=" + new Date().getTime(),
                    "data": jsonString,
                    "contentType": "application/json",
                    "success": function (data) {
                        alert(data.username + " " + data.password);
                    }
                });
            }
        </script>
    </head>
    <body>
        <input type="button" value="sendAjax1" onclick="javascript:test1()">
    </body>
</html>
```

（3）运行项目，单击按钮后，分别在控制台和前端输出相应的信息，如图 3-32 所示。

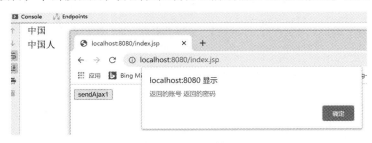

图 3-32　运行结果

3.3.18 在控制层返回 JSON 字符串

在 3.3.17 节中，实现了在 response 对象中返回 JSON 对象，本示例要在 response 对象中返回 JSON 字符串。其实返回 JSON 对象和 JSON 字符串都可以在前端进行处理，只是每个程序员的习惯不一样。

创建测试项目 returnjsonstring。

（1）创建实体类，代码如下：

```java
package com.ghy.www.entity;

import java.util.ArrayList;
import java.util.List;

public class Userinfo {
    private String username;
    private String password;
    private List xxxx = new ArrayList();

    public Userinfo() {
    }

    public String getUsername() {
        return username;
    }

    public void setUsername(String username) {
        this.username = username;
    }

    public String getPassword() {
        return password;
    }

    public void setPassword(String password) {
        this.password = password;
    }

    public List getXxxx() {
        return xxxx;
    }

    public void setXxxx(List xxxx) {
        this.xxxx = xxxx;
    }
}
```

（2）创建 JSP 文件，代码如下：

```jsp
<%@ page language="java" contentType="text/html; charset=utf-8"
    pageEncoding="utf-8" %>
<%@ taglib uri="http://java.sun.com/jsp/jstl/core" prefix="c" %>
<!DOCTYPE html>
<html>
    <head>
        <script src="jquery3.5.1.js">
        </script>
```

```html
        <script>
            function test1() {
                $.ajax({
                    "type": "post",
                    "url": "test1?t=" + new Date().getTime(),
                    "success": function (data) {
                        var jsonObject = JSON.parse(data);
                        alert(jsonObject.username + " " + jsonObject.password);

                        var listString = jsonObject.xxxx;
                        for (var i = 0; i < listString.length; i++) {
                            alert(listString[i]);
                        }
                    }
                });
            }
        </script>
    </head>
    <body>
        <input type="button" value="sendAjax1" onclick="javascript:test1()">
    </body>
</html>
```

（3）创建控制层，代码如下：

```java
package com.ghy.www.controller;

import com.fasterxml.jackson.core.JsonProcessingException;
import com.fasterxml.jackson.databind.ObjectMapper;
import com.ghy.www.entity.Userinfo;
import org.springframework.stereotype.Controller;
import org.springframework.web.bind.annotation.RequestMapping;
import org.springframework.web.bind.annotation.ResponseBody;

@Controller
public class TestController {
    @RequestMapping(value = "test1", produces = "text/html;charset=utf-8")
    @ResponseBody
    public String test1() throws JsonProcessingException {
        Userinfo returnUserinfo = new Userinfo();
        returnUserinfo.setUsername("返回的账号");
        returnUserinfo.setPassword("返回的密码");

        returnUserinfo.getXxxx().add("中国1");
        returnUserinfo.getXxxx().add("中国2");
        returnUserinfo.getXxxx().add("中国3");

        String jsonString = new ObjectMapper().writeValueAsString(returnUserinfo);
        return jsonString;
    }
}
```

（4）程序运行后，单击界面中的按钮，将会在前端正确地显示后台传递的数据。

3.3.19　使用 HttpServletResponse 对象输出响应字符

我们可以使用 HttpServletResponse 对象输出字符串，并把字符串作为 AJAX 请求的响应字符，从而在客户端进行进一步处理。

创建测试项目 responseprintstring。

（1）创建 JSP 文件，代码如下：

```jsp
<%@ page language="java" contentType="text/html; charset=utf-8"
    pageEncoding="utf-8" %>
<%@ taglib uri="http://java.sun.com/jsp/jstl/core" prefix="c" %>
<!DOCTYPE html>
<html>
    <head>
        <script src="jquery3.5.1.js">
        </script>
        <script>
            function test1() {
                $.ajax({
                    "type": "get",
                    "url": "test?t=" + new Date().getTime(),
                    "dataType": "json",
                    "success": function (data) {
                        alert(data.username + " " + data.password);
                    }
                });
            }
        </script>
    </head>
    <body>
        <input type="button" value="sendAjax1" onclick="javascript:test1()">
    </body>
</html>
```

（2）创建控制层，代码如下：

```java
package com.ghy.www.controller;

import com.fasterxml.jackson.databind.ObjectMapper;
import com.ghy.www.entity.Userinfo;
import org.springframework.stereotype.Controller;
import org.springframework.web.bind.annotation.RequestMapping;

import javax.servlet.http.HttpServletRequest;
import javax.servlet.http.HttpServletResponse;
import java.io.IOException;
import java.io.PrintWriter;

@Controller
public class TestController {
    @RequestMapping(value = "test")
    public void test1(HttpServletRequest request, HttpServletResponse response) throws IOException {
        Userinfo userinfo = new Userinfo();
        userinfo.setUsername("返回的账号");
        userinfo.setPassword("返回的密码");

        String returnJSONString = new ObjectMapper().writeValueAsString(userinfo);

        response.setCharacterEncoding("utf-8");
        response.setContentType("text/html");
```

```
        PrintWriter out = response.getWriter();
        out.print(returnJSONString);
        out.flush();
        out.close();
    }
}
```

(3)程序运行后,单击界面中的按钮,将会在前端正确地显示出后台传递的数据。

3.3.20 解决日期问题

在 Spring MVC 框架中,为了保持可扩展性,在默认情况下,对日期的处理是非常薄弱的,需要程序员自行扩展。

在中文环境中有 3 种常用的日期格式。

(1)2000-1-1。

(2)2000 年 1 月 1 日。

(3)2000/1/1。

Spring MVC 框架默认只支持"2000/1/1"这种日期格式,也就是可以直接将它传给控制层数据类型为 java.util.Date 的参数。如果想实现支持其他两种日期格式,那么需要自定义类型转换器。

创建测试项目 datetest。

(1)创建封装前端数据的实体类,代码如下:

```
package com.ghy.www.entity;

import java.util.Date;

public class Userinfo {
    private String username;
    private Date date;

    public Userinfo() {
    }

    public String getUsername() {
        return username;
    }

    public void setUsername(String username) {
        this.username = username;
    }

    public Date getDate() {
        return date;
    }

    public void setDate(Date date) {
        this.date = date;
    }
}
```

(2)创建将前端不同日期格式转换成后台 Date 日期对象的转换器,代码如下:

```
package com.ghy.www.convert;

import org.springframework.core.convert.converter.Converter;
```

```java
import org.springframework.stereotype.Component;

import java.text.SimpleDateFormat;
import java.util.Date;

@Component
public class DateConvert implements Converter<String, Date> {
    @Override
    public Date convert(String stringDate) {
        System.out.println("进入了DateConvert中的public Date convert(String stringDate)方法");
        SimpleDateFormat simpleDateFormat1 = new SimpleDateFormat("yyyy-MM-dd");
        SimpleDateFormat simpleDateFormat2 = new SimpleDateFormat("yyyy/MM/dd");
        SimpleDateFormat simpleDateFormat3 = new SimpleDateFormat("yyyy年MM月dd日");

        SimpleDateFormat[] formatArray = new SimpleDateFormat[]{simpleDateFormat1, simpleDateFormat2, simpleDateFormat3};
        boolean formatResult = false;
        for (int i = 0; i < formatArray.length; i++) {
            try {
                Date newDate = formatArray[i].parse(stringDate);
                formatResult = true;
                return newDate;
            } catch (Exception e) {
            }
        }
        if (formatResult == false) {
            System.err.println("stringDate=" + stringDate + ",日期格式错误!-------------------------");
        }
        return null;
    }
}
```

（3）创建控制层，代码如下：

```java
package com.ghy.www.controller;

import com.ghy.www.entity.Userinfo;
import org.springframework.stereotype.Controller;
import org.springframework.web.bind.annotation.RequestMapping;
import org.springframework.web.bind.annotation.ResponseBody;

import java.util.Date;

@Controller
public class TestController {
    @RequestMapping("/test1")
    public String test1(Date date) {
        System.out.println("test1 date=" + date);
        return "index.jsp";
    }

    @RequestMapping("/test2")
    public String test1(Userinfo userinfo) {
        System.out.println("test2 " + userinfo.getUsername() + " " + userinfo.getDate());
        return "index.jsp";
    }
```

```java
@RequestMapping("/test3")
@ResponseBody
public Userinfo test3() {
    Userinfo userinfo = new Userinfo();
    userinfo.setUsername("返回的账号");
    userinfo.setDate(new Date());
    return userinfo;
}

@RequestMapping("/test4")
@ResponseBody
public Date test4() {
    return new Date();
}
}
```

（4）在 resources 文件夹中创建配置文件 application.yml，其核心代码如下：

```yaml
spring:
  jackson:
    date-format: yyyy-MM-dd HH:mm:ss #Date 转换为 String 格式
```

（5）创建前端 JSP 文件，代码如下：

```jsp
<%@ page language="java" contentType="text/html; charset=utf-8"
    pageEncoding="utf-8" %>
<%@ taglib uri="http://java.sun.com/jsp/jstl/core" prefix="c" %>
<!DOCTYPE html>
<html>
    <head>
    </head>
    <body>
        test1:<br>
        <form action="test1" method="post">
            <input type="text" name="date" value="2000/1/1"><br>
            <input type="submit" value="submit">
        </form>
        <br>
        test1:<br>
        <form action="test1" method="post">
            <input type="text" name="date" value="2000-1-2"><br>
            <input type="submit" value="submit">
        </form>
        <br>
        test1:<br>
        <form action="test1" method="post">
            <input type="text" name="date" value="2000年1月3日"><br>
            <input type="submit" value="submit">
        </form>
        <br>
        <br>
        <br>
        test2:<br>
        <form action="test2" method="post">
            <input type="text" name="username" value="中国1"><br>
            <input type="text" name="date" value="2000/1/1"><br>
            <input type="submit" value="submit">
        </form>
```

```html
    <br>
    test2:
    <form action="test2" method="post">
        <input type="text" name="username" value="中国2"><br>
        <input type="text" name="date" value="2000-1-2"><br>
        <input type="submit" value="submit">
    </form>
    <br>
    test2:
    <form action="test2" method="post">
        <input type="text" name="username" value="中国3"><br>
        <input type="text" name="date" value="2000年1月3日"><br>
        <input type="submit" value="submit">
    </form>
    <br>
    <br>
    <br>
    error1:<br>
    <form action="test1" method="post">
        <input type="text" name="date" value="abc"><br>
        <input type="submit" value="submit">
    </form>
    <br/>
    error2:<br>
    <form action="test2" method="post">
        <input type="text" name="username" value="中国1"><br>
        <input type="text" name="date" value="abc"><br>
        <input type="submit" value="submit">
    </form>
</body>
</html>
```

（6）按顺序单击前端的按钮后，控制台输出的结果如下：

```
进入了DateConvert中的public Date convert(String stringDate)方法
test1 date=Sat Jan 01 00:00:00 CST 2000
进入了DateConvert中的public Date convert(String stringDate)方法
test1 date=Sun Jan 02 00:00:00 CST 2000
进入了DateConvert中的public Date convert(String stringDate)方法
test1 date=Mon Jan 03 00:00:00 CST 2000
进入了DateConvert中的public Date convert(String stringDate)方法
test2 中国1 Sat Jan 01 00:00:00 CST 2000
进入了DateConvert中的public Date convert(String stringDate)方法
test2 中国2 Sun Jan 02 00:00:00 CST 2000
进入了DateConvert中的public Date convert(String stringDate)方法
test2 中国3 Mon Jan 03 00:00:00 CST 2000
进入了DateConvert中的public Date convert(String stringDate)方法
test1 date=null
stringDate=abc，日期格式错误！------------------------
进入了DateConvert中的public Date convert(String stringDate)方法
test2 中国1 null
stringDate=abc，日期格式错误！------------------------
```

（7）执行test3和test4控制层后，控制台输出的结果如下：

```
{"username":"返回的账号","date":"2020-10-05 03:54:45"}
"2020-10-05 03:54:50"
```

3.3.21 单文件上传 1——使用 MultipartHttpServletRequest

Spring 5 MVC 可以实现文件上传。

创建测试项目 upload1。

（1）添加 commons-fileupload 和 commons-io 依赖代码：

```xml
<dependency>
    <groupId>commons-fileupload</groupId>
    <artifactId>commons-fileupload</artifactId>
    <version>1.4</version>
</dependency>

<dependency>
    <groupId>commons-io</groupId>
    <artifactId>commons-io</artifactId>
    <version>2.6</version>
</dependency>
```

（2）创建 JSP 文件，代码如下：

```jsp
<%@ page language="java" contentType="text/html; charset=utf-8"
    pageEncoding="utf-8" %>
<%@ taglib uri="http://java.sun.com/jsp/jstl/core" prefix="c" %>
<!DOCTYPE html>
<html>
    <head>
    </head>
    <body>
        <form action="upload" method="post" enctype="multipart/form-data">
            username:<input type="text" name="username">
            <br/>
            username:<input type="file" name="uploadFile">
            <br/>
            <input type="submit" value="submit">
        </form>
    </body>
</html>
```

（3）创建控制层，代码如下：

```java
package com.ghy.www.controller;

import org.apache.commons.io.FileUtils;
import org.springframework.stereotype.Controller;
import org.springframework.web.bind.annotation.RequestMapping;
import org.springframework.web.multipart.MultipartFile;
import org.springframework.web.multipart.MultipartHttpServletRequest;

import java.io.File;
import java.io.IOException;
import java.io.InputStream;

@Controller
public class TestController {
    @RequestMapping(value = "upload")
    public String loginMethod(MultipartHttpServletRequest request) throws IOException {
        String username = request.getParameter("username");
        System.out.println("username=" + username);
```

```
            MultipartFile file = request.getFile("uploadFile");
            String uploadFileName = file.getOriginalFilename();
            System.out.println("原始文件名:" + uploadFileName);

            InputStream fileStream = file.getInputStream();

            String uploadPath = request.getSession().getServletContext().getRealPath("/upload");

            System.out.println(uploadPath);

            File destination = new File(uploadPath, uploadFileName);
            FileUtils.copyInputStreamToFile(fileStream, destination);

            fileStream.close();

            return "index.jsp";
        }
    }
```

（4）在 application.yml 文件中添加如下配置代码：

```
spring:
  servlet:
    multipart:
      max-request-size: 2048MB
      max-file-size: 2048MB
```

通过 max-file-size 设置单个文件的大小，通过 max-request-size 设置上传的总数据大小。设置所有上传文件的总数据大小为 2GB。如果我们不在该配置文件中添加上面的配置，则在运行时会出现异常：

```
org.apache.tomcat.util.http.fileupload.impl.SizeLimitExceededException: the request was rejected because its size (1213502317) exceeds the configured maximum (10485760)
```

（5）程序运行后，成功实现文件上传。

3.3.22　单文件上传 2——使用 MultipartFile

创建测试项目 upload2。

（1）创建控制层，代码如下：

```
package com.ghy.www.controller;

import org.apache.commons.io.FileUtils;
import org.springframework.stereotype.Controller;
import org.springframework.web.bind.annotation.RequestMapping;
import org.springframework.web.multipart.MultipartFile;

import javax.servlet.http.HttpServletRequest;
import javax.servlet.http.HttpServletResponse;
import java.io.File;
import java.io.IOException;
import java.io.InputStream;

@Controller
public class TestController {
```

```java
@RequestMapping(value = "upload")
public String loginMethod(String username, MultipartFile uploadFile, HttpServletRequest request, HttpServletResponse response) throws IOException {
    System.out.println("username=" + username);

    String uploadFileName = uploadFile.getOriginalFilename();
    System.out.println("原始文件名: " + uploadFileName);

    InputStream fileStream = uploadFile.getInputStream();

    String uploadPath = request.getSession().getServletContext().getRealPath("/upload");
    System.out.println(uploadPath);

    File destination = new File(uploadPath, uploadFileName);
    FileUtils.copyInputStreamToFile(fileStream, destination);

    fileStream.close();
    return "index.jsp";
}
```

MultipartFile uploadFile 中的参数名 uploadFile 一定要和前端<input type="file" name="uploadFile">中文件域的 name 值一致。

（2）程序运行后，成功实现文件上传。

3.3.23　单文件上传 3——使用 MultipartFile 并结合实体类

创建测试项目 upload3。

（1）创建实体类，代码如下：

```java
package com.ghy.www.entity;

import org.springframework.web.multipart.MultipartFile;

public class Userinfo {
    private String username;
    private MultipartFile uploadFile;

    public Userinfo() {
    }

    public String getUsername() {
        return username;
    }

    public void setUsername(String username) {
        this.username = username;
    }

    public MultipartFile getUploadFile() {
        return uploadFile;
    }

    public void setUploadFile(MultipartFile uploadFile) {
        this.uploadFile = uploadFile;
```

}
}
```

（2）创建控制层，代码如下：

```java
package com.ghy.www.controller;

import com.ghy.www.entity.Userinfo;
import org.apache.commons.io.FileUtils;
import org.springframework.stereotype.Controller;
import org.springframework.web.bind.annotation.RequestMapping;
import org.springframework.web.multipart.MultipartFile;

import javax.servlet.http.HttpServletRequest;
import java.io.File;
import java.io.IOException;
import java.io.InputStream;

@Controller
public class TestController {
    @RequestMapping(value = "upload")
    public String loginMethod(Userinfo userinfo, HttpServletRequest request) throws IOException {
        System.out.println("username=" + userinfo.getUsername());

        MultipartFile uploadFile = userinfo.getUploadFile();

        String uploadFileName = uploadFile.getOriginalFilename();
        System.out.println("原始文件名：" + uploadFileName);

        InputStream fileStream = uploadFile.getInputStream();

        String uploadPath = request.getSession().getServletContext().getRealPath("/upload");

        System.out.println(uploadPath);

        File destination = new File(uploadPath, uploadFileName);
        FileUtils.copyInputStreamToFile(fileStream, destination);

        fileStream.close();

        return "index.jsp";
    }
}
```

（3）程序运行后，成功实现文件上传。

3.3.24 多文件上传 1——使用 MultipartHttpServletRequest

创建测试项目 upload4。

（1）创建 JSP 文件，代码如下：

```jsp
<%@ page language="java" contentType="text/html; charset=utf-8"
    pageEncoding="utf-8" %>
<%@ taglib uri="http://java.sun.com/jsp/jstl/core" prefix="c" %>
<!DOCTYPE html>
<html>
```
```

```html
<head>
</head>
<body>
    <form action="upload.spring" method="post" enctype="multipart/form-data">
        username:<input type="text" name="username">
        <br/>
        file1:<input type="file" name="uploadFile1">
        <br/>
        file2:<input type="file" name="uploadFile2">
        <br/>
        file3:<input type="file" name="uploadFile3">
        <br/>
        file4:<input type="file" name="uploadFile4">
        <br/>
        file5:<input type="file" name="uploadFile5">
        <br/>
        <input type="submit" value="submit">
        <br/>
    </form>
</body>
</html>
```

（2）创建控制层，代码如下：

```java
package com.ghy.www.controller;

import org.apache.commons.io.FileUtils;
import org.springframework.stereotype.Controller;
import org.springframework.web.bind.annotation.RequestMapping;
import org.springframework.web.multipart.MultipartFile;
import org.springframework.web.multipart.MultipartHttpServletRequest;

import java.io.File;
import java.io.IOException;
import java.io.InputStream;
import java.text.SimpleDateFormat;
import java.util.Date;
import java.util.Iterator;
import java.util.Map;

@Controller
public class TestController {
    @RequestMapping(value = "upload")
    public String loginMethod(MultipartHttpServletRequest request) throws IOException {

        String username = request.getParameter("username");
        System.out.println("username=" + username);

        SimpleDateFormat format = new SimpleDateFormat("yyyy-MM-dd");
        String uploadPath = request.getSession().getServletContext().getRealPath("/upload");

        Map<String, MultipartFile> fileMap = request.getFileMap();
        Iterator<String> iterator = fileMap.keySet().iterator();
        while (iterator.hasNext()) {
            String eachInputName = iterator.next();
            MultipartFile eachFile = fileMap.get(eachInputName);

            String eachFileName = eachFile.getOriginalFilename();
```

```
            InputStream eachFileStream = eachFile.getInputStream();

            String dateString = format.format(new Date());
            dateString = dateString + "_" + System.currentTimeMillis() + "_" + eachFileName;

            File destination = new File(uploadPath, dateString);
            FileUtils.copyInputStreamToFile(eachFileStream, destination);
            eachFileStream.close();
        }

        return "index.jsp";
    }
}
```

（3）程序运行后，成功实现文件上传。

3.3.25 多文件上传 2——使用 MultipartFile[]

创建测试项目 upload5。

（1）创建 JSP 文件，代码如下：

```
<%@ page language="java" contentType="text/html; charset=utf-8"
    pageEncoding="utf-8" %>
<%@ taglib uri="http://java.sun.com/jsp/jstl/core" prefix="c" %>
<!DOCTYPE html>
<html>
    <head>
    </head>
    <body>
        <form action="upload" method="post" enctype="multipart/form-data">
            username:<input type="text" name="username">
            <br/>
            file1:<input type="file" name="uploadFile">
            <br/>
            file2:<input type="file" name="uploadFile">
            <br/>
            file3:<input type="file" name="uploadFile">
            <br/>
            file4:<input type="file" name="uploadFile">
            <br/>
            file5:<input type="file" name="uploadFile">
            <br/>
            <input type="submit" value="submit">
            <br/>
        </form>
    </body>
</html>
```

（2）创建控制层，代码如下：

```
package com.ghy.www.controller;

import org.apache.commons.io.FileUtils;
import org.springframework.stereotype.Controller;
import org.springframework.web.bind.annotation.RequestMapping;
import org.springframework.web.multipart.MultipartFile;
```

```java
import javax.servlet.http.HttpServletRequest;
import javax.servlet.http.HttpServletResponse;
import java.io.File;
import java.io.IOException;
import java.io.InputStream;
import java.text.SimpleDateFormat;
import java.util.Date;

@Controller
public class TestController {
    @RequestMapping(value = "upload")
    public String loginMethod(String username, MultipartFile uploadFile[], HttpServletRequest request, HttpServletResponse response) throws IOException {

        System.out.println("username=" + username);

        SimpleDateFormat format = new SimpleDateFormat("yyyy-MM-dd");
        String uploadPath = request.getSession().getServletContext().getRealPath("/upload");

        System.out.println(uploadFile.length);

        for (int i = 0; i < uploadFile.length; i++) {
            MultipartFile eachFile = uploadFile[i];

            String eachFileName = eachFile.getOriginalFilename();
            InputStream eachFileStream = eachFile.getInputStream();

            String dateString = format.format(new Date());
            dateString = dateString + "_" + System.currentTimeMillis() + "_" + eachFileName;

            File destination = new File(uploadPath, dateString);
            FileUtils.copyInputStreamToFile(eachFileStream, destination);
            eachFileStream.close();
        }
        return "index.jsp";
    }
}
```

（3）程序运行后，成功实现文件上传。

3.3.26 多文件上传 3——使用 MultipartFile[]并结合实体类

创建测试项目 upload6。

（1）创建实体类，代码如下：

```java
package com.ghy.www.entity;

import org.springframework.web.multipart.MultipartFile;

public class Userinfo {
    private String username;
    private MultipartFile uploadFile[];

    public Userinfo() {
    }
```

```java
    public String getUsername() {
        return username;
    }

    public void setUsername(String username) {
        this.username = username;
    }

    public MultipartFile[] getUploadFile() {
        return uploadFile;
    }

    public void setUploadFile(MultipartFile[] uploadFile) {
        this.uploadFile = uploadFile;
    }
}
```

（2）创建 JSP 文件，代码如下：

```jsp
<%@ page language="java" contentType="text/html; charset=utf-8"
    pageEncoding="utf-8" %>
<%@ taglib uri="http://java.sun.com/jsp/jstl/core" prefix="c" %>
<!DOCTYPE html>
<html>
    <head>
    </head>
    <body>
        <form action="upload" method="post" enctype="multipart/form-data">
            username:<input type="text" name="username">
            <br/>
            file1:<input type="file" name="uploadFile">
            <br/>
            file2:<input type="file" name="uploadFile">
            <br/>
            file3:<input type="file" name="uploadFile">
            <br/>
            file4:<input type="file" name="uploadFile">
            <br/>
            file5:<input type="file" name="uploadFile">
            <br/>
            <input type="submit" value="submit">
            <br/>
        </form>
    </body>
</html>
```

（3）创建控制层，代码如下：

```java
package com.ghy.www.controller;

import com.ghy.www.entity.Userinfo;
import org.apache.commons.io.FileUtils;
import org.springframework.stereotype.Controller;
import org.springframework.web.bind.annotation.RequestMapping;
import org.springframework.web.multipart.MultipartFile;

import javax.servlet.http.HttpServletRequest;
```

```java
import java.io.File;
import java.io.IOException;
import java.io.InputStream;
import java.text.SimpleDateFormat;
import java.util.Date;

@Controller
public class TestController {
    @RequestMapping(value = "upload")
    public String loginMethod(Userinfo userinfo, HttpServletRequest request) throws IOException {
        System.out.println(userinfo.getUsername());
        SimpleDateFormat sdf = new SimpleDateFormat("yyyy_MM_dd_hh_mm_ss");
        MultipartFile[] files = userinfo.getUploadFile();
        for (int i = 0; i < files.length; i++) {
            MultipartFile file = files[i];
            String uploadFileName = file.getOriginalFilename();
            InputStream isRef = file.getInputStream();
            String targetDir = request.getSession().getServletContext().getRealPath("/upload");
            System.out.println(targetDir);
            String getDateString = sdf.format(new Date());
            File destination = new File(targetDir, getDateString + "_" + System.nanoTime() + "_" + uploadFileName);
            FileUtils.copyInputStreamToFile(isRef, destination);
            isRef.close();
        }
        return "index.jsp";
    }
}
```

（4）程序运行后，成功实现文件上传。

3.3.27 使用 AJAX 实现文件上传

创建测试项目 ajaxupload。

创建 JSP 文件，代码如下：

```jsp
<%@ page language="java" contentType="text/html; charset=utf-8"
    pageEncoding="utf-8" %>
<%@ taglib uri="http://java.sun.com/jsp/jstl/core" prefix="c" %>
<!DOCTYPE html>
<html>
    <head>
        <meta charset="utf-8">
        <title>
            Insert title here
        </title>
        <script type="text/javascript" src="jquery3.5.1.js"></script>
        <script type="text/javascript">
            function ajaxUpload() {
                var usernameValue = $("#username").val();
                var inputFile = $("#uploadFile")[0].files[0];

                var formData = new FormData();
                formData.append('username', usernameValue)
```

```
                formData.append('uploadFile', inputFile);

                $.ajax({
                    url: 'upload',
                    type: 'POST',
                    data: formData,
                    processData: false,
                    contentType: false,
                    success: function () {
                        alert("上传成功!");
                    }
                });
            }
        </script>
    </head>
    <body>
        <div>
            username:<input type="text" id="username"><br>
            file:<input type="file" id="uploadFile"><br>
            <input type="button" value="button" onclick="javascript:ajaxUpload()"><br>
        </div>
    </body>
</html>
```

创建控制层，代码如下：

```
package com.ghy.www.controller;

import org.apache.commons.io.FileUtils;
import org.springframework.stereotype.Controller;
import org.springframework.web.bind.annotation.RequestMapping;
import org.springframework.web.multipart.MultipartFile;
import org.springframework.web.multipart.MultipartHttpServletRequest;

import java.io.File;
import java.io.IOException;
import java.io.InputStream;

@Controller
public class TestController {
    @RequestMapping("/upload")
    public String upload(MultipartHttpServletRequest request) throws IOException {
        System.out.println(request.getParameter("username"));
        String uploadPath = request.getServletContext().getRealPath("/upload");
        System.out.println(uploadPath);

        MultipartFile uploadFile = request.getFile("uploadFile");
        String fileName = uploadFile.getOriginalFilename();
        InputStream inputStream = uploadFile.getInputStream();

        File newFile = new File(uploadPath, fileName);
        FileUtils.copyInputStreamToFile(inputStream, newFile);

        return "index.jsp";
    }
}
```

3.3.28 支持中文文件名的文件下载

创建测试项目 downloadfile。

（1）创建 JSP 文件，代码如下：

```jsp
<%@ page language="java" contentType="text/html; charset=utf-8"
    pageEncoding="utf-8" %>
<%@ taglib uri="http://java.sun.com/jsp/jstl/core" prefix="c" %>
<!DOCTYPE html>
<html>
    <head>
    </head>
    <body>
        <a href="downloadFile?fileName=<%=java.net.URLEncoder.encode(" 中国 ~!@#$%^&()_+{}[]; ',..rar", "utf-8")%>">中国         ~!@#$%^&()_+{}[];',..rar</a>
        <br/>
        <a href="downloadFile?fileName=<%=java.net.URLEncoder.encode("postTest.rar", "utf-8")%>">postTest.rar</a>
        <br/>
        <a href="downloadFile?fileName=<%=java.net.URLEncoder.encode("abc.rar", "utf-8")%>">abc.rar</a>
    </body>
</html>
```

（2）创建控制层，代码如下：

```java
package com.ghy.www.controller;

import org.apache.commons.io.IOUtils;
import org.springframework.stereotype.Controller;
import org.springframework.web.bind.annotation.RequestMapping;

import javax.servlet.ServletOutputStream;
import javax.servlet.http.HttpServletRequest;
import javax.servlet.http.HttpServletResponse;
import java.io.*;

@Controller
public class TestController {
    @RequestMapping(value = "downloadFile")
    public void testA(String fileName, HttpServletRequest request, HttpServletResponse response)
            throws UnsupportedEncodingException {
        try {
            System.out.println(fileName);
            String downPath = request.getSession().getServletContext().getRealPath("/");
            System.out.println(downPath + fileName);
            File downloadFile = new File(downPath + fileName);
            response.setContentType("application/octet-stream;");
            String useragent = request.getHeader("User-Agent").toLowerCase();
            System.out.println(useragent);
```

```
            if (useragent.contains("wow64")) {
               System.out.println("IE");
               // 1.IE 浏览器 UTF-8
               response.setHeader("Content-Disposition", "attachment;filename="
                      + java.net.URLEncoder.encode(fileName, "utf-8").replaceAll("\\+",
"%20"));
               // replaceAll("\\+", "%20")的作用是处理空格
            } else {
               // 2.其他浏览器
               System.out.println("NOT IE");
               response.setHeader("Content-Disposition", "attachment;filename*=utf-8'zh_
cn';filename="
                      + java.net.URLEncoder.encode(fileName, "utf-8").replaceAll("\\+",
"%20"));
            }
            response.setHeader("Content-Length", String.valueOf(downloadFile.length()));
            FileInputStream fis = new FileInputStream(downloadFile);
            ServletOutputStream out = response.getOutputStream();
            IOUtils.copyLarge(fis, out);
            out.flush();
            out.close();
            fis.close();
         } catch (FileNotFoundException e) {
            e.printStackTrace();
         } catch (IOException e) {
            e.printStackTrace();
         }
      }
   }
```

（3）程序运行后，成功下载中文文件名和英文文件名的文件。

3.3.29 使用@RestController 注解

使用@RestController 注解等同于一起使用@Controller 注解和@ResponseBody 注解。

创建测试项目 restcontroller。

创建控制层，代码如下：

```
package com.ghy.www.controller;

import org.springframework.web.bind.annotation.RequestMapping;
import org.springframework.web.bind.annotation.RestController;

@RestController
public class TestController {
   class Userinfo {
      private String username;
      private String password;

      public String getUsername() {
         return username;
      }

      public void setUsername(String username) {
```

```
            this.username = username;
        }

        public String getPassword() {
            return password;
        }

        public void setPassword(String password) {
            this.password = password;
        }
    }

    @RequestMapping(value = "test")
    public Userinfo printInfo() {
        Userinfo userinfo = new Userinfo();
        userinfo.setUsername("中国");
        userinfo.setPassword("中国人");
        return userinfo;
    }
}
```

执行控制层地址 test，浏览器输出的结果如图 3-33 所示。

图 3-33 运行结果

3.4 扩展技术

本节提供更多的 Spring MVC 示例，对开发软件项目起到辅助作用。

3.4.1 使用 prefix 和 suffix 简化返回的视图名称

在项目中，如果所有的 JSP 文件都存储在 JSP 文件夹中，则控制层转发到 JSP 文件需要填写完整的路径。示例代码如下：

```
@RequestMapping(value = "login")
public String loginMethod(String username, String password) {
    if (username.equals("a") && password.equals("aa")) {
        return "jsp/ok.jsp";
    } else {
        return "jsp/no.jsp";
    }
}
```

返回值都有"jsp/"字符串，比较烦琐，因此，我们可以简化返回的视图名称。
创建测试项目 shortviewpath。
（1）在 webapp 路径的 myview 文件夹中存在 showThisView.jsp 文件。
（2）配置文件 application.yml 中的配置代码如下：

```
spring:
  mvc:
    view:
      prefix: /myview/
      suffix: .jsp
```

属性 prefix 表示 JSP 文件存储在哪个路径，suffix 表示视图文件的扩展名。
（3）创建控制层，代码如下：

```
package com.ghy.www.controller;

import org.springframework.stereotype.Controller;
import org.springframework.web.bind.annotation.RequestMapping;

@Controller
public class TestController {
    @RequestMapping(value = "test")
    public String testController() {
        return "showThisView";
    }
}
```

（4）运行程序，控制层成功转发到 showThisView.jsp 文件。

3.4.2 控制层返回 List 对象及实体

在 Spring 5 MVC 中的控制层不但可以返回 Java 的数据类型，如返回 List 对象或 Userinfo 类的自定义实体等，而且可以将它们自动转发到 JSP 页面。本示例在项目 returnothertype 中进行。
（1）创建控制层，代码如下：

```
package com.ghy.www.controller;

import com.ghy.www.entity.Userinfo;
import org.springframework.stereotype.Controller;
import org.springframework.web.bind.annotation.RequestMapping;

import java.util.ArrayList;
import java.util.List;

@Controller
public class TestController {
    @RequestMapping(value = "listMethod")
    public List<String> listMethodXXXXXXXX() {
        List list = new ArrayList();
        list.add("中国1");
        list.add("中国2");
        list.add("中国3");
        list.add("中国4");
        return list;
    }

    @RequestMapping(value = "getUserinfo")
    public Userinfo getUserinfoXXXXXXXXX() {
        Userinfo userinfo = new Userinfo("100", "中国");
        return userinfo;
    }
}
```

（2）创建 JSP 文件 listMethod.jsp，其核心代码如下：

```
<%@ page import="java.util.Enumeration" %>
<%@ page language="java" contentType="text/html; charset=utf-8"
```

```
        pageEncoding="utf-8" %>
<%@ taglib uri="http://java.sun.com/jsp/jstl/core" prefix="c" %>
<!DOCTYPE html>
<html>
    <head>
        <meta charset="utf-8">
        <title>Insert title here</title>
    </head>
    <body>
        <%
            Enumeration enum1 = request.getAttributeNames();
            while (enum1.hasMoreElements()) {
                String key = (String) enum1.nextElement();
                out.println("key=" + key + "<br/>");
            } %>
        <br/>
        <br/>
        <c:forEach var="eachString" items="${stringList}">
            ${eachString}
            <br/>
        </c:forEach>
    </body>
</html>
```

（3）创建 JSP 文件，getUserinfo.jsp 核心代码如下：

```
<%@ page import="java.util.Enumeration" %>
<%@ page language="java" contentType="text/html; charset=utf-8"
        pageEncoding="utf-8" %>
<%@ taglib uri="http://java.sun.com/jsp/jstl/core" prefix="c" %>
<!DOCTYPE html>
<html>
    <head>
        <meta charset="utf-8">
        <title>Insert title here</title>
    </head>
    <body>
        <%
            Enumeration enum1 = request.getAttributeNames();
            while (enum1.hasMoreElements()) {
                String key = (String) enum1.nextElement();
                out.println("key=" + key + "<br/>");
            } %>
        <br/>
        <br/>
        ${userinfo.username}  ${userinfo.password}
    </body>
</html>
```

（4）配置文件 application.yml 中的配置更改如下：

```
spring:
  mvc:
    view:
      prefix: /
      suffix: .jsp
```

（5）部署项目运行程序，执行控制层地址 listMethod，程序运行结果如图 3-34 所示。

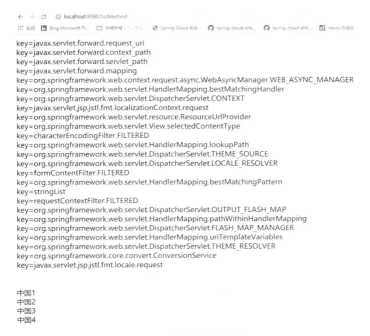

图 3-34　输出列表数据

（6）继续执行控制层地址 getUserinfo，程序运行结果如图 3-35 所示。

图 3-35　输出实体数据

如果自动放入 request 的 key 并不是自定义的，那么可以变成自定义的。

创建测试项目 returnothertyperename。

（1）更改控制层代码如下：

```java
package com.ghy.www.controller;

import com.ghy.www.entity.Userinfo;
import org.springframework.stereotype.Controller;
import org.springframework.web.bind.annotation.ModelAttribute;
import org.springframework.web.bind.annotation.RequestMapping;

import java.util.ArrayList;
import java.util.List;

@Controller
public class TestController {
    @ModelAttribute(name = "AAA")
    @RequestMapping(value = "listMethod")
    public List<String> listMethodXXXXXXXX() {
        List list = new ArrayList();
        list.add("中国1");
        list.add("中国2");
        list.add("中国3");
        list.add("中国4");
        return list;
    }

    @ModelAttribute(name = "BBB")
    @RequestMapping(value = "getUserinfo")
    public Userinfo getUserinfoXXXXXXXXX() {
        Userinfo userinfo = new Userinfo("100", "中国");
        return userinfo;
    }
}
```

（2）加入注解

```java
@ModelAttribute(name = "AAA")
```

相当于代码

```java
request.setAttribute("AAA",list);
```

将对象 list 以 key 为 AAA 放入 request 作用域中，那么前端 JSP 也需要改成

```
${AAA}
```

来取得对应的值。

（3）更改 JSP 文件后的 listMethod.jsp 核心代码如下：

```jsp
<%@ page import="java.util.Enumeration" %>
<%@ page language="java" contentType="text/html; charset=utf-8"
    pageEncoding="utf-8" %>
<%@ taglib uri="http://java.sun.com/jsp/jstl/core" prefix="c" %>
<!DOCTYPE html>
<html>
    <head>
        <meta charset="utf-8">
        <title>Insert title here</title>
    </head>
```

```jsp
    <body>
        <%
            Enumeration enum1 = request.getAttributeNames();
            while (enum1.hasMoreElements()) {
                String key = (String) enum1.nextElement();
                out.println("key=" + key + "<br/>");
            }
        %>
        <br/>
        <br/>
        <c:forEach var="eachString" items="${AAA}">
            ${eachString}
            <br/>
        </c:forEach>
    </body>
</html>
```

（4）更改 JSP 文件后的 getUserinfo.jsp 核心代码如下：

```jsp
<%@ page import="java.util.Enumeration" %>
<%@ page language="java" contentType="text/html; charset=utf-8"
    pageEncoding="utf-8" %>
<%@ taglib uri="http://java.sun.com/jsp/jstl/core" prefix="c" %>
<!DOCTYPE html>
<html>
    <head>
        <meta charset="utf-8">
        <title>Insert title here</title>
    </head>
    <body>
        <%
            Enumeration enum1 = request.getAttributeNames();
            while (enum1.hasMoreElements()) {
                String key = (String) enum1.nextElement();
                out.println("key=" + key + "<br/>");
            } %>
        <br/>
        <br/>
        ${BBB.username}___${BBB.password}
    </body>
</html>
```

（5）程序运行后，可以从 request 中取得自定义 attributeKey 对应的值。

3.4.3 实现国际化

使用国际化技术的优势是浏览器能根据设置的语言种类显示对应语言的信息，它是开发多语种软件项目的必备技术。

1. 在 JSP 文件中向国际化文本传入参数

创建测试项目 i18n-1。

（1）创建属性文件 myi18n.properties，其内容为空。
如果不创建该文件，运行时会出现如下异常：

```
No message found under code
```

（2）创建属性文件 myi18n_en_US.properties，其内容如下：

```
hasParamStatic=i am {0} ,age {1}
name=china
age=100
```

（3）创建属性文件 myi18n_zh_CN.properties，其内容如下：

```
hasParamStatic=我是 {0} ，年龄 {1}
name=中国
age=100
```

属性文件名中包含 zh 和 CN，以及 en 和 US。其中 zh 表示中文，CN 表示中国；en 表示英文，US 表示美国。要获取语言和国家（或地区）的代码，可通过如下程序：

```java
package com.ghy.www.test;

import java.util.Locale;

public class Test {
    public static void main(String[] args) {
        Locale[] localeArray = Locale.getAvailableLocales();
        for (int i = 0; i < localeArray.length; i++) {
            Locale locale = localeArray[i];
            System.out.println(locale.getCountry() + " " + locale.getLanguage());
        }
    }
}
```

（4）配置文件 application.yml 的内容如下：

```yaml
spring:
  messages:
    basename: myi18n
    encoding: UTF-8
    cache-duration: 0
```

（5）创建控制层，代码如下：

```java
package com.ghy.www.controller;

import org.springframework.stereotype.Controller;
import org.springframework.web.bind.annotation.RequestMapping;

@Controller
public class TestController {
    @RequestMapping(value = "test")
    public String test() {
        System.out.println("test run !");
        return "test.jsp";
    }
}
```

（6）创建 test.jsp 文件，代码如下：

```jsp
<%@ page language="java" contentType="text/html; charset=utf-8"
    pageEncoding="utf-8" %>
<%@ taglib uri="http://java.sun.com/jsp/jstl/core" prefix="c" %>
<%@ taglib uri="http://www.springframework.org/tags" prefix="spring" %>
<!DOCTYPE html>
```

```html
<html>
    <head>
        <meta charset="utf-8">
        <title>Insert title here</title>
    </head>
    <body>
        参数值非国际化的示例:
        <br/>
        <spring:message code="hasParamStatic">
            <spring:argument>
                <spring:message text="姓名非国际化">
                </spring:message>
            </spring:argument>
            <spring:argument>
                <spring:message text="年龄非国际化">
                </spring:message>
            </spring:argument>
        </spring:message>
        <br/>
        <br/>
        参数值国际化的示例:
        <br/>
        <spring:message code="hasParamStatic">
            <spring:argument>
                <spring:message code="name">
                </spring:message>
            </spring:argument>
            <spring:argument>
                <spring:message code="age">
                </spring:message>
            </spring:argument>
        </spring:message>
    </body>
</html>
```

(7) 运行项目，执行控制层地址 test，程序根据在浏览器中选择的语言显示对应语言的消息，效果如图 3-36 所示。

图 3-36　运行结果

(8) 在 Chrome 浏览器中切换语言，如图 3-37 所示。
可以将某种语言进行置顶，从而优先使用此语言。

注意：我们可以使用图 3-37 中的 ![icon] 确认使用的语言种类，也可以展开此项，然后对指定的语言调用"移到顶部"选项。

图 3-37 语言偏好排序设置

2. 服务端识别客户端使用不同语言的原理

服务端如何识别客户端（浏览器）使用的不同语言呢？

创建测试项目 i18n-2。

创建控制层，代码如下：

```
package com.ghy.www.controller;

import org.springframework.stereotype.Controller;
import org.springframework.web.bind.annotation.RequestMapping;

import javax.servlet.http.HttpServletRequest;
import java.util.Locale;

@Controller
public class TestController {
    @RequestMapping(value = "test")
    public String test(HttpServletRequest request) {
        Locale locale = request.getLocale();
        System.out.println("getLanguage=" + locale.getLanguage());
        return "test.jsp";
    }
}
```

当执行控制层地址 test，并设置浏览器使用不同的语言时，在服务端的控制台会输出指定的语言信息：

```
getLanguage=en
getLanguage=zh
```

语言信息 zh 和 en 是在请求（request）头中被发送给服务端的。

在设置浏览器使用英文时，请求头内容如下：

```
Accept-Language: en-US,en;q=0.9,zh;q=0.8,zh-CN;q=0.7
```

在设置浏览器使用中文时，请求头内容如下：

```
Accept-Language: zh,en-US;q=0.9,en;q=0.8,zh-CN;q=0.7
```

以上信息可以使用<F12>键进行截获。

服务端根据请求头中的语言信息就可以显示对应语言的文字。

3. 使用超链接实现语言的切换——使用 HttpSession

创建测试项目 i18n-3。

3.4 扩展技术

（1）创建配置类，代码如下：

```java
package com.ghy.www.config;

import org.springframework.context.annotation.Bean;
import org.springframework.context.annotation.Configuration;
import org.springframework.web.servlet.LocaleResolver;
import org.springframework.web.servlet.config.annotation.InterceptorRegistry;
import org.springframework.web.servlet.config.annotation.WebMvcConfigurer;
import org.springframework.web.servlet.i18n.LocaleChangeInterceptor;
import org.springframework.web.servlet.i18n.SessionLocaleResolver;

import java.util.Locale;

@Configuration
public class SpringConfig {
    //默认语言
    @Bean
    public LocaleResolver localeResolver() {
        SessionLocaleResolver localeResolver = new SessionLocaleResolver();
        localeResolver.setDefaultLocale(Locale.getDefault());
        return localeResolver;
    }

    //指定语言
    @Bean
    public WebMvcConfigurer localeInterceptor() {
        return new WebMvcConfigurer() {
            @Override
            public void addInterceptors(InterceptorRegistry registry) {
                LocaleChangeInterceptor localeInterceptor = new LocaleChangeInterceptor();
                localeInterceptor.setParamName("lang");
                registry.addInterceptor(localeInterceptor);
            }
        };
    }
}
```

（2）创建控制层，代码如下：

```java
package com.ghy.www.controller;

import org.springframework.stereotype.Controller;
import org.springframework.web.bind.annotation.RequestMapping;

import javax.servlet.http.HttpServletRequest;
import javax.servlet.http.HttpSession;
import java.util.Enumeration;

@Controller
public class TestController {
    @RequestMapping(value = "test")
    public String test(HttpSession session) {
        System.out.println("test run !");
        Enumeration enum1 = session.getAttributeNames();
        while (enum1.hasMoreElements()) {
            System.out.println("key=" + enum1.nextElement());
        }
        return "test.jsp";
    }
}
```

（3）创建 JSP 文件，代码如下：

```jsp
<%@ page language="java" contentType="text/html; charset=utf-8"
    pageEncoding="utf-8" %>
<%@ taglib uri="http://java.sun.com/jsp/jstl/core" prefix="c" %>
<%@ taglib uri="http://www.springframework.org/tags" prefix="spring" %>
<!DOCTYPE html>
<html>
    <head>
        <meta charset="utf-8">
        <title>Insert title here</title>
    </head>
    <body>
        <a href="?lang=zh_CN">中文</a>____<a href="?lang=en_US">en</a>
        <br/>
        <br/>
        参数值非国际化的示例：
        <br/>
        <spring:message code="hasParamStatic">
            <spring:argument>
                <spring:message text="姓名非国际化">
                </spring:message>
            </spring:argument>
            <spring:argument>
                <spring:message text="年龄非国际化">
                </spring:message>
            </spring:argument>
        </spring:message>
        <br/>
        <br/>
        参数值国际化的示例：
        <br/>
        <spring:message code="hasParamStatic">
            <spring:argument>
                <spring:message code="name">
                </spring:message>
            </spring:argument>
            <spring:argument>
                <spring:message code="age">
                </spring:message>
            </spring:argument>
        </spring:message>
    </body>
</html>
```

（4）运行项目，执行控制层地址 test，然后分别单击"中文"或"英文"的超链接，将会显示对应语言的消息，结果如图 3-38 所示。

图 3-38　运行结果

3.4 扩展技术

（5）该程序文件的代码如下：

```
@Bean
public LocaleResolver localeResolver() {
    SessionLocaleResolver localeResolver = new SessionLocaleResolver();
    localeResolver.setDefaultLocale(Locale.getDefault());
    return localeResolver;
}
```

经过分析，可以看出，语言信息保存在 HttpSession 中，key 的名称为：

org.springframework.web.servlet.i18n.SessionLocaleResolver.LOCALE

关闭当前浏览器，然后打开新的浏览器，并执行控制层地址 test，会产生新的 sessionId，导致语言的状态并没有被保留，还是使用默认的语言，这时可以将语言状态保存在 Cookie 中，这样就避免了语言状态保存在 HttpSession 中时丢失的问题。

4．使用超链接实现语言的切换——使用 Cookie

创建测试项目 i18n-4。

（1）创建配置类，代码如下：

```java
package com.ghy.www.config;

import org.springframework.context.annotation.Bean;
import org.springframework.context.annotation.Configuration;
import org.springframework.web.servlet.LocaleResolver;
import org.springframework.web.servlet.config.annotation.InterceptorRegistry;
import org.springframework.web.servlet.config.annotation.WebMvcConfigurer;
import org.springframework.web.servlet.i18n.CookieLocaleResolver;
import org.springframework.web.servlet.i18n.LocaleChangeInterceptor;

import java.util.Locale;

@Configuration
public class SpringConfig {
    //默认语言
    @Bean
    public LocaleResolver localeResolver() {
        CookieLocaleResolver localeResolver = new CookieLocaleResolver();
        localeResolver.setDefaultLocale(Locale.getDefault());
        localeResolver.setCookieMaxAge(36000);
        return localeResolver;
    }

    //指定语言
    @Bean
    public WebMvcConfigurer localeInterceptor() {
        return new WebMvcConfigurer() {
            @Override
            public void addInterceptors(InterceptorRegistry registry) {
                LocaleChangeInterceptor localeInterceptor = new LocaleChangeInterceptor();
                localeInterceptor.setParamName("lang");
                registry.addInterceptor(localeInterceptor);
            }
        };
    }
}
```

（2）运行项目，然后分别单击"中文"或"英文"的超链接，可以切换显示的语言种类，而且在打开新的浏览器进程时还是显示原有语言的信息，因为语言信息已经存储在 Cookie 中了。

5. 在控制层中处理国际化消息

这里将登录功能作为国际化测试的场景。在登录失败时，显示不同语言的提示信息。

创建测试项目 i18n-5。

（1）创建属性文件 myi18n_en_US.properties，其内容如下：

```
username=username
password=password
submit=submit

usernameisnull=username is null
passwordisnull=password is null
```

（2）创建属性文件 myi18n_zh_CN.properties，其内容如下：

```
username=账号
password=密码
submit=提交

usernameisnull=账号为空
passwordisnull=密码是空
```

（3）配置文件 application.yml 中的内容如下：

```
spring:
  messages:
    basename: myi18n
    encoding: UTF-8
    cache-duration: 0
```

（4）创建配置类，代码如下：

```java
package com.ghy.www.config;

import org.springframework.context.annotation.Bean;
import org.springframework.context.annotation.Configuration;
import org.springframework.web.servlet.LocaleResolver;
import org.springframework.web.servlet.config.annotation.InterceptorRegistry;
import org.springframework.web.servlet.config.annotation.WebMvcConfigurer;
import org.springframework.web.servlet.i18n.CookieLocaleResolver;
import org.springframework.web.servlet.i18n.LocaleChangeInterceptor;

import java.util.Locale;

@Configuration
public class SpringConfig {
    //默认语言
    @Bean
    public LocaleResolver localeResolver() {
        CookieLocaleResolver localeResolver = new CookieLocaleResolver();
        localeResolver.setDefaultLocale(Locale.getDefault());
        localeResolver.setCookieMaxAge(36000);
        return localeResolver;
    }
```

```java
        //指定语言
        @Bean
        public WebMvcConfigurer localeInterceptor() {
            return new WebMvcConfigurer() {
                @Override
                public void addInterceptors(InterceptorRegistry registry) {
                    LocaleChangeInterceptor localeInterceptor = new LocaleChangeInterceptor();
                    localeInterceptor.setParamName("lang");
                    registry.addInterceptor(localeInterceptor);
                }
            };
        }
    }
```

（5）创建控制层，代码如下：

```java
package com.ghy.www.controller;

import org.springframework.stereotype.Controller;
import org.springframework.web.bind.annotation.RequestMapping;
import org.springframework.web.servlet.mvc.support.RedirectAttributes;
import org.springframework.web.servlet.support.RequestContext;

import javax.servlet.http.HttpServletRequest;

@Controller
public class TestController {
    @RequestMapping(value = "login")
    public String login(String username, String password, HttpServletRequest request, RedirectAttributes attr) {
        System.out.println("login run !");

        RequestContext context = new RequestContext(request);

        if (username == null || "".equals(username)) {
            attr.addFlashAttribute("usernameisnull", context.getMessage("usernameisnull"));
        }
        if (password == null || "".equals(password)) {
            attr.addFlashAttribute("passwordisnull", context.getMessage("passwordisnull"));
        }
        return "redirect:/showLogin";
    }

    @RequestMapping(value = "showLogin")
    public String showLogin(HttpServletRequest request) {
        System.out.println("test run !");
        return "showLogin.jsp";
    }
}
```

（6）创建 JSP 文件，代码如下：

```jsp
<%@ page language="java" contentType="text/html; charset=utf-8"
    pageEncoding="utf-8" %>
<%@ taglib uri="http://java.sun.com/jsp/jstl/core" prefix="c" %>
<%@ taglib uri="http://www.springframework.org/tags" prefix="spring" %>
<!DOCTYPE html>
<html>
    <head>
```

```html
        <meta charset="utf-8">
        <title>Insert title here</title>
</head>
<body>
        <a href="?lang=zh_CN">中文</a>____-<a href="?lang=en_US">en</a>
        <br/>
        <br/>
        <br/>
        <form action="login" method="post">
            <spring:message code="username">
            </spring:message>
            ：<input type="text" name="username"/>${usernameisnull}
            <br/>
            <spring:message code="password">
            </spring:message>
            ：<input type="text" name="password"/>${passwordisnull}
            <br/>
            <input type="submit" value='<spring:message code="submit"></spring:message>'/>
        </form>
</body>
</html>
```

（7）运行项目，执行控制层地址 showLogin，在登录失败时，会根据不同语言显示对应语言的信息，结果如图 3-39 所示。

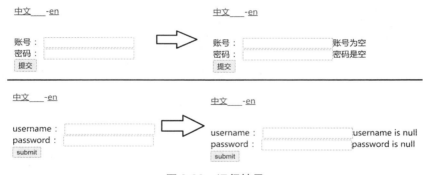

图 3-39　运行结果

3.4.4　处理异常

Spring 5 MVC 框架处理异常的方式有多种。

1．使用@ControllerAdvice 注解实现全局异常处理器

创建测试项目 exception1。
（1）创建异常处理类，代码如下：

```
package com.ghy.www.advice;

import com.ghy.www.exception.LoginException1;
import com.ghy.www.exception.LoginException2;
import org.springframework.web.bind.annotation.ExceptionHandler;
import org.springframework.web.bind.annotation.ResponseBody;
```

3.4 扩展技术

```java
import org.springframework.web.servlet.ModelAndView;

import java.sql.SQLException;
import java.util.HashMap;
import java.util.Map;

@org.springframework.web.bind.annotation.ControllerAdvice
public class ControllerAdvice {
    @ExceptionHandler(value = Exception.class)
    public ModelAndView processException(Exception ex) {
        ModelAndView mav = new ModelAndView();
        mav.addObject("ex", ex);
        mav.setViewName("errorPage.jsp");
        return mav;
    }

    @ExceptionHandler(value = SQLException.class)
    public ModelAndView processSQLException(SQLException ex) {
        ModelAndView mav = new ModelAndView();
        mav.addObject("ex", ex);
        mav.setViewName("sqlerror.jsp");
        return mav;
    }

    @ExceptionHandler(value = LoginException1.class)
    public ModelAndView processLoginException1(LoginException1 ex) {
        ModelAndView mav = new ModelAndView();
        mav.addObject("ex", ex);
        mav.setViewName("loginException1.jsp");
        return mav;
    }

    @ExceptionHandler(value = LoginException2.class)
    @ResponseBody
    public Map processLoginException2(LoginException2 ex) {
        Map map = new HashMap();
        map.put("controllerName", "UserinfoController");
        map.put("message", ex.getMessage());
        return map;
    }
}
```

（2）创建控制层，代码如下：

```java
package com.ghy.www.controller;

import com.ghy.www.exception.LoginException1;
import com.ghy.www.exception.LoginException2;
import org.springframework.stereotype.Controller;
import org.springframework.web.bind.annotation.RequestMapping;

import java.sql.SQLException;

@Controller
public class TestController {
    @RequestMapping(value = "test1")
    public String test1() {
        System.out.println("public String test1()");
```

```
        int i = 10;
        int j = 0;
        int result = i / j;
        return "index.jsp";
    }

    @RequestMapping(value = "test2")
    public String test2() throws Exception {
        System.out.println("public String test2()");
        if (1 == 1) {
            throw new SQLException("SQL 语句错误");
        }
        return "index.jsp";
    }

    @RequestMapping(value = "test3")
    public String test3() throws LoginException1 {
        System.out.println("public String test3()");
        if (1 == 1) {
            throw new LoginException1("账号××登录失败,请重新登录!");
        }
        return "index.jsp";
    }

    @RequestMapping(value = "test4")
    public String test4() throws LoginException2 {
        System.out.println("public String test4()");
        if (1 == 1) {
            throw new LoginException2("账号××登录失败,请重新登录!");
        }
        return "index.jsp";
    }
}
```

（3）创建异常类，代码如下：

```
package com.ghy.www.exception;

public class LoginException1 extends Exception {
    public LoginException1(String message) {
        super(message);
    }
}

package com.ghy.www.exception;

public class LoginException2 extends Exception {
    public LoginException2(String message) {
        super(message);
    }
}
```

（4）视图文件 errorPage.jsp 中的内容如下：

```
<body>
    到达了 errorPage.jsp 页面,异常信息为: ${ex.message}
</body>
```

loginException1.jsp 中的内容如下：

```
<body>
    到达了 loginException1.jsp 页面，异常信息为：${ex.message}
</body>
```

sqlerror.jsp 中的内容如下：

```
<body>
    到达了 sqlerror.jsp 页面，异常信息为：${ex.message}
</body>
```

（5）执行不同的 URL，在自定义的视图中显示异常信息，成功实现了全局异常处理器。

2．使用@RestControllerAdvice 注解实现全局异常处理器

@RestController 注解的作用是隐式地使用了@ResponseBody 注解。

创建测试项目 exception2。

（1）创建控制层，代码如下：

```java
package com.ghy.www.controller;

import com.ghy.www.exception.LoginException1;
import com.ghy.www.exception.LoginException2;
import org.springframework.web.bind.annotation.RequestMapping;
import org.springframework.web.bind.annotation.RestController;

import java.sql.SQLException;

@RestController
public class TestController {

    @RequestMapping(value = "test1")
    public String test1() {
        System.out.println("public String test1()");
        int i = 10;
        int j = 0;
        int result = i / j;
        return "responseBody1";
    }

    @RequestMapping(value = "test2")
    public String test2() throws Exception {
        System.out.println("public String test2()");
        if (1 == 1) {
            throw new SQLException("SQL 语句错误");
        }
        return "responseBody2";
    }

    @RequestMapping(value = "test3")
    public String test3() throws LoginException1 {
        System.out.println("public String test3()");
        if (1 == 1) {
            throw new LoginException1("账号××登录失败，请重新登录！");
        }
        return "responseBody3";
    }
```

```java
@RequestMapping(value = "test4")
public String test4() throws LoginException2 {
    System.out.println("public String test4()");
    if (1 == 1) {
        throw new LoginException2("账号××登录失败,请重新登录!");
    }
    return "responseBody4";
}
```

}

(2)创建异常处理器,代码如下:

```java
package com.ghy.www.advice;

import com.ghy.www.exception.LoginException1;
import com.ghy.www.exception.LoginException2;
import org.springframework.web.bind.annotation.ExceptionHandler;
import org.springframework.web.bind.annotation.RestControllerAdvice;
import org.springframework.web.servlet.ModelAndView;

import java.sql.SQLException;
import java.util.HashMap;
import java.util.Map;

@RestControllerAdvice
public class ControllerAdvice {
    @ExceptionHandler(value = Exception.class)
    public ModelAndView processException(Exception ex) {
        ModelAndView mav = new ModelAndView();
        mav.addObject("ex", ex);
        mav.setViewName("errorPage.jsp");
        return mav;
    }

    @ExceptionHandler(value = SQLException.class)
    public ModelAndView processSQLException(SQLException ex) {
        ModelAndView mav = new ModelAndView();
        mav.addObject("ex", ex);
        mav.setViewName("sqlerror.jsp");
        return mav;
    }

    @ExceptionHandler(value = LoginException1.class)
    public ModelAndView processLoginException1(LoginException1 ex) {
        ModelAndView mav = new ModelAndView();
        mav.addObject("ex", ex);
        mav.setViewName("loginException1.jsp");
        return mav;
    }

    @ExceptionHandler(value = LoginException2.class)
    public Map processLoginException2(LoginException2 ex) {
        Map map = new HashMap();
        map.put("controllerName", "UserinfoController");
        map.put("message", ex.getMessage());
        return map;
    }
}
```

3.4.5 方法的参数是 Model 数据类型

向 Model 类的对象中存储数据相当于向 request 作用域中存储数据。

创建测试项目 modeltest。

（1）创建控制层，代码如下：

```java
package com.ghy.www.controller;

import org.springframework.stereotype.Controller;
import org.springframework.ui.Model;
import org.springframework.web.bind.annotation.RequestMapping;

@Controller
public class TestController {
    @RequestMapping(value = "test")
    public String testMethod(Model model) {
        model.addAttribute("myKey1", "myValue1");
        return "index.jsp";
    }
}
```

（2）创建 JSP 文件，代码如下：

```jsp
<body>
    <%
        Enumeration enum1 = request.getAttributeNames();
        while (enum1.hasMoreElements()) {
            String key = "" + enum1.nextElement();
            out.print(key + "<br/>");
        } %>
    <br/>
    ${myKey1}
</body>
```

（3）程序运行结果如图 3-40 所示。

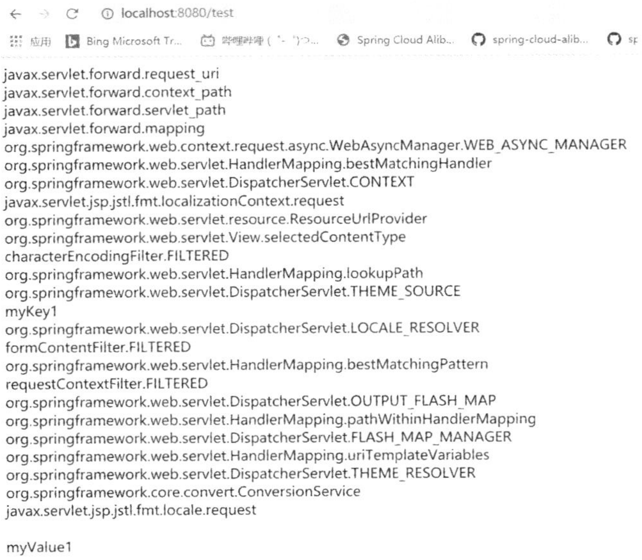

图 3-40　运行结果

3.4.6　方法的参数是 ModelMap 数据类型

向 ModelMap 类的对象中存储数据相当于向 request 作用域中存储数据。ModelMap 类比 Model 类的 API 丰富，两者的作用一样。

创建测试项目 modelmap1。

（1）创建控制层，代码如下：

```java
package com.ghy.www.controller;

import org.springframework.stereotype.Controller;
import org.springframework.ui.ModelMap;
import org.springframework.web.bind.annotation.RequestMapping;

import java.util.ArrayList;
import java.util.List;

@Controller
public class TestController {
    // ModelMap 类比 Model 类的对象的 API 丰富
    @RequestMapping(value = "test")
    public String testMethod(ModelMap modelMap) {
        List list = new ArrayList();
        list.add("中国1");
        list.add("中国2");
        list.add("中国3");
        list.add("中国4");
        modelMap.addAttribute("listStringKey", list);
        return "listString.jsp";
    }
}
```

（2）创建 JSP 文件，代码如下：

```jsp
<body>
    <%
        Enumeration enum1 = request.getAttributeNames();
        while (enum1.hasMoreElements()) {
            String key = (String) enum1.nextElement();
            out.print("key=" + key + "<br/>");
        } %>
    <br/>
    <c:forEach var="eachString" items="${listStringKey}">
        ${eachString}
        <br/>
    </c:forEach>
</body>
```

（3）程序运行结果如图 3-41 所示。

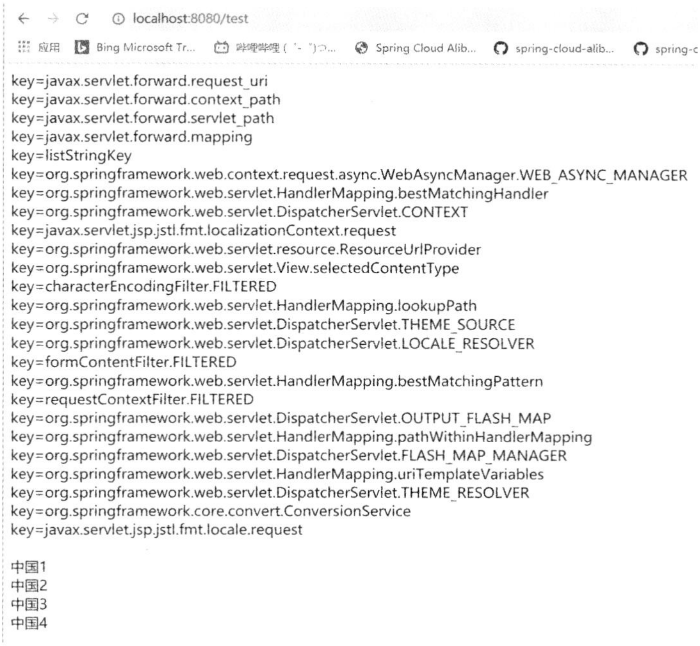
localhost:8080/test
应用
Bing Microsoft Tr...
Spring Cloud Alib...
spring-cloud-alib...
spring-c
key=javax.servlet.forward.request_uri
key=javax.servlet.forward.context_path
key=javax.servlet.forward.servlet_path
key=javax.servlet.forward.mapping
key=listStringKey
key=org.springframework.web.context.request.async.WebAsyncManager.WEB_ASYNC_MANAGER
key=org.springframework.web.servlet.HandlerMapping.bestMatchingHandler
key=org.springframework.web.servlet.DispatcherServlet.CONTEXT
key=javax.servlet.jsp.jstl.fmt.localizationContext.request
key=org.springframework.web.servlet.resource.ResourceUrlProvider
key=org.springframework.web.servlet.View.selectedContentType
key=characterEncodingFilter.FILTERED
key=org.springframework.web.servlet.HandlerMapping.lookupPath
key=org.springframework.web.servlet.DispatcherServlet.THEME_SOURCE
key=org.springframework.web.servlet.DispatcherServlet.LOCALE_RESOLVER
key=formContentFilter.FILTERED
key=org.springframework.web.servlet.HandlerMapping.bestMatchingPattern
key=requestContextFilter.FILTERED
key=org.springframework.web.servlet.DispatcherServlet.OUTPUT_FLASH_MAP
key=org.springframework.web.servlet.HandlerMapping.pathWithinHandlerMapping
key=org.springframework.web.servlet.DispatcherServlet.FLASH_MAP_MANAGER
key=org.springframework.web.servlet.HandlerMapping.uriTemplateVariables
key=org.springframework.web.servlet.DispatcherServlet.THEME_RESOLVER
key=org.springframework.core.convert.ConversionService
key=javax.servlet.jsp.jstl.fmt.locale.request
中国1
中国2
中国3
中国4

图 3-41 运行结果

3.4.7 方法的返回值是 ModelMap 数据类型

创建测试项目 modelmap2。

（1）创建控制层，代码如下：

```
package com.ghy.www.controller;

import org.springframework.stereotype.Controller;
import org.springframework.ui.ModelMap;
import org.springframework.web.bind.annotation.RequestMapping;

import java.util.ArrayList;
import java.util.List;

@Controller
public class TestController {
    @RequestMapping(value = "test")
    public ModelMap testMethod() {
        List list = new ArrayList();
        list.add("中国 1");
        list.add("中国 2");
        list.add("中国 3");
        list.add("中国 4");

        ModelMap map = new ModelMap();
        map.addAttribute("listString", list);

        return map;
    }
}
```

（2）配置文件 application.yml 中的代码如下：

```
spring:
  mvc:
    view:
      prefix: /
      suffix: .jsp
```

（3）创建 JSP 文件，代码如下：

```
<body>
    <%
        Enumeration enum1 = request.getAttributeNames();
        while (enum1.hasMoreElements()) {
            String key = (String) enum1.nextElement();
            out.println("key=" + key + "<br/>");
        } %>
    <br/>
    <c:forEach var="eachString" items="${listString}">
        ${eachString}
        <br/>
    </c:forEach>
</body>
```

（4）程序运行结果如图 3-42 所示。

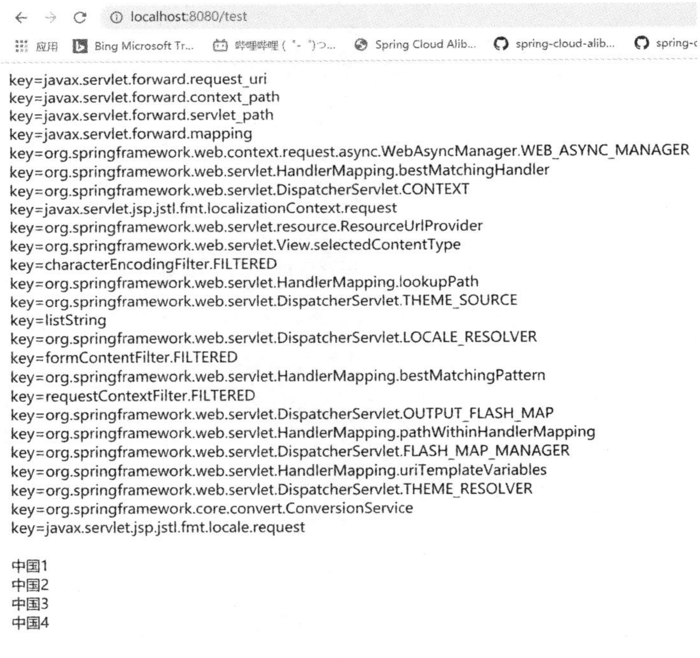

图 3-42　运行结果

3.4.8　方法的返回值是 ModelAndView 数据类型

返回 ModelMap 类的缺点是不能指定视图，而 ModelAndView 类的对象可以实现这一点。创建测试项目 modelandview1。

3.4 扩展技术

（1）创建控制层，代码如下：

```java
package com.ghy.www.controller;

import org.springframework.stereotype.Controller;
import org.springframework.web.bind.annotation.RequestMapping;
import org.springframework.web.servlet.ModelAndView;

@Controller
public class TestController {
    @RequestMapping(value = "test")
    public ModelAndView test() {
        ModelAndView view = new ModelAndView();
        view.setViewName("index.jsp");
        view.addObject("myKey", "大中国");
        return view;
    }
}
```

（2）创建 JSP 文件，代码如下：

```jsp
<body>
    <%
        Enumeration enum1 = request.getAttributeNames();
        while (enum1.hasMoreElements()) {
            String key = (String) enum1.nextElement();
            out.print("key=" + key + "<br/>");
        } %>
    <br/>
    ${myKey}
</body>
```

（3）程序运行结果如图 3-43 所示。

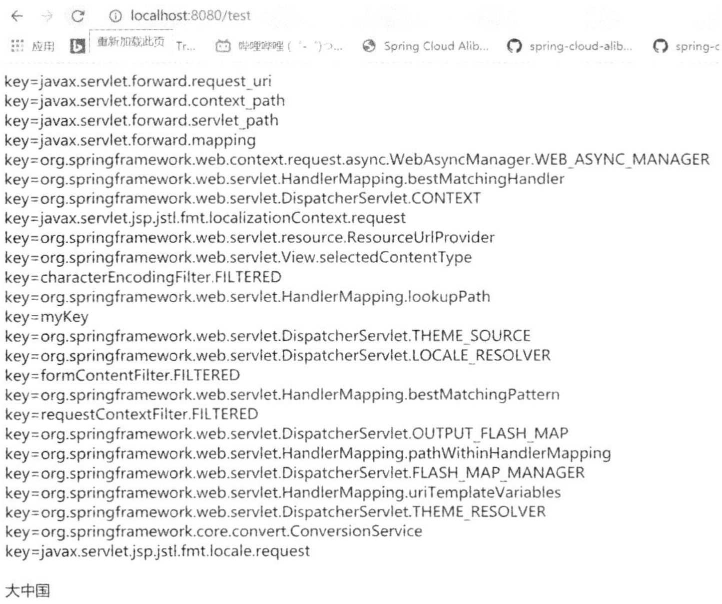

图 3-43 运行结果

3.4.9 方法的返回值是 ModelAndView 数据类型（实现重定向）

创建测试项目 modelandview2。
（1）创建控制层，代码如下：

```java
package com.ghy.www.controller;

import org.springframework.stereotype.Controller;
import org.springframework.web.bind.annotation.RequestMapping;
import org.springframework.web.servlet.ModelAndView;

@Controller
public class TestController {
    @RequestMapping(value = "test1")
    public ModelAndView test1() {
        System.out.println("test1");
        ModelAndView mav = new ModelAndView();
        mav.setViewName("index.jsp");
        return mav;
    }

    @RequestMapping(value = "test2")
    public ModelAndView test2() {
        System.out.println("test2");
        ModelAndView mav = new ModelAndView();
        mav.setViewName("redirect:/index.jsp");
        return mav;
    }
}
```

（2）先执行控制层地址 test1 实现转发操作，再执行控制层地址 test2 实现重定向操作。

3.4.10 使用@RequestAttribute 和@SessionAttribute 注解

使用@RequestAttribute 和@SessionAttribute 注解可以访问 request 和 session 作用域中的数据。
创建测试项目 request-session-attr。
（1）创建控制层，代码如下：

```java
package com.ghy.www.controller;

import org.springframework.stereotype.Controller;
import org.springframework.web.bind.annotation.RequestAttribute;
import org.springframework.web.bind.annotation.RequestMapping;
import org.springframework.web.bind.annotation.SessionAttribute;

import javax.servlet.http.HttpServletRequest;
import javax.servlet.http.HttpSession;

@Controller
public class TestController {
    @RequestMapping(value = "test")
    public String test(HttpServletRequest request, HttpSession session) {
        request.setAttribute("requestKey", "requestValue值1");
```

```java
            request.getSession().setAttribute("sessionKey", "sessionValue值2");
            return "printInfo";//转发到printInfo控制层，request中的数据得以保留
        }

        @RequestMapping(value = "printInfo")
        public String printInfo(@RequestAttribute("requestKey") String value1,
                        @SessionAttribute("sessionKey") String value2) {
            System.out.println("value1=" + value1);
            System.out.println("value2=" + value2);
            return "index.jsp";
        }
    }
```

（2）执行控制层地址 test 后会转发到网址 printInfo，控制台输出结果如下：

```
value1=requestValue值1
value2=sessionValue值2
```

3.4.11　使用@CookieValue 和@RequestHeader 注解

注解@CookieValue 用于取得 Cookie 值。

注解@RequestHeader 用于取得 request header 值。

创建测试项目 requestheader-cookie。

（1）创建控制层，代码如下：

```java
package com.ghy.www.controller;

import org.springframework.stereotype.Controller;
import org.springframework.web.bind.annotation.CookieValue;
import org.springframework.web.bind.annotation.RequestHeader;
import org.springframework.web.bind.annotation.RequestMapping;

import javax.servlet.http.Cookie;
import javax.servlet.http.HttpServletRequest;
import javax.servlet.http.HttpServletResponse;
import javax.servlet.http.HttpSession;

@Controller
public class TestController {
    @RequestMapping(value = "test")
    public String test(HttpServletRequest request, HttpServletResponse response, HttpSession session) {
        Cookie cookie = new Cookie("myCookieName", "我是Cookie值");
        cookie.setMaxAge(36000);
        response.addCookie(cookie);
        return "redirect:/printInfo";
    }

    @RequestMapping(value = "printInfo")
    public String printInfo(@CookieValue(name = "myCookieName") String cookieValue,
                        @RequestHeader(name = "Accept-Language") String acceptLanguage,
@RequestHeader(name = "Host") String host) {
        System.out.println("cookieValue=" + cookieValue);
```

```java
            System.out.println("Accept-Language=" + acceptLanguage);
            System.out.println("Host=" + host);
            return "index.jsp";
        }
    }
```

（2）执行控制层地址 test，控制台输出结果如下：

```
cookieValue=我是 Cookie 值
Accept-Language=zh,en-US;q=0.9,en;q=0.8,zh-CN;q=0.7
Host=localhost:8080
```

3.4.12　使用@SessionAttributes 注解

@SessionAttribute 注解常用于从 HttpSession 获取数据，@SessionAttributes 注解用于向 HttpSession 存放数据。

创建测试项目 sessionattributes。

（1）创建控制层，代码如下：

```java
package com.ghy.www.controller;

import org.springframework.stereotype.Controller;
import org.springframework.ui.Model;
import org.springframework.web.bind.annotation.RequestAttribute;
import org.springframework.web.bind.annotation.RequestMapping;
import org.springframework.web.bind.annotation.SessionAttribute;
import org.springframework.web.bind.annotation.SessionAttributes;

@Controller
@SessionAttributes(names = "myKey2")
public class TestController {
    @RequestMapping(value = "test1")
    public String test1(Model model) {
        model.addAttribute("myKey1", "我在 request 存在");
        model.addAttribute("myKey2", "我在 request 和 session 中都存在");
        return "test2";
    }

    @RequestMapping(value = "test2")
    public String test2(@RequestAttribute("myKey1") String a, @RequestAttribute("myKey2") String b,
                        @SessionAttribute(value = "myKey1", required = false) String c, @SessionAttribute("myKey2") String d) {
        System.out.println("request myKey1=" + a);
        System.out.println("request myKey2=" + b);
        System.out.println("session myKey1=" + c);
        System.out.println("session myKey2=" + d);
        return "test1.jsp";
    }
}
```

（2）创建 JSP 文件 test1.jsp，代码如下：

```
<body>
    request_keys:
```

```jsp
<br/>
<%
    Enumeration enum1 = request.getAttributeNames();
    while (enum1.hasMoreElements()) {
        String key = "" + enum1.nextElement();
        out.print(key + "<br/>");
    }
%>
<br/>
<br/> session_keys:
<br/>
<%
    Enumeration enum2 = request.getSession().getAttributeNames();
    while (enum2.hasMoreElements()) {
        String key = "" + enum2.nextElement();
        out.print(key + "<br/>");
    }
%>
<br/>
<br/> 从 request 获取值：myKey1:${requestScope.myKey1}
<br/> 从 request 获取值：myKey2:${requestScope.myKey2}
<br/> 从 session 获取值：myKey1:${sessionScope.myKey1}
<br/> 从 session 获取值：myKey2:${sessionScope.myKey2}
</body>
```

（3）先执行控制层 test1，然后执行控制层 test2，浏览器中输出的内容如图 3-44 所示。

```
request_keys:   ←
javax.servlet.forward.request_uri
javax.servlet.forward.context_path
javax.servlet.forward.servlet_path
javax.servlet.forward.mapping
org.springframework.web.context.request.async.WebAsyncManager.WEB_ASYNC_MANAGER
org.springframework.web.servlet.HandlerMapping.bestMatchingHandler
org.springframework.web.servlet.DispatcherServlet.CONTEXT
javax.servlet.jsp.jstl.fmt.localizationContext.request
org.springframework.web.servlet.resource.ResourceUrlProvider
org.springframework.web.servlet.View.selectedContentType
characterEncodingFilter.FILTERED
org.springframework.web.servlet.HandlerMapping.lookupPath
org.springframework.web.servlet.DispatcherServlet.THEME_SOURCE
org.springframework.validation.BindingResult.myKey2
myKey1   ←
org.springframework.web.servlet.DispatcherServlet.LOCALE_RESOLVER
myKey2   ←
formContentFilter.FILTERED
org.springframework.web.servlet.HandlerMapping.bestMatchingPattern
requestContextFilter.FILTERED
org.springframework.web.servlet.DispatcherServlet.OUTPUT_FLASH_MAP
org.springframework.web.servlet.HandlerMapping.pathWithinHandlerMapping
org.springframework.web.servlet.DispatcherServlet.FLASH_MAP_MANAGER
org.springframework.web.servlet.HandlerMapping.uriTemplateVariables
org.springframework.web.servlet.DispatcherServlet.THEME_RESOLVER
org.springframework.core.convert.ConversionService
javax.servlet.jsp.jstl.fmt.locale.request

session_keys:
myKey2

从request获取值：myKey1:我在request存在
从request获取值：myKey2:我在request存在
从session获取值：myKey1:
从session获取值：myKey2:我在request和session中都存在
```

图 3-44　运行结果

3.4.13 使用@ModelAttribute 注解实现作用域别名

创建测试项目 modelattribute。
（1）创建控制层，代码如下：

```java
package com.ghy.www.controller;

import org.springframework.stereotype.Controller;
import org.springframework.web.bind.annotation.ModelAttribute;
import org.springframework.web.bind.annotation.RequestMapping;

import java.util.ArrayList;
import java.util.List;

@Controller
public class TestController {
    @RequestMapping(value = "test")
    @ModelAttribute(name = "showList")
    public List test() {
        System.out.println("public String test(ModelMap map)");
        List list = new ArrayList();
        list.add("中国1");
        list.add("中国2");
        list.add("中国3");
        list.add("中国4");
        return list;
    }
}
```

@ModelAttribute(name = "showList")注解的作用是在 public List test()方法的返回值中放入 request，以设置一个别名 key。

（2）配置文件 application.yml 的代码如下：

```yml
spring:
  mvc:
    view:
      prefix: /
      suffix: .jsp
```

（3）创建 JSP 文件 test.jsp，代码如下：

```jsp
<body>
    <%
        Enumeration enum1 = request.getAttributeNames();
        while (enum1.hasMoreElements()) {
            String key = "" + enum1.nextElement();
            out.print(key + "<br/>");
        }
    %>
    <br/>
    <c:forEach var="eachString" items="${showList}">
        ${eachString}<br/>
    </c:forEach>
</body>
```

（4）执行控制层地址 test，浏览器中输出的内容如图 3-45 所示。

图 3-45　运行结果

3.4.14　在路径中添加通配符的功能

我们可以在访问映射路径中添加通配符。

创建测试项目 url-xing。

（1）创建控制层，核心代码如下：

```
package com.ghy.www.controller;

import org.springframework.stereotype.Controller;
import org.springframework.web.bind.annotation.RequestMapping;

import javax.servlet.http.HttpServletRequest;
import javax.servlet.http.HttpServletResponse;

@Controller
public class TestController {
    @RequestMapping(value = "findById_*")
    public String test(HttpServletRequest request, HttpServletResponse response) {
        String servletPath = request.getServletPath();
        servletPath = servletPath.substring(1);
        int beginIndex = servletPath.indexOf("_");
        servletPath = servletPath.substring(beginIndex + 1);
        System.out.println(servletPath);
        return "index.jsp";
    }
}
```

（2）通过如下 URL 进行访问：

```
findById_1
findById_100
```

控制台输出内容如下：

```
1
100
```

3.4.15　控制层返回 void 数据的情况

前面示例中大多数控制层返回的是 String 数据类型，表示转发到指定名称的 JSP 文件。控制层还可以返回 void 数据类型，存在以下两种情况。

（1）使用默认的 JSP 文件。

（2）通过 HttpServletResponse 输出。

创建测试项目 return-null。

创建控制层文件，其核心代码如下：

```
package com.ghy.www.controller;

import org.springframework.stereotype.Controller;
import org.springframework.web.bind.annotation.RequestMapping;

import javax.servlet.http.HttpServletRequest;
import javax.servlet.http.HttpServletResponse;
import java.io.IOException;
import java.io.PrintWriter;

@Controller
public class TestController {
    @RequestMapping(value = "test")
    public void test() {
        System.out.println("test run !");
        // test.jsp
    }

    // 如果方法存在 request 和 response 参数
    // 不进行转发操作
    @RequestMapping(value = "testHasParam")
    public void testHasParam(HttpServletRequest request, HttpServletResponse response) {
        System.out.println("testHasParam run !");
    }

    @RequestMapping(value = "getUsername")
    public void getUsername(HttpServletRequest request, HttpServletResponse response) {
        try {
            response.setCharacterEncoding("utf-8");
            response.setContentType("text/html;charset=utf-8");
            PrintWriter out = response.getWriter();
            out.print("中国");
            out.flush();
            out.close();
        } catch (IOException e) {
            e.printStackTrace();
```

            }
        }
}

（3）文件 test.jsp 的核心代码如下：

```
<body>
    test.jsp page!
</body>
```

（4）更改配置文件 application.yml 后，其核心代码如下：

```
spring:
  mvc:
    view:
      prefix: /
      suffix: .jsp
```

（5）部署并运行项目，执行控制层地址 test，程序运行结果如图 3-46 所示。

图 3-46　程序运行结果

（6）执行控制层地址 testHasParam，程序运行结果如图 3-47 所示。

（7）执行控制层地址 getUsername，程序运行结果如图 3-48 所示。

图 3-47　程序运行结果　　　　　　　　图 3-48　默认转发到 index.jsp 文件中

3.4.16　解决多人开发路径可能重复的问题

在开发 Java EE 项目时，分组开发、分工协作是软件公司常用的工作方式，这时就会出现一些问题，比如开发 A 模块时从前端登录，路径为 login，开发 B 模块时从后端登录，路径也为 login，这样的重复路径会导致出现错误。

创建测试项目 pathsame。

（1）开发 A 模块时的前端登录代码：

```java
package com.ghy.www.controllera;

import org.springframework.stereotype.Controller;
import org.springframework.web.bind.annotation.RequestMapping;

@Controller
public class TestControllerA {
    @RequestMapping(value = "login")
```

```java
    public String listStringMethod() {
        System.out.println("a login new");
        return "index.jsp";
    }
}
```

（2）开发 B 模块时的后台登录代码：

```java
package com.ghy.www.controllerb;

import org.springframework.stereotype.Controller;
import org.springframework.web.bind.annotation.RequestMapping;

@Controller
public class TestControllerB {
    @RequestMapping(value = "login")
    public String listStringMethod() {
        System.out.println("b login new");
        return "index.jsp";
    }
}
```

（3）启动 Tomcat 时出现异常，信息如下：

```
org.springframework.beans.factory.BeanCreationException: Error creating bean with name
'requestMappingHandlerMapping' defined in class path resource [org/springframework/boot/
autoconfigure/web/servlet/WebMvcAutoConfiguration$EnableWebMvcConfiguration.class]: Invocation
of init method failed; nested exception is java.lang.IllegalStateException: Ambiguous mapping.
Cannot map 'testControllerB' method
    com.ghy.www.controllerb.TestControllerB#listStringMethod()
to { /login}: There is already 'testControllerA' bean method
    com.ghy.www.controllera.TestControllerA#listStringMethod() mapped.
```

从提示的出错信息可以看到，路径/login 已经被注册，不能重复注册，那么，如果遇到这种情况该怎么办呢？

（4）针对上述情况，解决办法就是限定各模块的访问路径。

将 A 模块的控制层代码更改如下：

```java
package com.ghy.www.controllera;

import org.springframework.stereotype.Controller;
import org.springframework.web.bind.annotation.RequestMapping;

@Controller
@RequestMapping("/a")
public class TestControllerA {
    @RequestMapping(value = "login")
    public String listStringMethod() {
        System.out.println("a login new");
        return "../index.jsp";
    }
}
```

将 B 模块的控制层代码更改如下：

```java
package com.ghy.www.controllerb;

import org.springframework.stereotype.Controller;
import org.springframework.web.bind.annotation.RequestMapping;
```

```
@Controller
@RequestMapping("/b")
public class TestControllerB {
    @RequestMapping(value = "login")
    public String listStringMethod() {
        System.out.println("b login new");
        return "../index.jsp";
    }
}
```

如果在控制层类的上方使用@RequestMapping 注解，表示首先定义了相对的父路径，然后在类的方法上定义的路径就是相对于类级别的。

（5）重新启动 Tomcat 时没有出现异常。
（6）执行控制层地址/a/login 和/b/login 后，成功执行不同模块相同控制层路径对应的方法。
（7）通过在控制层类的上方加入@RequestMapping("/a")注解

```
@Controller
@RequestMapping("/a")
public class TestControllerA {
```

可以在 Spring MVC 中进行模块化开发。

现在的状态虽然正确实现了多人开发，但控制层中的代码 return "../index.jsp";看起来不太美观，因为如果路径过多，../就会出现很多，该如何解决呢？在 application.yml 配置文件中加入如下配置即可：

```
spring:
  mvc:
    view:
      prefix: /
```

该配置的功能就是限定默认访问资源的路径是/（根路径），也就是相对于 Web 的路径。我们将控制层代码更改如下：

```
return "index.jsp";
```

（8）再次执行控制层地址/a/login 和/b/login，就能在控制台中正确输出我们想要的字符串。

3.4.17 使用@PathVariable 注解

@PathVariable 注解可以将 URL 中的参数和参数值的写法由?name=value 转换为内嵌在 URL 地址中。

创建测试项目 path-variable。
（1）创建 A 模块的控制层，代码如下：

```
package com.ghy.www.controller;

import org.springframework.stereotype.Controller;
import org.springframework.web.bind.annotation.PathVariable;
import org.springframework.web.bind.annotation.RequestMapping;

@Controller
public class TestControllerA {
```

```java
    @RequestMapping(value = "findUserinfo1/{userId}")
    public String findUserinfo1(@PathVariable("userId") String xxxxxx) {
        System.out.println(xxxxxx);
        return "index.jsp";
    }

    @RequestMapping(value = "findUserinfo2/{userId}")
    public String findUserinfo2(@PathVariable String userId) {
        System.out.println(userId);
        return "index.jsp";
    }

    @RequestMapping(value = "findUserinfo3/username/{username}/age/{age}")
    public String findUserinfo2(@PathVariable String username, @PathVariable String age) {
        System.out.println(username + " " + age);
        return "index.jsp";
    }
}
```

（2）创建 B 模块的控制层，代码如下：

```java
package com.ghy.www.controller;

import org.springframework.stereotype.Controller;
import org.springframework.web.bind.annotation.PathVariable;
import org.springframework.web.bind.annotation.RequestMapping;

@Controller
@RequestMapping(value = "findUserinfo4/username/{username}")
public class TestControllerB {
    @RequestMapping(value = "address/{address}")
    public String findUserinfo1(@PathVariable String username, @PathVariable String address) {
        System.out.println(username + " " + address);
        return "index.jsp";
    }
}
```

（3）部署项目，执行控制层地址/findUserinfo1/123，控制台输出结果如下：

```
123
```

（4）执行控制层地址/findUserinfo2/456，控制台输出结果如下：

```
456
```

（5）执行控制层地址/findUserinfo3/username/abc/age/123，控制台输出结果如下：

```
abc 123
```

（6）执行控制层地址/findUserinfo4/username/abc/address/bj，控制台输出结果如下：

```
abc bj
```

3.4.18　通过 URL 参数访问指定的业务方法

如果我们想要在访问同一个 URL 地址的同时以传递参数的方式来调用指定控制层中指定的业务方法，使用@RequestMapping 注解很容易实现。

创建测试项目 urlparam-method。

（1）创建控制层，代码如下：

```
package com.ghy.www.controller;

import org.springframework.stereotype.Controller;
import org.springframework.web.bind.annotation.RequestMapping;

@Controller
public class TestController {
    @RequestMapping(value = "listInfo", params = "type=A")
    public String listInfoAAA() {
        System.out.println("AAA");
        return "index.jsp";
    }

    @RequestMapping(value = "listInfo", params = "type=B")
    public String listInfoBBB() {
        System.out.println("BBB");
        return "index.jsp";
    }
}
```

（2）控制层中有两个业务方法，如何调用这两个业务方法呢？

执行控制层地址：

```
/listInfo?type=A
```

和：

```
/listInfo?type=B
```

3.4.19 使用@GetMapping、@PostMapping、@PutMapping 和@DeleteMapping 注解

在 HTTP 中，对于提交类型，除 get 和 post 外，还可以使用常见的 put 和 delete，在 Spring 5 MVC 中分别对应@GetMapping、@PostMapping、@PutMapping 和@DeleteMapping 注解。

@GetMapping 注解的作用是查询。

@PostMapping 注解的作用是添加。

@PutMapping 注解的作用是更新。

@DeleteMapping 注解的作用是删除。

在后面的章节中，我们使用两种方式来测试这 4 个注解的使用，分别是使用表单<form>和 AJAX。

1．使用<form>

创建测试项目 form-methodtype。

（1）创建控制层，代码如下：

```
package com.ghy.www.controller;

import org.springframework.stereotype.Controller;
import org.springframework.web.bind.annotation.DeleteMapping;
import org.springframework.web.bind.annotation.GetMapping;
```

```java
import org.springframework.web.bind.annotation.PostMapping;
import org.springframework.web.bind.annotation.PutMapping;

@Controller
public class TestController {
    @GetMapping(value = "get")
    public String get(String username) {
        System.out.println("get username=" + username);
        return "index.jsp";
    }

    @PostMapping(value = "post")
    public String post(String username) {
        System.out.println("post username=" + username);
        return "redirect:/get?username=getValue";
    }

    @PutMapping(value = "put")
    public String put(String username) {
        System.out.println("put username=" + username);
        return "redirect:/get?username=getValue";
    }

    @DeleteMapping(value = "delete")
    public String delete(String username) {
        System.out.println("delete username=" + username);
        return "redirect:/get?username=getValue";
    }
}
```

（2）创建 JSP 文件，代码如下：

```html
<body>
    <form action="get" method="get">
        username:<input type="text" name="username" value="getValue">
        <br/>
        <input type="submit" value="get">
    </form>
    <form action="post" method="post">
        username:<input type="text" name="username" value="postValue">
        <br/>
        <input type="submit" value="post">
    </form>
    <form action="put" method="post">
        <input type="hidden" name="_method" value="put"/>
        <br/>
        username:<input type="text" name="username" value="putValue">
        <br/>
        <input type="submit" value="put">
    </form>
    <form action="delete" method="post">
        <input type="hidden" name="_method" value="delete"/>
        <br/>
        username:<input type="text" name="username" value="deleteValue">
        <br/>
        <input type="submit" value="delete">
    </form>
</body>
```

在执行控制层的方法 put 和 delete 对应的 HTML 中，一定要在<form>中添加<input type="hidden" name="_method" value="delete" />隐藏域，因为如果使用 Spring 的标签 form

```
<springForm:form action="put" method="put">
    <input type="text" name="age" value="123"/>
    <input type="submit" value="put"/>
</springForm:form>
```

程序运行后，Spring 的标签 form 会自动生成该隐藏域，服务器根据隐藏域中的值来调用指定的方法。

（3）在 application.yml 文件中配置 filter（过滤器），代码如下：

```
spring:
  mvc:
    hiddenmethod:
      filter:
        enabled: true
```

（4）运行项目，执行 index.jsp 文件，按顺序单击 4 个按钮，控制台输出结果如下：

```
get username=getValue
post username=postValue
get username=getValue
put username=putValue
get username=getValue
delete username=deleteValue
get username=getValue
```

2. 使用 AJAX

创建测试项目 ajax-methodtype。

（1）创建控制层，代码如下：

```java
package com.ghy.www.controller;

import org.springframework.stereotype.Controller;
import org.springframework.web.bind.annotation.*;

@Controller
public class TestController {
    @GetMapping(value = "get")
    @ResponseBody
    public String get(String username) {
        System.out.println("get username=" + username);
        return "getReturn";
    }

    @PostMapping(value = "post")
    @ResponseBody
    public String post(String username) {
        System.out.println("post username=" + username);
        return "postReturn";
    }

    @PutMapping(value = "put")
    @ResponseBody
    public String put(String username) {
```

```java
        System.out.println("put username=" + username);
        return "putReturn";
    }

    @DeleteMapping(value = "delete")
    @ResponseBody
    public String delete(String username) {
        System.out.println("delete username=" + username);
        return "deleteReturn";
    }
}
```

(2) 创建 JSP 文件，代码如下：

```jsp
<%@ page language="java" contentType="text/html; charset=utf-8"
    pageEncoding="utf-8" %>
<%@ taglib uri="http://java.sun.com/jsp/jstl/core" prefix="c" %>
<%@ taglib uri="http://www.springframework.org/tags" prefix="spring" %>
<!DOCTYPE html>
<html>
    <head>
        <script src="jquery3.5.1.js">
        </script>
        <script>
            function getMethod() {
                $.ajax({
                    "type": "get",
                    "url": "get?t=" + new Date().getTime(),
                    "data": {
                        "username": "getValue"
                    },
                    success: function (data) {
                        alert(data);
                    }
                });
            }

            function postMethod() {
                $.ajax({
                    "type": "post",
                    "url": "post?t=" + new Date().getTime(),
                    "data": {
                        "username": "postValue"
                    },
                    success: function (data) {
                        alert(data);
                    }
                });
            }

            function putMethod() {
                $.ajax({
                    "type": "post",
                    "url": "put?t=" + new Date().getTime(),
                    "data": {
                        "_method": "put",
                        "username": "putValue"
                    },
                    success: function (data) {
                        alert(data);
                    }
```

```
                });
            }
            function deleteMethod() {
                $.ajax({
                    "type": "post",
                    "url": "delete?t=" + new Date().getTime(),
                    "data": {
                        "_method": "delete",
                        "username": "deleteValue"
                    },
                    success: function (data) {
                        alert(data);
                    }
                });
            }
        </script>
    </head>
    <body>
        <input type="button" value="get" onclick="getMethod()"/>
        <br/>
        <input type="button" value="post" onclick="postMethod()"/>
        <br/>
        <input type="button" value="put" onclick="putMethod()"/>
        <br/>
        <input type="button" value="delete" onclick="deleteMethod()"/>
    </body>
</html>
```

（3）在 application.yml 文件中配置 filter（过滤器），代码如下：

```
spring:
  mvc:
    hiddenmethod:
      filter:
        enabled: true
```

（4）运行项目，执行 index.jsp 文件，按顺序单击 4 个按钮，控制台输出结果如下：

```
get username=getValue
post username=postValue
put username=putValue
delete username=deleteValue
```

在浏览器中成功接收从服务端返回的数据。

3.4.20 使用拦截器

使用 Spring MVC 中的拦截器可以实现在执行控制层方法之前完成一些功能的处理，如权限验证、执行时间统计等工作，这点和 AOP 有些类似，但拦截器比 AOP 更适合处理一些与 Web 对象有关的功能，这些 Web 对象可以是 request、response、session、application 和 cookie 等，AOP 对于使用这些对象有些不太适合。

创建测试项目 handlerinterceptor-test。

创建控制层，代码如下：

```
package com.ghy.www.controller;

import org.springframework.stereotype.Controller;
```

```java
import org.springframework.web.bind.annotation.RequestMapping;
import org.springframework.web.bind.annotation.ResponseBody;
import org.springframework.web.servlet.ModelAndView;

@Controller
public class TestController {
    @RequestMapping(value = "/test1")
    @ResponseBody
    public String test1() {
        System.out.println("public String test1()");
        System.out.println();
        return "i am test1 string";
    }

    @RequestMapping(value = "/test2")
    public ModelAndView test2() {
        System.out.println("public String test2()");
        System.out.println();
        ModelAndView view = new ModelAndView();
        view.setViewName("welcome.jsp");
        return view;
    }

    @RequestMapping(value = "/test3")
    public ModelAndView test3() {
        System.out.println("public String test3()");
        System.out.println();
        String username = null;
        username.toString();// 出现异常
        ModelAndView view = new ModelAndView();
        view.setViewName("welcome.jsp");
        return view;
    }
}
```

创建拦截器类，代码如下：

```java
package com.ghy.www.myinterceptor;

import org.springframework.stereotype.Component;
import org.springframework.web.servlet.HandlerInterceptor;
import org.springframework.web.servlet.ModelAndView;

import javax.servlet.http.HttpServletRequest;
import javax.servlet.http.HttpServletResponse;

@Component
public class MyHandlerInterceptor implements HandlerInterceptor {
    // 在调用控制层方法之前执行
    @Override
    public boolean preHandle(HttpServletRequest request, HttpServletResponse response, Object handler)
            throws Exception {
        System.out.println("preHandle: ");
        System.out.println("request=" + request + " getServletPath=" + request.getServletPath());
        System.out.println("response=" + response);
        System.out.println("handler=" + handler.getClass().getName());
        System.out.println();
        return true;
    }
```

3.4 扩展技术

```java
        // 调用控制层方法没有出现异常之后执行
        // 调用控制层方法出现异常之后不执行
        @Override
        public void postHandle(HttpServletRequest request, HttpServletResponse response, Object handler,
                               ModelAndView modelAndView) throws Exception {
            HandlerInterceptor.super.postHandle(request, response, handler, modelAndView);
            System.out.println("postHandle: ");
            System.out.println("request=" + request + " getServletPath=" + request.getServletPath());
            System.out.println("response=" + response);
            System.out.println("handler=" + handler.getClass().getName());
            System.out.println("modelAndView=" + modelAndView);
            System.out.println();
        }

        // 不管调用控制层方法是否出现异常都执行
        @Override
        public void afterCompletion(HttpServletRequest request, HttpServletResponse response, Object handler, Exception ex)
                throws Exception {
            HandlerInterceptor.super.afterCompletion(request, response, handler, ex);
            System.out.println("afterCompletion: ");
            System.out.println("request=" + request + " getServletPath=" + request.getServletPath());
            System.out.println("response=" + response);
            System.out.println("handler=" + handler.getClass().getName());
            System.out.println("ex=" + ex);
        }
    }
```

创建配置类，代码如下：

```java
package com.ghy.www.config;

import com.ghy.www.myinterceptor.MyHandlerInterceptor;
import org.springframework.beans.factory.annotation.Autowired;
import org.springframework.context.annotation.Configuration;
import org.springframework.web.servlet.config.annotation.InterceptorRegistry;
import org.springframework.web.servlet.config.annotation.WebMvcConfigurer;

@Configuration
public class WebMvcConfiguration implements WebMvcConfigurer {
    @Autowired
    MyHandlerInterceptor myHandlerInterceptor;

    @Override
    public void addInterceptors(InterceptorRegistry registry) {
        registry.addInterceptor(myHandlerInterceptor).addPathPatterns("/**").excludePathPatterns("/**.jsp");
    }
}
```

执行控制层地址 test1，输出结果如下：

```
preHandle:
request=org.apache.catalina.connector.RequestFacade@51792bd getServletPath=/test1
response=org.apache.catalina.connector.ResponseFacade@509f2859
handler=org.springframework.web.method.HandlerMethod
```

```
public String test1()

postHandle:
request=org.apache.catalina.connector.RequestFacade@51792bd getServletPath=/test1
response=org.apache.catalina.connector.ResponseFacade@509f2859
handler=org.springframework.web.method.HandlerMethod
modelAndView=null

afterCompletion:
request=org.apache.catalina.connector.RequestFacade@51792bd getServletPath=/test1
response=org.apache.catalina.connector.ResponseFacade@509f2859
handler=org.springframework.web.method.HandlerMethod
ex=null
```

执行控制层地址 test2，输出结果如下：

```
preHandle:
request=org.apache.catalina.connector.RequestFacade@51792bd getServletPath=/test2
response=org.apache.catalina.connector.ResponseFacade@509f2859
handler=org.springframework.web.method.HandlerMethod

public String test2()

postHandle:
request=org.apache.catalina.connector.RequestFacade@51792bd getServletPath=/test2
response=org.apache.catalina.connector.ResponseFacade@509f2859
handler=org.springframework.web.method.HandlerMethod
modelAndView=ModelAndView: reference to view with name 'welcome.jsp'; model is {}

afterCompletion:
request=org.apache.catalina.connector.RequestFacade@51792bd getServletPath=/test2
response=org.apache.catalina.connector.ResponseFacade@509f2859
handler=org.springframework.web.method.HandlerMethod
ex=null
```

执行控制层地址 test3，输出结果如下：

```
preHandle:
request=org.apache.catalina.connector.RequestFacade@7338edda getServletPath=/test3
response=org.apache.catalina.connector.ResponseFacade@53976913
handler=org.springframework.web.method.HandlerMethod

public String test3()

afterCompletion:
request=org.apache.catalina.connector.RequestFacade@7338edda getServletPath=/test3
response=org.apache.catalina.connector.ResponseFacade@53976913
handler=org.springframework.web.method.HandlerMethod
ex=java.lang.NullPointerException
 2020-10-09 14:18:23.647 ERROR 796 --- [nio-8080-exec-2] o.a.c.c.C.[.[./].[dispatcher
Servlet]    : Servlet.service() for servlet [dispatcherServlet] in context with path [] threw
exception [Request processing failed; nested exception is java.lang.NullPointerException] with
root cause

java.lang.NullPointerException: null
    at com.ghy.www.controller.TestController.test3(TestController.java:32) ~[classes/:na]
    at sun.reflect.NativeMethodAccessorImpl.invoke0(Native Method) ~[na:1.8.0_251]
    at sun.reflect.NativeMethodAccessorImpl.invoke(NativeMethodAccessorImpl.java:62) ~[na:1.8.0_251]
```

        at   sun.reflect.DelegatingMethodAccessorImpl.invoke(DelegatingMethodAccessorImpl.java:43)
~[na:1.8.0_251]
        at java.lang.reflect.Method.invoke(Method.java:498) ~[na:1.8.0_251]
        at org.springframework.web.method.support.InvocableHandlerMethod.doInvoke(InvocableHandler
Method.java:190) ~[spring-web-5.2.9.RELEASE.jar:5.2.9.RELEASE]
        at org.springframework.web.method.support.InvocableHandlerMethod.invokeForRequest(Invocable
HandlerMethod.java:138) ~[spring-web-5.2.9.RELEASE.jar:5.2.9.RELEASE]
        at org.springframework.web.servlet.mvc.method.annotation.ServletInvocableHandlerMethod.
invokeAndHandle(ServletInvocableHandlerMethod.java:105) ~[spring-webmvc-5.2.9.RELEASE.jar:
5.2.9.RELEASE]
        at org.springframework.web.servlet.mvc.method.annotation.RequestMappingHandlerAdapter.
invokeHandlerMethod(RequestMappingHandlerAdapter.java:878) ~[spring-webmvc-5.2.9.RELEASE.jar:
5.2.9.RELEASE]
        at org.springframework.web.servlet.mvc.method.annotation.RequestMappingHandlerAdapter.
handleInternal(RequestMappingHandlerAdapter.java:792) ~[spring-webmvc-5.2.9.RELEASE.jar:5.2.9.
RELEASE]
        at org.springframework.web.servlet.mvc.method.AbstractHandlerMethodAdapter.handle(Abstract
HandlerMethodAdapter.java:87) ~[spring-webmvc-5.2.9.RELEASE.jar:5.2.9.RELEASE]
        at org.springframework.web.servlet.DispatcherServlet.doDispatch(DispatcherServlet.java:
1040) ~[spring-webmvc-5.2.9.RELEASE.jar:5.2.9.RELEASE]
        at org.springframework.web.servlet.DispatcherServlet.doService(DispatcherServlet.java:
943) ~[spring-webmvc-5.2.9.RELEASE.jar:5.2.9.RELEASE]
        at org.springframework.web.servlet.FrameworkServlet.processRequest(FrameworkServlet.java:
1006) ~[spring-webmvc-5.2.9.RELEASE.jar:5.2.9.RELEASE]
        at org.springframework.web.servlet.FrameworkServlet.doGet(FrameworkServlet.java:898) ~[spring-
webmvc-5.2.9.RELEASE.jar:5.2.9.RELEASE]
        at javax.servlet.http.HttpServlet.service(HttpServlet.java:626) ~[tomcat-embed-core-9.0.38.
jar:4.0.FR]
        at org.springframework.web.servlet.FrameworkServlet.service(FrameworkServlet.java:883)
~[spring-webmvc-5.2.9.RELEASE.jar:5.2.9.RELEASE]
        at javax.servlet.http.HttpServlet.service(HttpServlet.java:733) ~[tomcat-embed-core-9.0.
38.jar:4.0.FR]
        at org.apache.catalina.core.ApplicationFilterChain.internalDoFilter(ApplicationFilter
Chain.java:231) ~[tomcat-embed-core-9.0.38.jar:9.0.38]
        at org.apache.catalina.core.ApplicationFilterChain.doFilter(ApplicationFilterChain.java:
166) ~[tomcat-embed-core-9.0.38.jar:9.0.38]
        at org.apache.tomcat.websocket.server.WsFilter.doFilter(WsFilter.java:53) ~[tomcat-embed-
websocket-9.0.38.jar:9.0.38]
        at org.apache.catalina.core.ApplicationFilterChain.internalDoFilter(ApplicationFilter
Chain.java:193) ~[tomcat-embed-core-9.0.38.jar:9.0.38]
        at org.apache.catalina.core.ApplicationFilterChain.doFilter(ApplicationFilterChain.java:
166) ~[tomcat-embed-core-9.0.38.jar:9.0.38]
        at org.springframework.web.filter.RequestContextFilter.doFilterInternal(RequestContext
Filter.java:100) ~[spring-web-5.2.9.RELEASE.jar:5.2.9.RELEASE]
        at org.springframework.web.filter.OncePerRequestFilter.doFilter(OncePerRequestFilter.
java:119) ~[spring-web-5.2.9.RELEASE.jar:5.2.9.RELEASE]
        at org.apache.catalina.core.ApplicationFilterChain.internalDoFilter(ApplicationFilter
Chain.java:193) ~[tomcat-embed-core-9.0.38.jar:9.0.38]
        at org.apache.catalina.core.ApplicationFilterChain.doFilter(ApplicationFilterChain.java:
166) ~[tomcat-embed-core-9.0.38.jar:9.0.38]
        at org.springframework.web.filter.FormContentFilter.doFilterInternal(FormContentFilter.
java:93) ~[spring-web-5.2.9.RELEASE.jar:5.2.9.RELEASE]
        at org.springframework.web.filter.OncePerRequestFilter.doFilter(OncePerRequestFilter.
java:119) ~[spring-web-5.2.9.RELEASE.jar:5.2.9.RELEASE]
        at org.apache.catalina.core.ApplicationFilterChain.internalDoFilter(ApplicationFilter
Chain.java:193) ~[tomcat-embed-core-9.0.38.jar:9.0.38]
        at org.apache.catalina.core.ApplicationFilterChain.doFilter(ApplicationFilterChain.java:
166) ~[tomcat-embed-core-9.0.38.jar:9.0.38]

```
        at org.springframework.web.filter.CharacterEncodingFilter.doFilterInternal(Character
EncodingFilter.java:201) ~[spring-web-5.2.9.RELEASE.jar:5.2.9.RELEASE]
        at org.springframework.web.filter.OncePerRequestFilter.doFilter(OncePerRequestFilter.
java:119) ~[spring-web-5.2.9.RELEASE.jar:5.2.9.RELEASE]
        at org.apache.catalina.core.ApplicationFilterChain.internalDoFilter(ApplicationFilter
Chain.java:193) ~[tomcat-embed-core-9.0.38.jar:9.0.38]
        at org.apache.catalina.core.ApplicationFilterChain.doFilter(ApplicationFilterChain.java:
166) ~[tomcat-embed-core-9.0.38.jar:9.0.38]
        at org.apache.catalina.core.StandardWrapperValve.invoke(StandardWrapperValve.java:202)
~[tomcat-embed-core-9.0.38.jar:9.0.38]
        at  org.apache.catalina.core.StandardContextValve.invoke(StandardContextValve.java:97)
[tomcat-embed-core-9.0.38.jar:9.0.38]
        at org.apache.catalina.authenticator.AuthenticatorBase.invoke(AuthenticatorBase.java:541)
[tomcat-embed-core-9.0.38.jar:9.0.38]
        at org.apache.catalina.core.StandardHostValve.invoke(StandardHostValve.java:143) [tomcat-
embed-core-9.0.38.jar:9.0.38]
        at org.apache.catalina.valves.ErrorReportValve.invoke(ErrorReportValve.java:92) [tomcat-
embed-core-9.0.38.jar:9.0.38]
        at org.apache.catalina.core.StandardEngineValve.invoke(StandardEngineValve.java:78) [tomcat-
embed-core-9.0.38.jar:9.0.38]
        at org.apache.catalina.connector.CoyoteAdapter.service(CoyoteAdapter.java:343) [tomcat-
embed-core-9.0.38.jar:9.0.38]
        at org.apache.coyote.http11.Http11Processor.service(Http11Processor.java:374) [tomcat-
embed-core-9.0.38.jar:9.0.38]
        at org.apache.coyote.AbstractProcessorLight.process(AbstractProcessorLight.java:65) [tomcat-
embed-core-9.0.38.jar:9.0.38]
        at org.apache.coyote.AbstractProtocol$ConnectionHandler.process(AbstractProtocol.java:868)
[tomcat-embed-core-9.0.38.jar:9.0.38]
        at org.apache.tomcat.util.net.NioEndpoint$SocketProcessor.doRun(NioEndpoint.java:1590)
[tomcat-embed-core-9.0.38.jar:9.0.38]
        at org.apache.tomcat.util.net.SocketProcessorBase.run(SocketProcessorBase.java:49) [tomcat-
embed-core-9.0.38.jar:9.0.38]
        at java.util.concurrent.ThreadPoolExecutor.runWorker(ThreadPoolExecutor.java:1149) [na:1.
8.0_251]
        at java.util.concurrent.ThreadPoolExecutor$Worker.run(ThreadPoolExecutor.java:624) [na:1.
8.0_251]
        at org.apache.tomcat.util.threads.TaskThread$WrappingRunnable.run(TaskThread.java:61)
[tomcat-embed-core-9.0.38.jar:9.0.38]
        at java.lang.Thread.run(Thread.java:748) [na:1.8.0_251]

    preHandle:
    request=org.apache.catalina.core.ApplicationHttpRequest@4543e12d getServletPath=/error
    response=org.apache.catalina.connector.ResponseFacade@53976913
    handler=org.springframework.web.method.HandlerMethod

    postHandle:
    request=org.apache.catalina.core.ApplicationHttpRequest@4543e12d getServletPath=/error
    response=org.apache.catalina.connector.ResponseFacade@53976913
    handler=org.springframework.web.method.HandlerMethod
    modelAndView=ModelAndView [view="error"; model={timestamp=Fri Oct 09 14:18:23 CST 2020,
status=500, error=Internal Server Error, message=, path=/test3}]

    afterCompletion:
    request=org.apache.catalina.core.ApplicationHttpRequest@4543e12d getServletPath=/error
    response=org.apache.catalina.connector.ResponseFacade@53976913
    handler=org.springframework.web.method.HandlerMethod
    ex=null
```

当把拦截器类 MyHandlerInterceptor 的 preHandle()方法的返回值改成 return false 时，完整代码如下：

```java
// 在调用控制层方法之前执行
@Override
public boolean preHandle(HttpServletRequest request, HttpServletResponse response,
Object handler)
        throws Exception {
    System.out.println("preHandle: ");
    System.out.println("request=" + request + " getServletPath=" + request.getServletPath());
    System.out.println("response=" + response);
    System.out.println("handler=" + handler.getClass().getName());
    System.out.println();
    return false;
}
```

执行完控制层方法后，不再执行后面的方法，执行流程就停止了，执行控制层地址 test1，输出结果如下：

```
preHandle:
request=org.apache.catalina.connector.RequestFacade@5e2af0a6 getServletPath=/test1
response=org.apache.catalina.connector.ResponseFacade@1be8450a
handler=org.springframework.web.method.HandlerMethod
```

3.4.21　Spring 5 MVC 应用 AOP 切面

创建 Java 项目 springmvc-aop。

（1）创建数据访问层，代码如下：

```java
package com.ghy.www.dao;

import com.ghy.www.entity.Userinfo;
import org.springframework.stereotype.Repository;

@Repository
public class UserinfoDAO {
    public Userinfo getUserinfoById(int userId) {
        Userinfo userinfo = new Userinfo();
        userinfo.setUsername("大中国");
        return userinfo;
    }
}
```

（2）创建业务层，代码如下：

```java
package com.ghy.www.service;

import com.ghy.www.dao.UserinfoDAO;
import com.ghy.www.entity.Userinfo;
import org.springframework.beans.factory.annotation.Autowired;
import org.springframework.stereotype.Service;

@Service
public class UserinfoService {
    @Autowired
    private UserinfoDAO userinfoDAO;

    public String getUsername(int userId) {
```

```java
        Userinfo userinfo = userinfoDAO.getUserinfoById(userId);
        return userinfo.getUsername();
    }
}
```

(3)创建切面类,代码如下:

```java
package com.ghy.www.myaspect;

import org.aspectj.lang.ProceedingJoinPoint;
import org.aspectj.lang.annotation.Around;
import org.aspectj.lang.annotation.Aspect;
import org.aspectj.lang.annotation.Before;
import org.springframework.stereotype.Component;

@Component
@Aspect
public class MyAspect {
    @Around(value = "execution(* com.ghy.www.controller.TestController.*(..))")
    public Object aroundMethod(ProceedingJoinPoint point) throws Throwable {
        System.out.println("begin time");
        Object returnValue = point.proceed();
        System.out.println("  end time");
        return returnValue;
    }

    @Before(value = "execution(* com.ghy.www.service.UserinfoService.*(..))")
    public void beforeMethod() {
        System.out.println("before " + System.currentTimeMillis());
    }
}
```

(4)创建控制层,代码如下:

```java
package com.ghy.www.controller;

import com.ghy.www.service.UserinfoService;
import org.springframework.beans.factory.annotation.Autowired;
import org.springframework.stereotype.Controller;
import org.springframework.web.bind.annotation.RequestMapping;

@Controller
public class TestController {
    @Autowired
    private UserinfoService userinfoService;

    @RequestMapping("test")
    public String test() {
        System.out.println("public String test()");
        System.out.println(userinfoService.getUsername(10000));
        return "test.jsp";
    }
}
```

(5)部署项目并执行程序,在控制台输出的结果说明切面成功应用到控制层和业务层上,结果如下:

```
begin time
public String test()
before 1602224889552
大中国
  end time
```

第 4 章 MyBatis 3 核心技术之必备技能

本章目标
（1）掌握 ORM 概念
（2）掌握 ORM 映射原理
（3）掌握泛型 DAO
（4）掌握 MyBatis 核心对象的生命周期
（5）使用 Mapper 接口操作数据库
（6）在 IntelliJ IDEA 中手动搭建 MyBatis 开发环境

4.1 ORM 简介

MyBatis 是一个基于"ORM 映射"的框架，ORM 的全称是对象关系映射（Object Relational Mapping）。

对象（Object）就是 Java 中的对象，关系（Relational）就是数据库中的数据表。基于"ORM 映射"的框架使数据在对象和关系之间进行双向转换，将对象中的数据转换成数据表中的一行，或者将数据表中的一行转换成一个对象，如图 4-1 所示。MyBatis 可以实现双向转换的过程。

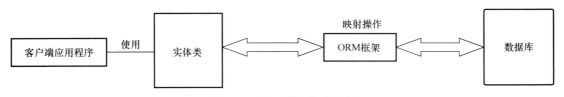

图 4-1 ORM 框架的主流程

使用 ORM 框架后，程序员不再直接使用 JDBC 对象访问数据库，而是以面向对象的方式使用实体类，对实体类进行的增删改查操作都会由 ORM 框架转换成对数据库的增删改查操作。

ORM 映射的细节体现如下 3 个方面。
（1）1 个类对应 1 个表。
（2）1 个类的对象对应表中的 1 行。

(3)1个类中的属性对应 1 个表中的列。

ORM 映射关系如图 4-2 所示。

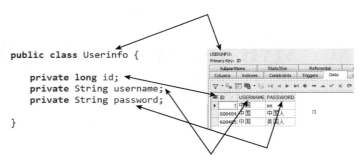

图 4-2　ORM 映射关系

ORM 框架核心技术的原理是 JDBC+反射。

4.2　MyBatis 的优势

MyBatis 是现阶段操作数据库的主流 ORM 框架，此框架的主要作用就是更加便捷地操作数据库。MyBatis 具有很多优势，和程序员密切相关的优势主要体现在以下 7 个方面。

（1）行（ROW）与实体类（Entity）双向转换：可以将数据表中的行与实体类进行互相转换，比如将 ResultSet 对象返回的数据自动封装进实体类或 List 中，或将实体类中的数据转换成数据表中新的一行。

（2）SQL 语句与 Java 文件分离：可以把 SQL 语句写入 XML 文件中，其目的是将 SQL 语句与 Java 文件进行分离，有利于代码的后期维护，也使代码的分层更加明确。

（3）允许对 SQL 语句进行自定义优化：由于 MyBatis 使用 SQL 语句对数据库进行操作，因此可以单独对慢 SQL 语句进行优化，以提高操作效率，而 Hibernate 却做不到这点，所以 MyBatis 相比 Hibernate 就有了很大的优势，这也是现阶段大部分软件公司逐步用 MyBatis 替换 Hibernate 的主要原因。

（4）减化 DAO 层代码：使用传统的 JDBC 开发方式时，需要编写必要的 DAO 层代码以对数据库进行操作，但是这样的代码编写在软件开发的过程中非常不便，因为多个 DAO 类中的大部分 JDBC 代码都是冗余的，而 MyBatis 解决了这个问题。使用 MyBatis 做查询时可以自动将数据表中的数据记录封装到实体类或 Map 中，再将它们放入 List 进行返回，对于如此常见且有利于提高开发效率的功能，MyBatis 完全进行了封装，不需要程序员编写底层的实现代码，从此观点来看，使用 MyBatis 开发软件非常方便快捷，去除了大量冗余的 JDBC 代码，提高了开发效率。

（5）半自动化所带来的灵活性：MyBatis 是"半自动化"的"ORM 映射"框架，但它应该算作 SQL 映射框架（SQL mapper framework）。将 MyBatis 称为"半自动化"的 ORM 框架是因为 MyBatis 操作数据库时还是使用原始的 SQL 语句，这些 SQL 语句还需要程序员自行设计。MyBatis 在使用方式上和"全自动化"的 ORM 框架 Hibernate 有着非常大的差异，因为 MyBatis 以 SQL 语句为映射基础，而 Hibernate 完全基于实体类与表进行映射，基本上属于全自动化的

ORM 映射框架，但正是 MyBatis 的半自动化 ORM 映射框架的特性，可以将 SQL 语句灵活多变的特点融入项目开发中。

（6）支持 XML 或注解（Annotation）的方式进行 ORM 映射：MyBatis 可以使用 XML 或注解的方式将数据表中的记录映射成一个 Map 或一个实体类对象，但 MyBatis 官方还是推荐使用 XML 方式，因为使用注解方式编写 SQL 语句时，如果 SQL 语句非常复杂，会难以进行代码的后期维护，另外使用这种方式把 SQL 语句和 Java 代码放在了一起，没有分离。

（7）功能丰富：MyBatis 还可以实现定义 SQL 段落、调用存储过程、数据缓存和高级映射等功能。

4.3 使用 JDBC+反射技术实现泛型 DAO

MyBatis 实现行与实体类双向转换的原理是基于反射和 JDBC 技术。MyBatis 只是对 JDBC 技术进行了轻量级的封装，使程序员更方便地操作数据库。

为了使读者对 MyBatis 核心功能的原理有更加细致的了解，本节实现了行与实体类的双向转换。

创建 maven-archetype-quickstart 类型的 maven 项目 entity-row。

添加依赖配置：

```xml
<dependency>
    <groupId>cn.easyproject</groupId>
    <artifactId>ojdbc6</artifactId>
    <version>11.2.0.4</version>
</dependency>
```

创建获得 Connection（连接）对象，代码如下：

```java
package com.ghy.www.tools;

import java.sql.Connection;
import java.sql.DriverManager;
import java.sql.SQLException;

public class GetConnection {
    public static Connection getConnection() throws ClassNotFoundException, SQLException {
        String driverName = "oracle.jdbc.OracleDriver";
        String url = "jdbc:oracle:thin:@localhost:1521:orcl";
        String username = "y2";
        String password = "123123";
        Class.forName(driverName);
        Connection conn = DriverManager.getConnection(url, username, password);
        return conn;
    }
}
```

创建实体类 Userinfo，核心代码如下：

```java
package com.ghy.www.entity;

public class Userinfo {
```

```java
    private long id;
    private String username;
    private String password;

    public Userinfo() {
    }

    public Userinfo(long id, String username, String password) {
        super();
        this.id = id;
        this.username = username;
        this.password = password;
    }

    public long getId() {
        return id;
    }

    public void setId(long id) {
        this.id = id;
    }

    public String getUsername() {
        return username;
    }

    public void setUsername(String username) {
        this.username = username;
    }

    public String getPassword() {
        return password;
    }

    public void setPassword(String password) {
        this.password = password;
    }
}
```

创建泛型 DAO 类 BaseDAO，代码如下：

```java
package com.ghy.www.dao;

import com.ghy.www.tools.GetConnection;

import java.lang.reflect.Field;
import java.sql.Connection;
import java.sql.PreparedStatement;
import java.sql.ResultSet;
import java.sql.SQLException;
import java.util.ArrayList;
import java.util.List;

public class BaseDAO<T> {
    // 此方法模拟了 MyBatis 的 save() 方法
    // MyBatis 框架内部的核心和本示例基本一样，使用的技术就是 JDBC+反射
    public void save(T t)
```

4.3 使用 JDBC+反射技术实现泛型 DAO

```java
            throws IllegalArgumentException, IllegalAccessException, ClassNotFoundException,
SQLException {
        String sql = "insert into ";
        String colName = "";
        String colParam = "";
        String begin = "(";
        String end = ")";
        Class classRef = t.getClass();
        String tableName = classRef.getSimpleName().toLowerCase();
        sql = sql + tableName;
        List values = new ArrayList();
        Field[] fieldArray = classRef.getDeclaredFields();
        for (int i = 0; i < fieldArray.length; i++) {
            Field eachField = fieldArray[i];
            eachField.setAccessible(true);
            String eachFieldName = eachField.getName();
            Object eachValue = eachField.get(t);
            colName = colName + "," + eachFieldName;
            colParam = colParam + ",?";
            values.add(eachValue);
        }
        colName = colName.substring(1);
        colParam = colParam.substring(1);
        sql = sql + begin + colName + end;
        sql = sql + " values" + begin + colParam + end;
        System.out.println(sql);
        for (int i = 0; i < values.size(); i++) {
            System.out.println(values.get(i));
        }

        Connection conn = GetConnection.getConnection();
        PreparedStatement ps = conn.prepareStatement(sql);
        for (int i = 0; i < values.size(); i++) {
            ps.setObject(i + 1, values.get(i));
        }
        ps.executeUpdate();
        ps.close();
        conn.close();
    }

    // 此方法模拟了 MyBatis 的 get()方法
    public T get(Class<T> classObject, long id)
            throws InstantiationException, IllegalAccessException, ClassNotFoundException,
SQLException {
        T t = null;
        String sql = "select * from " + classObject.getSimpleName().toLowerCase() + " where id=?";
        Connection conn = GetConnection.getConnection();
        PreparedStatement ps = conn.prepareStatement(sql);
        ps.setLong(1, id);
        ResultSet rs = ps.executeQuery();
        while (rs.next()) {
            t = classObject.newInstance();
            Field[] fieldArray = classObject.getDeclaredFields();
            for (int i = 0; i < fieldArray.length; i++) {
                Field eachField = fieldArray[i];
                eachField.setAccessible(true);
                String fieldName = eachField.getName();
```

```java
            Object value = rs.getObject(fieldName);
            if (value.getClass().getTypeName().equals("java.math.BigDecimal")) {
                long longValue = Long.parseLong("" + value);
                eachField.set(t, longValue);
            } else {
                eachField.set(t, value);
            }
        }
    }
    rs.close();
    ps.close();
    conn.close();
    return t;
}

// 此方法模拟了 MyBatis 的 update()方法
public void update(T t)
        throws IllegalArgumentException, IllegalAccessException, ClassNotFoundException, SQLException {
    String sql = "update " + t.getClass().getSimpleName().toLowerCase() + " set ";
    String whereSQL = " where id=?";
    String colName = "";
    Class classRef = t.getClass();
    List values = new ArrayList();
    Field[] fieldArray = classRef.getDeclaredFields();
    long idValue = 0;
    for (int i = 0; i < fieldArray.length; i++) {
        Field eachField = fieldArray[i];
        eachField.setAccessible(true);
        String eachFieldName = eachField.getName();
        if (!eachFieldName.equals("id")) {
            Object eachValue = eachField.get(t);
            colName = colName + "," + eachFieldName + "=?";
            values.add(eachValue);
        } else {
            Object eachValue = eachField.get(t);
            idValue = Long.parseLong(eachValue.toString());
        }
    }
    values.add(idValue);

    colName = colName.substring(1);
    sql = sql + colName + whereSQL;
    System.out.println(sql);
    for (int i = 0; i < values.size(); i++) {
        System.out.println(values.get(i));
    }

    Connection conn = GetConnection.getConnection();
    PreparedStatement ps = conn.prepareStatement(sql);
    for (int i = 0; i < values.size(); i++) {
        ps.setObject(i + 1, values.get(i));
    }
    ps.executeUpdate();
    ps.close();
    conn.close();
}
```

4.3　使用 JDBC+反射技术实现泛型 DAO

```java
// 此方法模拟了MyBatis的delete()方法
public void delete(T t) throws IllegalArgumentException, IllegalAccessException, ClassNotFoundException,
        SQLException, NoSuchFieldException, SecurityException {
    String sql = "delete from " + t.getClass().getSimpleName().toLowerCase();
    String whereSQL = " where id=?";
    Class classRef = t.getClass();
    List values = new ArrayList();
    Field idField = classRef.getDeclaredField("id");
    idField.setAccessible(true);
    Object object = idField.get(t);
    values.add(object);

    sql = sql + whereSQL;
    System.out.println(sql);
    for (int i = 0; i < values.size(); i++) {
        System.out.println(values.get(i));
    }

    Connection conn = GetConnection.getConnection();
    PreparedStatement ps = conn.prepareStatement(sql);
    for (int i = 0; i < values.size(); i++) {
        ps.setObject(i + 1, values.get(i));
    }
    ps.executeUpdate();
    ps.close();
    conn.close();
}
```

增加记录，代码如下：

```java
package com.ghy.www.test1;

import com.ghy.www.dao.BaseDAO;
import com.ghy.www.entity.Userinfo;

import java.sql.SQLException;

public class Insert {
    public static void main(String[] args)
            throws IllegalArgumentException, IllegalAccessException, ClassNotFoundException, SQLException {
        Userinfo userinfo = new Userinfo();
        userinfo.setId(1000L);
        userinfo.setUsername("中国");
        userinfo.setPassword("中国人");
        BaseDAO<Userinfo> dao = new BaseDAO<>();
        dao.save(userinfo);
    }
}
```

查询记录，代码如下：

```java
package com.ghy.www.test1;

import com.ghy.www.dao.BaseDAO;
```

```java
import com.ghy.www.entity.Userinfo;

import java.sql.SQLException;

public class Select {
    public static void main(String[] args)
            throws InstantiationException, IllegalAccessException, ClassNotFoundException, SQLException {
        BaseDAO<Userinfo> dao = new BaseDAO<>();
        Userinfo userinfo = dao.get(Userinfo.class, 1000L);
        System.out.println(userinfo.getId() + " " + userinfo.getUsername() + " " + userinfo.getPassword());
    }
}
```

修改记录，代码如下：

```java
package com.ghy.www.test1;

import com.ghy.www.dao.BaseDAO;
import com.ghy.www.entity.Userinfo;

import java.sql.SQLException;

public class Update {
    public static void main(String[] args)
            throws InstantiationException, IllegalAccessException, ClassNotFoundException, SQLException {
        BaseDAO<Userinfo> dao = new BaseDAO<>();
        Userinfo userinfo = dao.get(Userinfo.class, 1000L);
        userinfo.setUsername("xxx");
        userinfo.setPassword("xxxxxx");
        dao.update(userinfo);
    }
}
```

删除记录，代码如下：

```java
package com.ghy.www.test1;

import com.ghy.www.dao.BaseDAO;
import com.ghy.www.entity.Userinfo;

import java.sql.SQLException;

public class Delete {
    public static void main(String[] args) throws InstantiationException, IllegalAccessException,
            ClassNotFoundException, SQLException, IllegalArgumentException, NoSuchFieldException, SecurityException {
        BaseDAO<Userinfo> dao = new BaseDAO<>();
        Userinfo userinfo = dao.get(Userinfo.class, 1000L);
        dao.delete(userinfo);
    }
}
```

使用 JDBC 结合反射技术将数据表中的行和实体类进行双向转换，是 ORM 框架核心功能的底层原理，MyBatis 实现 ORM 映射的原理也是 JDBC+反射技术。

4.4 三大核心对象的介绍

MyBatis 中的三大核心对象是 SqlSessionFactoryBuilder、SqlSessionFactory 和 SqlSession。这三者之间的创建关系是：SqlSessionFactoryBuilder 创建 SqlSessionFactory，SqlSession Factory 创建 SqlSession。

这三者的作用如下。

（1）SqlSessionFactoryBuilder：读取 mybatis-config.xml 文件中的信息，根据此信息产生 SqlSession Factory。

（2）SqlSessionFactory：保存 MyBatis 全局信息，可以根据 SqlSessionFactory 产生 SqlSession。

（3）SqlSession：实现 CURD 操作。

使用 SqlSessionFactoryBuilder 类创建 SqlSessionFactory 对象的方式可以来自一个 XML 配置文件，还可以来自一个实例化的 Configuration 对象。使用 XML 方式创建 SqlSessionFactory 对象在使用上比较广泛，也是官方推荐的。

4.5 三大核心对象的生命周期

对象的生命周期也就是对象从创建到销毁的过程，但在此过程中，如果实现的代码质量不高，那么很容易造成程序错误或代码效率的降低。

（1）SqlSessionFactoryBuilder 对象可以被 Java 虚拟机（JVM）实例化、使用或者销毁。一旦使用 SqlSessionFactoryBuilder 对象创建了 SqlSessionFactory，就不需要 SqlSessionFactoryBuilder 类了，也就是不需要保持此类对象的状态，可以任由 JVM 销毁，因此 SqlSessionFactoryBuilder 对象的最佳使用范围是在方法内部，也就是说，可以在方法内部声明 SqlSessionFactoryBuilder 对象来创建 SqlSessionFactory 对象。

（2）SqlSessionFactory 对象由 SqlSessionFactoryBuilder 对象创建而来。一旦 SqlSessionFactory 类的实例被创建，该实例就应该在应用程序执行期间一直存在，而不需要每次操作数据库时都重新创建。因为创建 SqlSessionFactory 对象比较耗时，所以我们应用它的最佳方式就是自己写一个单例模式或使用第三方的 Spring 框架实现单例模式，从而对 SqlSessionFactory 对象进行有效的管理。SqlSessionFactory 对象是线程安全的。

（3）SqlSession 对象由 SqlSessionFactory 类创建而来。需要注意的是，每个线程都应该有其自己的 SqlSession 实例，所以常和 ThreadLocal 联合使用。SqlSession 的实例不能共享，因为它是线程不安全的。千万不要在 Servlet 中声明 SglSession 对象的实例变量，这是由于 Servlet 是单例的，声明该对象的实例变量会造成线程的安全问题，也不能将 SqlSession 对象放在一个类的静态字段甚至是实例字段中，同时不能将 SqlSession 对象放在 HttpSession 会话或 ServletContext 上下文中。在接收到 HTTP 请求后，可以打开一个 SqlSession 对象操作数据库，在响应之前需要关闭 SqlSession。关闭 SqlSession 很重要，应该确保使用 finally 块来关闭它。下面的示例就是一个确保正常关闭 SqlSession 对象的基本模式，代码如下：

```
SqlSession sqlSession = ……;
try {
```

```
        // sqlSession curd code
        sqlSession.commit();
    } catch (Exception e) {
        sqlSession.rollback();
        e.printStackTrace();
    } finally {
        sqlSession.close();
    }
```

注意：由于本书的示例全部在 Spring Boot 环境中进行测试，因此不再需要配置文件 mybatis-config.xml，可以被 application.yml 配置文件所取代。另外，三大对象的生命周期也由 Spring Boot 进行托管，不需要程序员自己进行 commit()、rollback()和 close()操作，但还是有必要了解三大核心对象生命周期的相关知识。

4.6 使用 MyBatis Generator 插件：单模块

使用 MyBatis 对数据库进行 CURD 操作时需要必备的 SQL 映射文件和实体类，但这两个文件的代码内容比较繁杂，尤其是如果数据表的字段很多，则手写代码的工作量会非常大，此种情况也存在于 Hibernate 中。为了加快开发效率，需要使用 MyBatis Generator 插件。该插件的主要功能就是根据数据表结构生成对应的 SQL 映射文件和实体类。

在 IDEA 中安装 MyBatis Generator 插件来实现对数据表的逆向操作，搜索 mybatis 后的结果如图 4-3 所示。

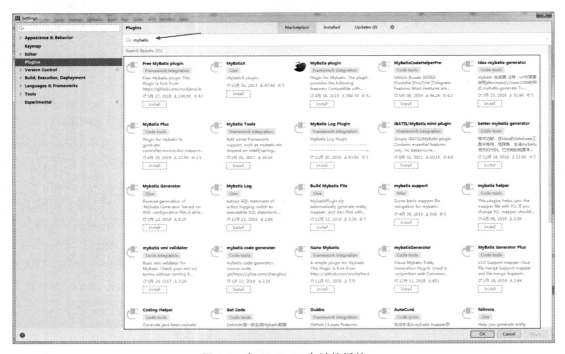

图 4-3　与 MyBatis 有关的插件

但可惜的是，这些搜索出的与 MyBatis Generator 有关的插件并不是 MyBatis 官方提供的，官方并没有提供 IDEA 的 MyBatis Generator 插件。图 4-3 中显示的插件是个人 Java 爱好者对 MyBatis Generator 插件的封装，非官方版本，所以为了稳定性，建议不要安装这些插件，这些插件会因个人原因停止更新而造成 bug 遗留。

本章使用 MyBatis 官方提供的 MyBatis Generator 插件实现对数据表的逆向操作。官方 MyBatis Generator 插件以 Maven 插件 plugin 的方式提供。

为了演示示例的丰富性，我们会在 IDEA 中的两种模块（Modules）版本的项目中使用 MyBatis Generator 插件。

（1）单模块版本。

（2）多模块版本。

还有一种使用方式是使用两个项目进行搭配开发，一个项目为 MyBatis Generator 插件逆向的项目 A，另一个项目 B 是运行的项目，将 MyBatis Generator 插件逆向项目 A 中的代码复制到另一个项目 B 中运行。

4.6.1 操作 Oracle 数据库

本节使用 MyBatis Generator 插件逆向的代码操作 Oracle 数据库。

创建 maven-archetype-webapp 类型的项目 idea-onemodules-oracle，然后搭建基本的 Spring MVC+Spring Boot 开发环境。

1. 编辑 pom.xml 配置文件

添加依赖配置，代码如下：

```xml
<parent>
    <groupId>org.springframework.boot</groupId>
    <artifactId>spring-boot-starter-parent</artifactId>
    <version>2.3.4.RELEASE</version>
</parent>

<properties>
    <project.build.sourceEncoding>UTF-8</project.build.sourceEncoding>
    <maven.compiler.source>1.8</maven.compiler.source>
    <maven.compiler.target>1.8</maven.compiler.target>
</properties>

<dependencies>
    <dependency>
        <groupId>junit</groupId>
        <artifactId>junit</artifactId>
        <version>4.11</version>
        <scope>test</scope>
    </dependency>

    <dependency>
        <groupId>org.springframework.boot</groupId>
        <artifactId>spring-boot-starter-web</artifactId>
    </dependency>

    <dependency>
        <groupId>javax.servlet</groupId>
```

```xml
        <artifactId>jstl</artifactId>
    </dependency>

    <dependency>
        <groupId>org.apache.tomcat.embed</groupId>
        <artifactId>tomcat-embed-jasper</artifactId>
        <scope>provided</scope>
    </dependency>

    <dependency>
        <groupId>org.mybatis.spring.boot</groupId>
        <artifactId>mybatis-spring-boot-starter</artifactId>
        <version>2.1.3</version>
    </dependency>

    <dependency>
        <groupId>cn.easyproject</groupId>
        <artifactId>ojdbc6</artifactId>
        <version>11.2.0.4</version>
    </dependency>
</dependencies>
```

我们还需要在 pom.xml 配置文件中添加 MyBatis Generator 插件,代码如下:

```xml
<build>
    <plugins>
        <!--mybatis-generator 插件-->
        <plugin>
            <groupId>org.mybatis.generator</groupId>
            <artifactId>mybatis-generator-maven-plugin</artifactId>
            <version>1.4.0</version>
            <configuration>
                <verbose>true</verbose>
                <overwrite>true</overwrite>
                <configurationFile>src/main/resources/generatorConfig.xml</configurationFile>
            </configuration>
        </plugin>
    </plugins>
</build>
```

2. 创建 generatorConfig.xml 配置文件

在 resources 文件夹中创建 generatorConfig.xml 配置文件,代码如下:

```xml
<?xml version="1.0" encoding="UTF-8"?>
<!DOCTYPE generatorConfiguration PUBLIC "-//mybatis.org//DTD MyBatis Generator Configuration 1.0//EN"
    "http://mybatis.org/dtd/mybatis-generator-config_1_0.dtd">
<generatorConfiguration>
    <classPathEntry location="C:\mvn_repository\cn\easyproject\ojdbc6\11.2.0.4\ojdbc6-11.2.0.4.jar"/>
    <context id="context1" targetRuntime="MyBatis3Simple">
        <jdbcConnection
            connectionURL="jdbc:oracle:thin:@localhost:1521:orcl"
            driverClass="oracle.jdbc.OracleDriver" password="123123" userId="y2"/>
        <javaModelGenerator targetPackage="com.ghy.www.entity"
            targetProject="src/main/java"/>
```

4.6 使用 MyBatis Generator 插件：单模块

```xml
            <sqlMapGenerator targetPackage="com.ghy.www.sqlmapping"
                    targetProject="src/main/java"/>
            <javaClientGenerator targetPackage="com.ghy.www.sqlmapping"
                    targetProject="src/main/java" type="XMLMAPPER"/>
            <table schema="y2" tableName="userinfo">
                <generatedKey column="id"
                        sqlStatement="select idauto.nextval from dual" identity="false"/>
            </table>
        </context>
</generatorConfiguration>
```

注意：使用代码显式地指定 JDBC 驱动的位置：

```xml
<classPathEntry location="C:\mvn_repository\cn\easyproject\ojdbc6\11.2.0.4\ojdbc6-11.
    2.0.4.jar"/>
```

3. 执行逆向操作

单击如图 4-4 所示的菜单项，执行逆向操作。

逆向操作后的项目结构如图 4-5 所示。

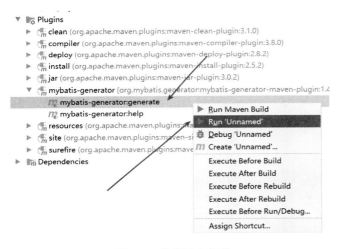

图 4-4　执行逆向操作

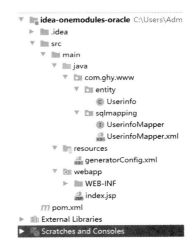

图 4-5　项目结构

4. 添加 @Mapper 注解

在 UserinfoMapper.java 文件中添加 @Mapper 注解，如图 4-6 所示。

图 4-6　添加注解

5. 创建 application.yml 配置文件

在 resources 文件夹中创建 application.yml 配置文件,代码如下:

```yaml
spring:
  datasource:
    driver-class-name: oracle.jdbc.OracleDriver
    hikari:
      auto-commit: true
      connection-test-query: SELECT 1 from dual
      connection-timeout: 30000
      idle-timeout: 30000
      max-lifetime: 1800000
      maximum-pool-size: 15
      minimum-idle: 5
      pool-name: MyHikariCP
    password: 123123
    type: com.zaxxer.hikari.HikariDataSource
    url: jdbc:oracle:thin:@localhost:1521:orcl
    username: y2

server:
  servlet:
    encoding:
      charset: utf-8
      enabled: true
      force: true
```

6. 创建业务类

创建业务类,代码如下:

```java
package com.ghy.www.service;

import com.ghy.www.entity.Userinfo;
import com.ghy.www.sqlmapping.UserinfoMapper;
import org.springframework.beans.factory.annotation.Autowired;
import org.springframework.stereotype.Service;

@Service
public class UserinfoServiceA {

    @Autowired
    private UserinfoMapper userinfoMapper;

    public void insertUserinfo1() {
        Userinfo userinfo = new Userinfo();
        userinfo.setUsername("中国");
        userinfo.setPassword("中国人");
        userinfoMapper.insert(userinfo);
    }
}
package com.ghy.www.service;
```

```
import com.ghy.www.entity.Userinfo;
import com.ghy.www.sqlmapping.UserinfoMapper;
import org.springframework.beans.factory.annotation.Autowired;
import org.springframework.stereotype.Service;

@Service
public class UserinfoServiceB {
    @Autowired
    private UserinfoMapper userinfoMapper;

    public void insertUserinfo2() {
        Userinfo userinfo = new Userinfo();
        userinfo.setUsername("中国");
        userinfo.setPassword("中国人中国人中国人中国人中国人中国人中国人中国人中国人中国人中国人中国人中国人中国人中国人中国人中国人中国人中国人中国人中国人中国人中国人中国人中国人中国人中国人中国人中国人中国人中国人中国人中国人中国人中国人中国人中国人中国人中国人中国人中国人中国人中国人中国人中国人中国人");
        userinfoMapper.insert(userinfo);
    }
}
```

7．创建控制层

创建控制层，代码如下：

```
package com.ghy.www.controller;

import com.ghy.www.service.UserinfoServiceA;
import com.ghy.www.service.UserinfoServiceB;
import org.springframework.beans.factory.annotation.Autowired;
import org.springframework.stereotype.Controller;
import org.springframework.transaction.annotation.Transactional;
import org.springframework.web.bind.annotation.RequestMapping;

import javax.servlet.http.HttpServletRequest;
import javax.servlet.http.HttpServletResponse;

@Controller
@Transactional
public class TestController {

    @Autowired
    private UserinfoServiceA userinfoServiceA;

    @Autowired
    private UserinfoServiceB userinfoServiceB;

    @RequestMapping("test1")
    public void test1(HttpServletRequest request, HttpServletResponse response) {
        userinfoServiceA.insertUserinfo1();
        userinfoServiceA.insertUserinfo1();
    }

    @RequestMapping("test2")
    public void test2(HttpServletRequest request, HttpServletResponse response) {
```

```
            userinfoServiceA.insertUserinfo1();
            userinfoServiceB.insertUserinfo2();
    }
}
```

8. 创建运行类

创建运行类，代码如下：

```
package com.ghy.www;

import org.springframework.boot.SpringApplication;
import org.springframework.boot.autoconfigure.SpringBootApplication;

@SpringBootApplication
public class Application {
    public static void main(String[] args) {
        SpringApplication.run(Application.class, args);
    }
}
```

9. 运行项目时出现异常

执行运行类后，再执行如下网址：

http://localhost:8080/test1

控制台输出异常的信息如下：

```
org.apache.ibatis.binding.BindingException: Invalid bound statement (not found): com.ghy.www.sqlmapping.UserinfoMapper.insert
```

出现异常的原因是 UserinfoMapper.xml 文件消失了，如图 4-7 所示。

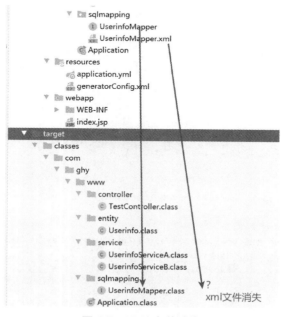

图 4-7　XML 文件消失

10. 处理异常

在 pom.xml 文件中添加配置，代码如下：

```xml
<build>
    <resources>
        <resource>
            <directory>src/main/java</directory>
            <includes>
                <include>**/*.xml</include>
                <include>**/*.properties</include>
            </includes>
        </resource>
    </resources>
</build>
```

添加配置代码后一定要执行 Reload project 菜单项，如图 4-8 所示。

图 4-8　执行 Reload project 菜单项

重启项目，出现了 UserinfoMapper.xml 文件，如图 4-9 所示。

11. 执行成功

再次执行如下网址：

http://localhost:8080/test1

成功添加了两条新的记录，如图 4-10 所示。

执行如下网址：

http://localhost:8080/test2

控制台输出异常信息：

```
java.sql.SQLException: ORA-12899: 列 "Y2"."USERINFO"."PASSWORD" 的值太大 (实际值: 294, 最大值: 50)
```

该异常信息提示 PASSWORD 存储的值过大。userinfo 数据表中保持两条记录，如图 4-11 所示。这说明 Spring Boot 在内部默认处理了回滚事务，非常方便。

图 4-10　成功添加两条记录　　　　图 4-11　保持两条记录

图 4-9　XML 文件出现

4.6.2　操作 MySQL 数据库

创建 maven-archetype-webapp 类型的项目 idea-onemodules-mysql，然后搭建基本的 Spring MVC+Spring Boot 开发环境。

1. 编辑 pom.xml 配置文件

添加依赖配置，代码如下：

```xml
<parent>
    <groupId>org.springframework.boot</groupId>
    <artifactId>spring-boot-starter-parent</artifactId>
    <version>2.3.4.RELEASE</version>
</parent>

<properties>
    <project.build.sourceEncoding>UTF-8</project.build.sourceEncoding>
    <maven.compiler.source>1.8</maven.compiler.source>
    <maven.compiler.target>1.8</maven.compiler.target>
</properties>

<dependencies>
    <dependency>
        <groupId>junit</groupId>
        <artifactId>junit</artifactId>
        <version>4.11</version>
        <scope>test</scope>
    </dependency>

    <dependency>
        <groupId>org.springframework.boot</groupId>
        <artifactId>spring-boot-starter-web</artifactId>
    </dependency>

    <dependency>
        <groupId>javax.servlet</groupId>
        <artifactId>jstl</artifactId>
    </dependency>

    <dependency>
        <groupId>org.apache.tomcat.embed</groupId>
        <artifactId>tomcat-embed-jasper</artifactId>
        <scope>provided</scope>
    </dependency>

    <dependency>
        <groupId>org.mybatis.spring.boot</groupId>
        <artifactId>mybatis-spring-boot-starter</artifactId>
        <version>2.1.3</version>
    </dependency>

    <dependency>
        <groupId>mysql</groupId>
        <artifactId>mysql-connector-java</artifactId>
        <version>8.0.20</version>
    </dependency>
</dependencies>
```

我们还需要在 pom.xml 配置文件中添加配置，代码如下：

```xml
<build>
    <plugins>
        <!--mybatis-generator 插件-->
        <plugin>
            <groupId>org.mybatis.generator</groupId>
            <artifactId>mybatis-generator-maven-plugin</artifactId>
            <version>1.4.0</version>
            <configuration>
                <verbose>true</verbose>
                <overwrite>true</overwrite>
                <configurationFile>src/main/resources/generatorConfig.xml</configurationFile>
            </configuration>
        </plugin>
    </plugins>
    <resources>
        <resource>
            <directory>src/main/java</directory>
            <includes>
                <include>**/*.xml</include>
                <include>**/*.properties</include>
            </includes>
        </resource>
    </resources>
</build>
```

2. 创建 generatorConfig.xml 配置文件

在 resources 文件夹中创建 generatorConfig.xml 配置文件，代码如下：

```xml
<?xml version="1.0" encoding="UTF-8"?>
<!DOCTYPE generatorConfiguration PUBLIC "-//mybatis.org//DTD MyBatis Generator Configuration 1.0//EN"
        "http://mybatis.org/dtd/mybatis-generator-config_1_0.dtd">
<generatorConfiguration>
    <classPathEntry location="C:\mvn_repository\mysql\mysql-connector-java\8.0.20\mysql-connector-java-8.0.20.jar"/>
    <context id="context1" targetRuntime="MyBatis3Simple">
        <jdbcConnection
                connectionURL="jdbc:mysql://localhost:3306/y2?serverTimezone=Asia/Shanghai"
                driverClass="com.mysql.cj.jdbc.Driver" password="123123" userId="root"/>
        <javaModelGenerator targetPackage="com.ghy.www.entity"
                    targetProject="src/main/java"/>
        <sqlMapGenerator targetPackage="com.ghy.www.sqlmapping"
                    targetProject="src/main/java"/>
        <javaClientGenerator targetPackage="com.ghy.www.sqlmapping"
                    targetProject="src/main/java" type="XMLMAPPER"/>
        <table schema="y2" tableName="userinfo">
        </table>
    </context>
</generatorConfiguration>
```

3. 执行逆向操作

单击如图 4-12 所示的菜单项，执行逆向操作。

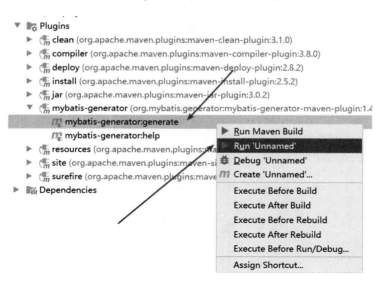

图 4-12　执行逆向操作

逆向操作后的项目结构如图 4-13 所示。

4．添加@Mapper 注解

在 UserinfoMapper.java 文件中添加@Mapper 注解，如图 4-14 所示。

图 4-13　项目结构

图 4-14　添加注解

5．创建 application.yml 配置文件

在 resources 文件夹中创建 application.yml 配置文件，代码如下：

```
spring:
  datasource:
    driver-class-name: com.mysql.cj.jdbc.Driver
```

```yaml
    hikari:
      auto-commit: true
      connection-test-query: SELECT 1
      connection-timeout: 30000
      idle-timeout: 30000
      max-lifetime: 1800000
      maximum-pool-size: 15
      minimum-idle: 5
      pool-name: MyHikariCP
    password: 123123
    type: com.zaxxer.hikari.HikariDataSource
    url: jdbc:mysql://localhost:3306/y2?serverTimezone=Asia/Shanghai
    username: root

server:
  servlet:
    encoding:
      charset: utf-8
      enabled: true
      force: true
```

6. 创建业务类

创建业务类，代码如下：

```java
package com.ghy.www.service;

import com.ghy.www.entity.Userinfo;
import com.ghy.www.sqlmapping.UserinfoMapper;
import org.springframework.beans.factory.annotation.Autowired;
import org.springframework.stereotype.Service;

@Service
public class UserinfoServiceA {
    @Autowired
    private UserinfoMapper userinfoMapper;

    public void insertUserinfo1() {
        Userinfo userinfo = new Userinfo();
        userinfo.setUsername("中国");
        userinfo.setPassword("中国人");
        userinfoMapper.insert(userinfo);
    }
}

package com.ghy.www.service;

import com.ghy.www.entity.Userinfo;
import com.ghy.www.sqlmapping.UserinfoMapper;
import org.springframework.beans.factory.annotation.Autowired;
import org.springframework.stereotype.Service;

@Service
public class UserinfoServiceB {
    @Autowired
    private UserinfoMapper userinfoMapper;
```

```java
    public void insertUserinfo1() {
        Userinfo userinfo = new Userinfo();
        userinfo.setUsername("中国");
        userinfo.setPassword("中国人中国人中国人中国人中国人中国人中国人中国人中国人中国人中国人中国人中国人中国人中国人中国人中国人中国人中国人中国人中国人中国人中国人中国人中国人中国人中国人中国人中国人中国人中国人中国人");
        userinfoMapper.insert(userinfo);
    }
}
```

7. 创建控制层

创建控制层，代码如下：

```java
package com.ghy.www.controller;

import com.ghy.www.service.UserinfoServiceA;
import com.ghy.www.service.UserinfoServiceB;
import org.springframework.beans.factory.annotation.Autowired;
import org.springframework.stereotype.Controller;
import org.springframework.transaction.annotation.Transactional;
import org.springframework.web.bind.annotation.RequestMapping;

import javax.servlet.http.HttpServletRequest;
import javax.servlet.http.HttpServletResponse;

@Controller
@Transactional
public class TestController {

    @Autowired
    private UserinfoServiceA userinfoServiceA;

    @Autowired
    private UserinfoServiceB userinfoServiceB;

    @RequestMapping("test1")
    public void test1(HttpServletRequest request, HttpServletResponse response) {
        userinfoServiceA.insertUserinfo1();
        userinfoServiceA.insertUserinfo1();
    }

    @RequestMapping("test2")
    public void test2(HttpServletRequest request, HttpServletResponse response) {
        userinfoServiceA.insertUserinfo1();
        userinfoServiceB.insertUserinfo2();
    }
}
```

8. 创建运行类

创建运行类，代码如下：

```java
package com.ghy.www;

import org.springframework.boot.SpringApplication;
```

```
import org.springframework.boot.autoconfigure.SpringBootApplication;

@SpringBootApplication
public class Application {
    public static void main(String[] args) {
        SpringApplication.run(Application.class, args);
    }
}
```

执行运行类后,再执行如下网址:

http://localhost:8080/test1

成功添加了两条新的记录,如图 4-15 所示。

执行如下网址:

http://localhost:8080/test2

控制台输出异常信息:

```
com.mysql.cj.jdbc.exceptions.MysqlDataTruncation: Data truncation: Data too long for column 'password' at row 1
```

提示 password 存储的值过大。userinfo 数据表中保持两条记录,如图 4-16 所示。说明 Spring Boot 在内部默认处理了回滚事务,非常方便。

图 4-15　成功添加两条记录　　　　　图 4-16　保持两条记录

> 注意:以上两个项目并没有创建 DAO 数据访问层,建议在实际的项目中添加 DAO 层。DAO 和 sqlmapping 的作用不同,sqlmapping 便于操作数据库,而 DAO 是对 sqlmapping 操作数据库的二次封装。

4.7　使用 MyBatis Generator 插件:多模块

4.6 节实现的是在单模块中搭建 MyBatis 开发环境,但在实际的软件项目中,为了实现代码的分层和模块的隔离,大多使用多模块(Modules)的项目结构来组织项目中的源代码。

4.7.1　操作 Oracle 数据库

新建名称为 idea_moreModules_oracle 的 Empty Project 项目,如图 4-17 所示。
然后添加以下 4 个模块。
(1)MyBatisGenerator,项目类型为 maven-archetype-quickstart。
(2)DAO,项目类型为 maven-archetype-quickstart。
(3)Service,项目类型为 maven-archetype-quickstart。
(4)Web,项目类型为 maven-archetype-webapp。
模块的存储位置在父项目所在的文件夹中,如图 4-18 所示。

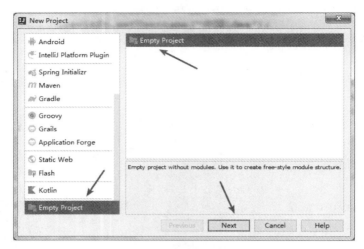

图 4-17　创建 Empty Project 项目

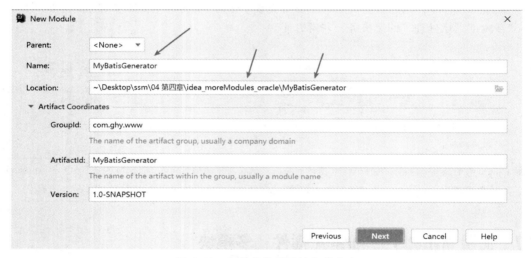

图 4-18　存放在父项目的文件夹中

模块列表如图 4-19 所示。项目结构如图 4-20 所示。

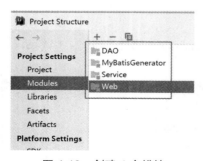

图 4-19　创建 4 个模块

图 4-20　项目结构

1. 搭建 MyBatisGenerator 模块环境

在 pom.xml 配置文件中添加代码如下：

```xml
<parent>
    <groupId>org.springframework.boot</groupId>
    <artifactId>spring-boot-starter-parent</artifactId>
    <version>2.3.4.RELEASE</version>
</parent>

<properties>
    <project.build.sourceEncoding>UTF-8</project.build.sourceEncoding>
    <maven.compiler.source>1.8</maven.compiler.source>
    <maven.compiler.target>1.8</maven.compiler.target>
</properties>

<dependencies>
    <dependency>
        <groupId>junit</groupId>
        <artifactId>junit</artifactId>
        <version>4.11</version>
        <scope>test</scope>
    </dependency>

    <dependency>
        <groupId>org.mybatis.spring.boot</groupId>
        <artifactId>mybatis-spring-boot-starter</artifactId>
        <version>2.1.3</version>
    </dependency>

    <dependency>
        <groupId>cn.easyproject</groupId>
        <artifactId>ojdbc6</artifactId>
        <version>11.2.0.4</version>
    </dependency>
</dependencies>
```

添加插件和资源配置，代码如下：

```xml
<build>
    <plugins>
        <!--mybatis-generator 插件-->
        <plugin>
            <groupId>org.mybatis.generator</groupId>
            <artifactId>mybatis-generator-maven-plugin</artifactId>
            <version>1.4.0</version>
            <configuration>
                <verbose>true</verbose>
                <overwrite>true</overwrite>
                <configurationFile>src/main/resources/generatorConfig.xml</configurationFile>
            </configuration>
        </plugin>
    </plugins>
    <resources>
        <resource>
            <directory>src/main/java</directory>
            <includes>
                <include>**/*.xml</include>
```

```
            <include>**/*.properties</include>
        </includes>
    </resource>
  </resources>
</build>
```

创建 generatorConfig.xml 配置文件,可以参考 4.6.1 节,然后执行逆向操作。
对 UserinfoMapper.java 接口使用@Mapper 注解。

2. 搭建 DAO 模块环境

创建 DAO 类,代码如下:

```
package com.ghy.www.dao;

import com.ghy.www.entity.Userinfo;
import com.ghy.www.sqlmapping.UserinfoMapper;
import org.springframework.beans.factory.annotation.Autowired;
import org.springframework.stereotype.Repository;

@Repository
public class UserinfoDAO {
    @Autowired
    private UserinfoMapper userinfoMapper;

    public void insert(Userinfo userinfo) {
        userinfoMapper.insert(userinfo);
    }
}
```

3. 搭建 Service 模块环境

创建业务类 A,代码如下:

```
package com.ghy.www.service;

import com.ghy.www.dao.UserinfoDAO;
import com.ghy.www.entity.Userinfo;
import org.springframework.beans.factory.annotation.Autowired;
import org.springframework.stereotype.Service;

@Service
public class UserinfoServiceA {

    @Autowired
    private UserinfoDAO userinfoDAO;

    public void insertUserinfo1() {
        Userinfo userinfo = new Userinfo();
        userinfo.setUsername("中国");
        userinfo.setPassword("中国人");
        userinfoDAO.insert(userinfo);
    }
}
```

创建业务类 B,代码如下:

```
package com.ghy.www.service;

import com.ghy.www.dao.UserinfoDAO;
```

```java
import com.ghy.www.entity.Userinfo;
import org.springframework.beans.factory.annotation.Autowired;
import org.springframework.stereotype.Service;

@Service
public class UserinfoServiceB {
    @Autowired
    private UserinfoDAO userinfoDAO;

    public void insertUserinfo2() {
        Userinfo userinfo = new Userinfo();
        userinfo.setUsername("中国");
        userinfo.setPassword("中国人中国人中国人中国人中国人中国人中国人中国人中国人中国人中国人中国人中国人中国人中国人中国人中国人中国人中国人中国人中国人中国人中国人中国人中国人中国人中国人中国人中国人中国人中国人中国人");
        userinfoDAO.insert(userinfo);
    }
}
```

4. 搭建 Web 模块环境

创建控制层，代码如下：

```java
package com.ghy.www.controller;

import com.ghy.www.service.UserinfoServiceA;
import com.ghy.www.service.UserinfoServiceB;
import org.springframework.beans.factory.annotation.Autowired;
import org.springframework.stereotype.Controller;
import org.springframework.transaction.annotation.Transactional;
import org.springframework.web.bind.annotation.RequestMapping;

import javax.servlet.http.HttpServletRequest;
import javax.servlet.http.HttpServletResponse;

@Controller
@Transactional
public class TestController {

    @Autowired
    private UserinfoServiceA userinfoServiceA;

    @Autowired
    private UserinfoServiceB userinfoServiceB;

    @RequestMapping("test1")
    public void test1(HttpServletRequest request, HttpServletResponse response) {
        userinfoServiceA.insertUserinfo1();
        userinfoServiceA.insertUserinfo1();
    }

    @RequestMapping("test2")
    public void test2(HttpServletRequest request, HttpServletResponse response) {
        userinfoServiceA.insertUserinfo1();
        userinfoServiceB.insertUserinfo2();
```

    }
}

创建 application.yml 配置文件，可以参考 4.6.1 节。

5．在 Web 模块创建启动类

创建启动类，代码如下：

```
package com.ghy.www;

import org.springframework.boot.SpringApplication;
import org.springframework.boot.autoconfigure.SpringBootApplication;

@SpringBootApplication
public class Application {
    public static void main(String[] args) {
        SpringApplication.run(Application.class, args);
    }
}
```

6．执行两个 URL

执行网址：

http://localhost:8080/test1
成功添加了两条记录。
执行网址：
http://localhost:8080/test2
正确实现了回滚。

4.7.2　操作 MySQL 数据库

创建名称为 idea_moreModules_mysql 的 Empty Project 项目，如图 4-21 所示。

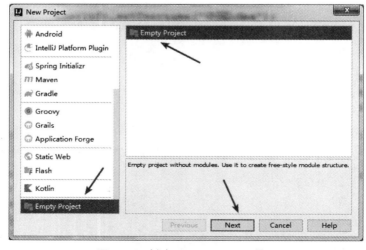

图 4-21　创建 Empty Project 项目

然后添加以下 4 个模块。
（1）MyBatisGenerator，项目类型为 maven-archetype-quickstart。
（2）DAO，项目类型为 maven-archetype-quickstart。
（3）Service，项目类型为 maven-archetype-quickstart。
（4）Web，项目类型为 maven-archetype-webapp。
模块的存储位置是在父项目所在的文件夹中。

1. 搭建 MyBatisGenerator 模块环境

在 pom.xml 配置文件中添加代码如下：

```xml
<parent>
    <groupId>org.springframework.boot</groupId>
    <artifactId>spring-boot-starter-parent</artifactId>
    <version>2.3.4.RELEASE</version>
</parent>

<properties>
    <project.build.sourceEncoding>UTF-8</project.build.sourceEncoding>
    <maven.compiler.source>1.8</maven.compiler.source>
    <maven.compiler.target>1.8</maven.compiler.target>
</properties>

<dependencies>
    <dependency>
        <groupId>junit</groupId>
        <artifactId>junit</artifactId>
        <version>4.11</version>
        <scope>test</scope>
    </dependency>

    <dependency>
        <groupId>org.mybatis.spring.boot</groupId>
        <artifactId>mybatis-spring-boot-starter</artifactId>
        <version>2.1.3</version>
    </dependency>

    <dependency>
        <groupId>mysql</groupId>
        <artifactId>mysql-connector-java</artifactId>
        <version>8.0.20</version>
    </dependency>
</dependencies>
```

添加插件和资源配置，代码如下：

```xml
<build>
    <plugins>
        <!--mybatis-generator 插件-->
        <plugin>
            <groupId>org.mybatis.generator</groupId>
            <artifactId>mybatis-generator-maven-plugin</artifactId>
            <version>1.4.0</version>
            <configuration>
                <verbose>true</verbose>
```

```xml
                <overwrite>true</overwrite>
                <configurationFile>src/main/resources/generatorConfig.xml</configurationFile>
            </configuration>
        </plugin>
    </plugins>
    <resources>
        <resource>
            <directory>src/main/java</directory>
            <includes>
                <include>**/*.xml</include>
                <include>**/*.properties</include>
            </includes>
        </resource>
    </resources>
</build>
```

创建 generatorConfig.xml 配置文件, 可以参考 4.6.2 节, 然后执行逆向操作。对 UserinfoMapper.java 接口使用 @Mapper 注解。

2. 搭建 DAO 模块环境

创建 DAO 类, 代码如下:

```java
package com.ghy.www.dao;

import com.ghy.www.entity.Userinfo;
import com.ghy.www.sqlmapping.UserinfoMapper;
import org.springframework.beans.factory.annotation.Autowired;
import org.springframework.stereotype.Repository;

@Repository
public class UserinfoDAO {
    @Autowired
    private UserinfoMapper userinfoMapper;

    public void insert(Userinfo userinfo) {
        userinfoMapper.insert(userinfo);
    }
}
```

3. 搭建 Service 模块环境

创建业务类 A, 代码如下:

```java
package com.ghy.www.service;

import com.ghy.www.dao.UserinfoDAO;
import com.ghy.www.entity.Userinfo;
import org.springframework.beans.factory.annotation.Autowired;
import org.springframework.stereotype.Service;

@Service
public class UserinfoServiceA {

    @Autowired
    private UserinfoDAO userinfoDAO;
```

```java
    public void insertUserinfo1() {
        Userinfo userinfo = new Userinfo();
        userinfo.setUsername("中国");
        userinfo.setPassword("中国人");
        userinfoDAO.insert(userinfo);
    }
}
```

创建业务类 B，代码如下：

```java
package com.ghy.www.service;

import com.ghy.www.dao.UserinfoDAO;
import com.ghy.www.entity.Userinfo;
import org.springframework.beans.factory.annotation.Autowired;
import org.springframework.stereotype.Service;

@Service
public class UserinfoServiceB {
    @Autowired
    private UserinfoDAO userinfoDAO;

    public void insertUserinfo2() {
        Userinfo userinfo = new Userinfo();
        userinfo.setUsername("中国");
        userinfo.setPassword("中国人中国人中国人中国人中国人中国人中国人中国人中国人中国人中国人中国人中国人中国人中国人中国人中国人中国人中国人中国人中国人中国人中国人中国人中国人中国人中国人中国人中国人中国人中国人中国人中国人");
        userinfoDAO.insert(userinfo);
    }
}
```

4．搭建 Web 模块环境

创建控制层，代码如下：

```java
package com.ghy.www.controller;

import com.ghy.www.service.UserinfoServiceA;
import com.ghy.www.service.UserinfoServiceB;
import org.springframework.beans.factory.annotation.Autowired;
import org.springframework.stereotype.Controller;
import org.springframework.transaction.annotation.Transactional;
import org.springframework.web.bind.annotation.RequestMapping;

import javax.servlet.http.HttpServletRequest;
import javax.servlet.http.HttpServletResponse;

@Controller
@Transactional
public class TestController {

    @Autowired
    private UserinfoServiceA userinfoServiceA;

    @Autowired
    private UserinfoServiceB userinfoServiceB;
```

```
    @RequestMapping("test1")
    public void test1(HttpServletRequest request, HttpServletResponse response) {
        userinfoServiceA.insertUserinfo1();
        userinfoServiceA.insertUserinfo1();
    }

    @RequestMapping("test2")
    public void test2(HttpServletRequest request, HttpServletResponse response) {
        userinfoServiceA.insertUserinfo1();
        userinfoServiceB.insertUserinfo2();
    }
}
```

创建 application.yml 配置文件,可以参考 4.6.2 节。

5. 在 Web 模块创建启动类

创建启动类,代码如下:

```
package com.ghy.www;

import org.springframework.boot.SpringApplication;
import org.springframework.boot.autoconfigure.SpringBootApplication;

@SpringBootApplication
public class Application {
    public static void main(String[] args) {
        SpringApplication.run(Application.class, args);
    }
}
```

6. 执行两个 URL

执行网址:

http://localhost:8080/test1

成功添加了两条记录。

执行网址:

http://localhost:8080/test2

正确实现了回滚。

4.8 自建环境使用 Mapper 接口操作 Oracle-MySQL 数据库

前面介绍的是使用 MyBatis Generator 插件生成的实体类和 SQL 映射文件来操作数据库,并不能从基础上掌握 MyBatis 框架的使用。本节将从自搭建开发环境开始,使用 Mapper 接口实现经典功能 CURD,并且针对 Oracle 和 MySQL 这样的主流数据库。

Spring Boot 整合 MyBatis 时,主流的使用方式会采用面向接口编程,也就是"接口-SQL 映射"功能,使程序员完全面向 Mapper 接口进行编程,在代码规范上再上一个台阶。

4.8.1 接口-SQL 映射的对应关系

"接口-SQL 映射"的对应关系如图 4-22 所示。

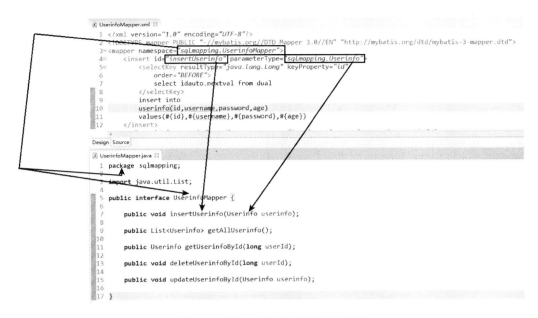

图 4-22　接口-SQL 映射的对应关系

下面介绍一下“接口-SQL 映射”的原理。

SQL 映射文件 UserinfoMapper.xml 中的 namespace 属性值 sqlmapping.UserinfoMapper 表示该映射对应的是 sqlmapping 包中的 UserinfoMapper 接口，而<insert>标签的 id 属性值 insertUserinfo 是 UserinfoMapper 接口中的 public void insertUserinfo(Userinfo userinfo)方法，<insert>标签的 parameterType 属性值 sqlmapping.Userinfo 是 public void insertUserinfo (Userinfo userinfo)方法的参数类型，只要它们一一对应，便能实现“接口-SQL 映射”，这样程序员完全能够以面向接口的方式设计软件。

4.8.2　针对 Oracle 的 CURD

本节将演示使用 Mapper 接口对 Oracle 数据库进行 CURD 操作。

1. 准备开发环境

创建 userinfo 数据表，表结构如图 4-23 所示。

ID	Column Name	Data Type	Size	Byte/Char	Precision	Scale	PK	Not Null	Virtual	Default / Virtual
1	ID	NUMBER	0		18	0	☑	☑	☐	
2	USERNAME	VARCHAR2	50	Byte	0	0	☐	☐	☐	
3	PASSWORD	VARCHAR2	50	Byte	0	0	☐	☐	☐	
4	AGE	NUMBER	0		18	0	☐	☐	☐	
5	INSERTDATE	DATE	0		0	0	☐	☐	☐	

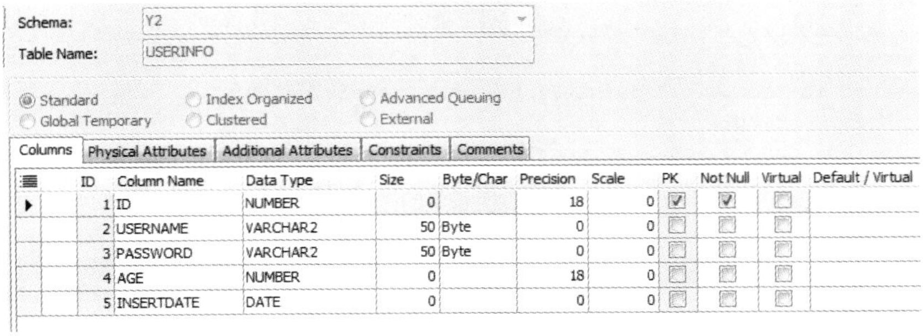

图 4-23　userinfo 数据表结构

创建名称为 mybatis_mapper_curd_oracle 的 Empty Project 项目，并创建多 Maven 模块的 Spring Boot+Spring MVC+MyBatis 开发环境。

完成后的项目结构如图 4-24 所示。

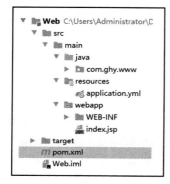

图 4-24　项目结构

2. 搭建 MyBatisGenerator 模块环境

创建逆向配置文件 generatorConfig.xml，代码如下：

```xml
<?xml version="1.0" encoding="UTF-8"?>
<!DOCTYPE generatorConfiguration PUBLIC "-//mybatis.org//DTD MyBatis Generator Configuration 1.0//EN"
        "http://mybatis.org/dtd/mybatis-generator-config_1_0.dtd">
<generatorConfiguration>
    <classPathEntry location="C:\mvn_repository\cn\easyproject\ojdbc6\11.2.0.4\ojdbc6-11.2.0.4.jar"/>
    <context id="context1" targetRuntime="MyBatis3Simple">
        <jdbcConnection
                connectionURL="jdbc:oracle:thin:@localhost:1521:orcl"
                driverClass="oracle.jdbc.OracleDriver" password="123123" userId="y2"/>
        <javaModelGenerator targetPackage="com.ghy.www.entity"
                            targetProject="src/main/java"/>
        <table schema="y2" tableName="userinfo">
            <generatedKey column="id"
                          sqlStatement="select idauto.nextval from dual" identity="false"/>
        </table>
```

```
        </context>
</generatorConfiguration>
```

添加@Mapper 注解，代码如下：

```
@Mapper
public interface IUserinfoMapping {
    public void insert(Userinfo userinfo);
```

3. 搭建 DAO 模块环境

创建 DAO 类，代码如下：

```
package com.ghy.www.dao;

import com.ghy.www.entity.Userinfo;
import com.ghy.www.sqlmapping.IUserinfoMapping;
import org.springframework.beans.factory.annotation.Autowired;
import org.springframework.stereotype.Repository;

import java.util.List;

@Repository
public class UserinfoDAO {
    @Autowired
    private IUserinfoMapping userinfoMapping;

    public void insert(Userinfo userinfo) {
        userinfoMapping.insert(userinfo);
    }

    public void deleteById(long userId) {
        userinfoMapping.deleteById(userId);
    }

    public void updateById(Userinfo userinfo) {
        userinfoMapping.updateById(userinfo);
    }

    public List<Userinfo> selectAll() {
        return userinfoMapping.selectAll();
    }

    public Userinfo selectById(long userId) {
        return userinfoMapping.selectById(userId);
    }
}
```

4. 搭建 Service 模块环境

创建业务类 A，代码如下：

```
package com.ghy.www.service;

import com.ghy.www.dao.UserinfoDAO;
import com.ghy.www.entity.Userinfo;
import org.springframework.beans.factory.annotation.Autowired;
import org.springframework.stereotype.Service;
```

```
import java.util.Date;

@Service
public class UserinfoServiceA {
    @Autowired
    private UserinfoDAO userinfoDAO;

    public void insertUserinfo1() {
        Userinfo userinfo = new Userinfo();
        userinfo.setUsername("中国");
        userinfo.setPassword("中国人");
        userinfo.setAge(100L);
        userinfo.setInsertdate(new Date());
        userinfoDAO.insert(userinfo);
    }
}
```

创建业务类 B，代码如下：

```
package com.ghy.www.service;

import com.ghy.www.dao.UserinfoDAO;
import com.ghy.www.entity.Userinfo;
import org.springframework.beans.factory.annotation.Autowired;
import org.springframework.stereotype.Service;

import java.util.Date;

@Service
public class UserinfoServiceB {
    @Autowired
    private UserinfoDAO userinfoDAO;

    public void insertUserinfo2() {
        Userinfo userinfo = new Userinfo();
        userinfo.setUsername("中国");
        userinfo.setPassword("中国人中国人中国人中国人中国人中国人中国人中国人中国人中国人中国人中国人中国人中国人中国人中国人中国人中国人中国人中国人中国人中国人中国人中国人中国人中国人中国人中国人中国人中国人中国人中国人中国人中国人中国人中国人中国人");
        userinfo.setAge(100L);
        userinfo.setInsertdate(new Date());
        userinfoDAO.insert(userinfo);
    }
}
```

创建业务类 UserinfoService，代码如下：

```
package com.ghy.www.service;

import com.ghy.www.dao.UserinfoDAO;
import com.ghy.www.entity.Userinfo;
import org.springframework.beans.factory.annotation.Autowired;
import org.springframework.stereotype.Service;

import java.util.List;

@Service
public class UserinfoService {
    @Autowired
```

```java
    private UserinfoDAO userinfoDAO;

    public void insert(Userinfo userinfo) {
        userinfoDAO.insert(userinfo);
    }

    public void deleteById(long userId) {
        userinfoDAO.deleteById(userId);
    }

    public void updateById(Userinfo userinfo) {
        userinfoDAO.updateById(userinfo);
    }

    public List<Userinfo> selectAll() {
        return userinfoDAO.selectAll();
    }

    public Userinfo selectById(long userId) {
        return userinfoDAO.selectById(userId);
    }
}
```

5. 搭建 Web 模块环境

创建控制层 1，代码如下：

```java
package com.ghy.www.controller;

import com.ghy.www.service.UserinfoServiceA;
import com.ghy.www.service.UserinfoServiceB;
import org.springframework.beans.factory.annotation.Autowired;
import org.springframework.stereotype.Controller;
import org.springframework.transaction.annotation.Transactional;
import org.springframework.web.bind.annotation.RequestMapping;

import javax.servlet.http.HttpServletRequest;
import javax.servlet.http.HttpServletResponse;

@Controller
@Transactional
public class TestController1 {

    @Autowired
    private UserinfoServiceA userinfoServiceA;

    @Autowired
    private UserinfoServiceB userinfoServiceB;

    @RequestMapping("test1")
    public void test1(HttpServletRequest request, HttpServletResponse response) {
        userinfoServiceA.insertUserinfo1();
        userinfoServiceA.insertUserinfo1();
    }

    @RequestMapping("test2")
    public void test2(HttpServletRequest request, HttpServletResponse response) {
```

```
        userinfoServiceA.insertUserinfo1();
        userinfoServiceB.insertUserinfo2();
    }
}
```

创建控制层 2,代码如下:

```java
package com.ghy.www.controller;

import com.ghy.www.entity.Userinfo;
import com.ghy.www.service.UserinfoService;
import org.springframework.beans.factory.annotation.Autowired;
import org.springframework.stereotype.Controller;
import org.springframework.transaction.annotation.Transactional;
import org.springframework.web.bind.annotation.RequestMapping;

import javax.servlet.http.HttpServletRequest;
import javax.servlet.http.HttpServletResponse;
import java.util.Date;
import java.util.List;

@Controller
@Transactional
public class TestController2 {
    @Autowired
    private UserinfoService userinfoService;

    @RequestMapping("insert")
    public void insert(HttpServletRequest request, HttpServletResponse response) {
        Userinfo userinfo = new Userinfo();
        userinfo.setUsername("中国");
        userinfo.setPassword("中国人");
        userinfo.setAge(100L);
        userinfo.setInsertdate(new Date());
        userinfoService.insert(userinfo);
        System.out.println("生成的id值为: " + userinfo.getId());
    }

    @RequestMapping("deleteById")
    public void deleteById(int userId, HttpServletRequest request, HttpServletResponse response) {
        userinfoService.deleteById(userId);
    }

    @RequestMapping("selectById")
    public void selectById(int userId, HttpServletRequest request, HttpServletResponse response) {
        Userinfo userinfo = userinfoService.selectById(userId);
        System.out.println(userinfo.getId() + " " + userinfo.getUsername() + " " + userinfo.getPassword() + " " + userinfo.getAge() + " " + userinfo.getInsertdate());
    }

    @RequestMapping("selectAll")
    public void selectAll(HttpServletRequest request, HttpServletResponse response) {
        List<Userinfo> listUserinfo = userinfoService.selectAll();
        for (int i = 0; i < listUserinfo.size(); i++) {
            Userinfo userinfo = listUserinfo.get(i);
```

4.8 自建环境使用 Mapper 接口操作 Oracle-MySQL 数据库

```java
            System.out.println(userinfo.getId() + " " + userinfo.getUsername() + " " +
userinfo.getPassword() + " " + userinfo.getAge() + " " + userinfo.getInsertdate());
        }
    }

    @RequestMapping("updateById")
    public void updateById(int userId, HttpServletRequest request, HttpServletResponse response) {
        Userinfo userinfo = userinfoService.selectById(userId);

        userinfo.setUsername("x");
        userinfo.setPassword("xx");
        userinfo.setAge(200L);
        userinfo.setInsertdate(new Date());
        userinfoService.updateById(userinfo);
    }
}
```

创建启动类，代码如下：

```java
package com.ghy.www;

import org.springframework.boot.SpringApplication;
import org.springframework.boot.autoconfigure.SpringBootApplication;

@SpringBootApplication
public class Application {
    public static void main(String[] args) {
        SpringApplication.run(Application.class, args);
    }
}
```

创建配置文件 application.yml，代码如下：

```yaml
spring:
  datasource:
    driver-class-name: oracle.jdbc.OracleDriver
    hikari:
      auto-commit: true
      connection-test-query: SELECT 1 from dual
      connection-timeout: 30000
      idle-timeout: 30000
      max-lifetime: 1800000
      maximum-pool-size: 15
      minimum-idle: 5
      pool-name: MyHikariCP
    password: 123123
    type: com.zaxxer.hikari.HikariDataSource
    url: jdbc:oracle:thin:@localhost:1521:orcl
    username: y2

server:
  servlet:
    encoding:
      charset: utf-8
      enabled: true
      force: true
```

6. 核心代码解释

在 SQL 映射文件 IUserinfoMapping.xml 中添加如下配置代码：

```
<insert id="insert" parameterType="com.ghy.www.entity.Userinfo">
    <selectKey resultType="java.lang.Long" keyProperty="id"
             order="BEFORE">
        select idauto.nextval from dual
    </selectKey>
    insert into
    userinfo(id,username,password,age,insertDate)
    values(#{id},#{username},#{password},#{age},#{insertdate})
</insert>
```

其中的配置代码如下：

```
    <selectKey resultType="java.lang.Long" keyProperty="id"
             order="BEFORE">
        select idauto.nextval from dual
    </selectKey>
```

标签<selectKey>的 order="BEFORE"属性的含义是 select 语句比 insert 语句先执行，resultType 属性值为“java.lang.Long”表示将序列返回的数字类型转成 Long 类型，keyProperty="id"的作用是将这个 Long 值放入 parameterType 的 Userinfo 的 id 属性中。此段配置代码的主要功能是根据序列对象生成一个主键 id 值，将 id 值放入 Userinfo 对象的 id 属性中，然后执行 insert 语句将记录插入数据表里。使用序列对象生成的 id 值还可以在代码中获取，也就是插入一条记录后再使用程序代码可以从 Userinfo 对象的 id 属性中获取刚才插入记录的 id 值。

属性 parameterType 定义参数类型，属性 resultType 定义返回值的类型。

至此，自建开发环境并使用 Mapper 接口针对 Oracle 数据库的 CURD 操作结束。

4.8.3　针对 MySQL 的 CURD

本节将演示使用 Mapper 接口对 MySQL 数据库进行 CURD 操作。

1. 准备开发环境

创建 userinfo 数据表，表结构如图 4-25 所示。

Column Name	Datatype	PK	NN	UQ	B	UN	ZF	AI	G	Default/Expression
id	INT	☑	☑	☐	☐	☐	☐	☑	☐	
username	VARCHAR(45)	☐	☐	☐	☐	☐	☐	☐	☐	NULL
password	VARCHAR(45)	☐	☐	☐	☐	☐	☐	☐	☐	NULL
age	INT	☐	☐	☐	☐	☐	☐	☐	☐	NULL
insertdate	DATE	☐	☐	☐	☐	☐	☐	☐	☐	NULL
		☐	☐	☐	☐	☐	☐	☐	☐	

图 4-25　userinfo 数据表结构

创建名称为 mybatis_mapper_curd_mysql 的 Empty Project 项目，并创建多 Maven 模块的 Spring Boot+Spring MVC+MyBatis 开发环境。

完成后的项目结构如图 4-26 所示。

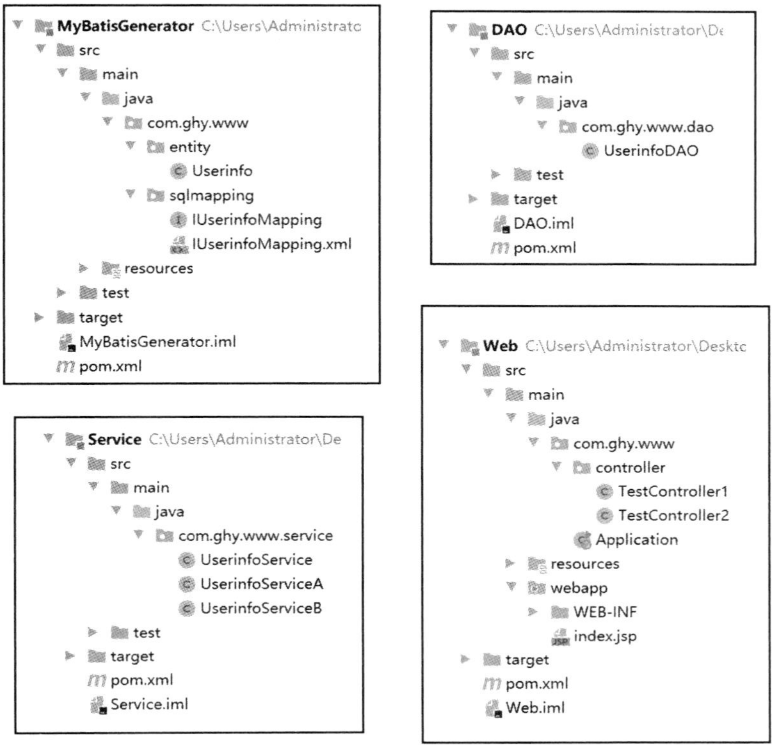

4.8　自建环境使用 Mapper 接口操作 Oracle-MySQL 数据库　　**245**

图 4-26　项目结构

2. 搭建 MyBatisGenerator 模块环境

创建逆向配置文件 generatorConfig.xml，代码如下：

```xml
<?xml version="1.0" encoding="UTF-8"?>
<!DOCTYPE generatorConfiguration PUBLIC "-//mybatis.org//DTD MyBatis Generator Configuration 1.0//EN"
        "http://mybatis.org/dtd/mybatis-generator-config_1_0.dtd">
<generatorConfiguration>
    <classPathEntry location="C:\mvn_repository\mysql\mysql-connector-java\8.0.20\mysql-connector-java-8.0.20.jar"/>
    <context id="context1" targetRuntime="MyBatis3Simple">
        <jdbcConnection
                connectionURL="jdbc:mysql://localhost:3306/y2?serverTimezone=Asia/Shanghai"
                driverClass="com.mysql.cj.jdbc.Driver" password="123123" userId="root"/>
        <javaModelGenerator targetPackage="com.ghy.www.entity"
                    targetProject="src/main/java"/>
        <table schema="y2" tableName="userinfo">
        </table>
    </context>
</generatorConfiguration>
```

添加@Mapper 注解，代码如下：

```java
@Mapper
public interface IUserinfoMapping {
    public void insert(Userinfo userinfo);
```

3. 搭建 DAO 模块环境

创建 DAO 类，代码如下：

```java
package com.ghy.www.dao;

import com.ghy.www.entity.Userinfo;
import com.ghy.www.sqlmapping.IUserinfoMapping;
import org.springframework.beans.factory.annotation.Autowired;
import org.springframework.stereotype.Repository;

import java.util.List;

@Repository
public class UserinfoDAO {
    @Autowired
    private IUserinfoMapping userinfoMapping;

    public void insert(Userinfo userinfo) {
        userinfoMapping.insert(userinfo);
    }

    public void deleteById(long userId) {
        userinfoMapping.deleteById(userId);
    }

    public void updateById(Userinfo userinfo) {
        userinfoMapping.updateById(userinfo);
    }

    public List<Userinfo> selectAll() {
        return userinfoMapping.selectAll();
    }

    public Userinfo selectById(long userId) {
        return userinfoMapping.selectById(userId);
    }
}
```

4. 搭建 Service 模块环境

创建业务类 A，代码如下：

```java
package com.ghy.www.service;

import com.ghy.www.dao.UserinfoDAO;
import com.ghy.www.entity.Userinfo;
import org.springframework.beans.factory.annotation.Autowired;
import org.springframework.stereotype.Service;

import java.util.Date;

@Service
public class UserinfoServiceA {
    @Autowired
    private UserinfoDAO userinfoDAO;
```

```java
    public void insertUserinfo1() {
        Userinfo userinfo = new Userinfo();
        userinfo.setUsername("中国");
        userinfo.setPassword("中国人");
        userinfo.setAge(100);
        userinfo.setInsertdate(new Date());
        userinfoDAO.insert(userinfo);
    }
}
```

创建业务类 B,代码如下:

```java
package com.ghy.www.service;

import com.ghy.www.dao.UserinfoDAO;
import com.ghy.www.entity.Userinfo;
import org.springframework.beans.factory.annotation.Autowired;
import org.springframework.stereotype.Service;

import java.util.Date;

@Service
public class UserinfoServiceB {
    @Autowired
    private UserinfoDAO userinfoDAO;

    public void insertUserinfo2() {
        Userinfo userinfo = new Userinfo();
        userinfo.setUsername("中国");
        userinfo.setPassword("中国人中国人中国人中国人中国人中国人中国人中国人中国人中国人中国人中国人中国人中国人中国人中国人中国人中国人中国人中国人中国人中国人中国人中国人中国人中国人中国人中国人中国人中国人中国人中国人");
        userinfo.setAge(100);
        userinfo.setInsertdate(new Date());
        userinfoDAO.insert(userinfo);
    }
}
```

创建业务类 UserinfoService,代码如下:

```java
package com.ghy.www.service;

import com.ghy.www.dao.UserinfoDAO;
import com.ghy.www.entity.Userinfo;
import org.springframework.beans.factory.annotation.Autowired;
import org.springframework.stereotype.Service;

import java.util.List;

@Service
public class UserinfoService {
    @Autowired
    private UserinfoDAO userinfoDAO;

    public void insert(Userinfo userinfo) {
        userinfoDAO.insert(userinfo);
    }

    public void deleteById(long userId) {
        userinfoDAO.deleteById(userId);
```

```java
    }

    public void updateById(Userinfo userinfo) {
        userinfoDAO.updateById(userinfo);
    }

    public List<Userinfo> selectAll() {
        return userinfoDAO.selectAll();
    }

    public Userinfo selectById(long userId) {
        return userinfoDAO.selectById(userId);
    }
}
```

5. 搭建 Web 模块环境

创建控制层 1,代码如下:

```java
package com.ghy.www.controller;

import com.ghy.www.service.UserinfoServiceA;
import com.ghy.www.service.UserinfoServiceB;
import org.springframework.beans.factory.annotation.Autowired;
import org.springframework.stereotype.Controller;
import org.springframework.transaction.annotation.Transactional;
import org.springframework.web.bind.annotation.RequestMapping;

import javax.servlet.http.HttpServletRequest;
import javax.servlet.http.HttpServletResponse;

@Controller
@Transactional
public class TestController1 {

    @Autowired
    private UserinfoServiceA userinfoServiceA;

    @Autowired
    private UserinfoServiceB userinfoServiceB;

    @RequestMapping("test1")
    public void test1(HttpServletRequest request, HttpServletResponse response) {
        userinfoServiceA.insertUserinfo1();
        userinfoServiceA.insertUserinfo1();
    }

    @RequestMapping("test2")
    public void test2(HttpServletRequest request, HttpServletResponse response) {
        userinfoServiceA.insertUserinfo1();
        userinfoServiceB.insertUserinfo2();
    }
}
```

创建控制层 2,代码如下:

```java
package com.ghy.www.controller;

import com.ghy.www.entity.Userinfo;
```

4.8 自建环境使用 Mapper 接口操作 Oracle-MySQL 数据库

```java
import com.ghy.www.service.UserinfoService;
import org.springframework.beans.factory.annotation.Autowired;
import org.springframework.stereotype.Controller;
import org.springframework.transaction.annotation.Transactional;
import org.springframework.web.bind.annotation.RequestMapping;

import javax.servlet.http.HttpServletRequest;
import javax.servlet.http.HttpServletResponse;
import java.util.Date;
import java.util.List;

@Controller
@Transactional
public class TestController2 {
    @Autowired
    private UserinfoService userinfoService;

    @RequestMapping("insert")
    public void insert(HttpServletRequest request, HttpServletResponse response) {
        Userinfo userinfo = new Userinfo();
        userinfo.setUsername("中国");
        userinfo.setPassword("中国人");
        userinfo.setAge(100);
        userinfo.setInsertdate(new Date());
        userinfoService.insert(userinfo);
        System.out.println("生成的 id 值为: " + userinfo.getId());
    }

    @RequestMapping("deleteById")
    public void deleteById(int userId, HttpServletRequest request, HttpServletResponse response) {
        userinfoService.deleteById(userId);
    }

    @RequestMapping("selectById")
    public void selectById(int userId, HttpServletRequest request, HttpServletResponse response) {
        Userinfo userinfo = userinfoService.selectById(userId);
        System.out.println(userinfo.getId() + " " + userinfo.getUsername() + " " + userinfo.getPassword() + " " + userinfo.getAge() + " " + userinfo.getInsertdate());
    }

    @RequestMapping("selectAll")
    public void selectAll(HttpServletRequest request, HttpServletResponse response) {
        List<Userinfo> listUserinfo = userinfoService.selectAll();
        for (int i = 0; i < listUserinfo.size(); i++) {
            Userinfo userinfo = listUserinfo.get(i);
            System.out.println(userinfo.getId() + " " + userinfo.getUsername() + " " + userinfo.getPassword() + " " + userinfo.getAge() + " " + userinfo.getInsertdate());
        }
    }

    @RequestMapping("updateById")
    public void updateById(int userId, HttpServletRequest request, HttpServletResponse response) {
        Userinfo userinfo = userinfoService.selectById(userId);
        userinfo.setUsername("x");
        userinfo.setPassword("xx");
```

```java
            userinfo.setAge(200);
            userinfo.setInsertdate(new Date());
            userinfoService.updateById(userinfo);
    }
}
```

创建启动类，代码如下：

```java
package com.ghy.www;

import org.springframework.boot.SpringApplication;
import org.springframework.boot.autoconfigure.SpringBootApplication;

@SpringBootApplication
public class Application {
    public static void main(String[] args) {
        SpringApplication.run(Application.class, args);
    }
}
```

创建配置文件 application.yml，代码如下：

```yaml
spring:
  datasource:
    driver-class-name: com.mysql.cj.jdbc.Driver
    hikari:
      auto-commit: true
      connection-test-query: SELECT 1
      connection-timeout: 30000
      idle-timeout: 30000
      max-lifetime: 1800000
      maximum-pool-size: 15
      minimum-idle: 5
      pool-name: MyHikariCP
    password: 123123
    type: com.zaxxer.hikari.HikariDataSource
    url: jdbc:mysql://localhost:3306/y2?serverTimezone=Asia/Shanghai
    username: root

server:
  servlet:
    encoding:
      charset: utf-8
      enabled: true
      force: true
```

至此，自建开发环境并使用 Mapper 接口针对 MySQL 数据库的 CURD 操作结束。

4.9 向 Mapper 接口传入参数类型

向 Mapper 接口传入参数的常见类型有如下 5 种。
（1）传入简单数据类型。
（2）传入复杂数据类型。
（3）传入 Map 数据类型。
（4）传入简单数组/复杂数组数据类型。

（5）传入 List<Long/Entity/Map>数据类型。

创建新的项目 mybatis_mapper_parameterType 来测试上述 5 种情况。

创建 SQL 映射文件 IUserinfoMapping.xml，代码如下：

```xml
<?xml version="1.0" encoding="UTF-8" ?>
<!DOCTYPE mapper PUBLIC "-//mybatis.org//DTD Mapper 3.0//EN" "mybatis-3-mapper.dtd">
<mapper namespace="com.ghy.www.sqlmapping.IUserinfoMapping">
    <select id="test1" resultType="com.ghy.www.entity.Userinfo">
        select *
        from userinfo where
        id=#{id}
    </select>

    <select id="test11" resultType="com.ghy.www.entity.Userinfo">
        select *
        from userinfo where
        id=#{param1} or id=#{param2}
    </select>

    <select id="test2" resultType="com.ghy.www.entity.Userinfo">
        select *
        from userinfo where
        id=#{id}
    </select>

    <select id="test22" resultType="com.ghy.www.entity.Userinfo">
        select *
        from userinfo where
        id=#{param1.id} or id=#{param2.id}
    </select>

    <select id="test3" resultType="com.ghy.www.entity.Userinfo">
        select *
        from userinfo where
        id=#{findId}
    </select>

    <select id="test41" resultType="com.ghy.www.entity.Userinfo">
        select *
        from userinfo where
        id=#{array[0]} or
        id=#{array[1]} or
        id=#{array[2]}
    </select>

    <select id="test42" resultType="com.ghy.www.entity.Userinfo">
        select *
        from userinfo where
        id=#{array[0].id} or
        id=#{array[1].id} or
        id=#{array[2].id}
    </select>

    <select id="test43" resultType="com.ghy.www.entity.Userinfo">
        select *
        from userinfo where
        id=#{param1[0]} or
```

```xml
            id=#{param1[1]} or
            id=#{param2[0].id} or id=#{param2[1].id}
    </select>

    <select id="test51" resultType="com.ghy.www.entity.Userinfo">
        select *
        from userinfo where
        id=#{list[0]} or
        id=#{list[1]}
    </select>

    <select id="test52" resultType="com.ghy.www.entity.Userinfo">
        select *
        from userinfo where
        id=#{list[0].id} or
        id=#{list[1].id}
    </select>

    <select id="test53" resultType="com.ghy.www.entity.Userinfo">
        select *
        from userinfo where
        id=#{list[0].myKey1} or
        id=#{list[1].myKey2}
    </select>
</mapper>
```

创建 Mapper 接口，代码如下：

```java
package com.ghy.www.sqlmapping;

import com.ghy.www.entity.Userinfo;
import org.apache.ibatis.annotations.Mapper;

import java.util.List;
import java.util.Map;

@Mapper
public interface IUserinfoMapping {
    public Userinfo test1(long id);

    public List<Userinfo> test11(long id1, long id2);

    public Userinfo test2(Userinfo userinfo);

    public List<Userinfo> test22(Userinfo userinfo1, Userinfo userinfo2);

    public Userinfo test3(Map map);

    public List<Userinfo> test41(long[] idArray);

    public List<Userinfo> test42(Userinfo[] userinfoArray);

    public List<Userinfo> test43(long[] idArray, Userinfo[] userinfoArray);

    public List<Userinfo> test51(List<Long> idList);

    public List<Userinfo> test52(List<Userinfo> idList);
```

```
    public List<Userinfo> test53(List<Map> idList);
}
```

4.10 从 SQL 映射取得返回值类型

常见的从 SQL 映射取得的返回值类型有如下 3 种。
（1）返回简单数据类型。
（2）返回复杂数据类型。
（3）返回 Map 数据类型。
在项目 mybatis_mapper_resultType 中测试上述 3 种情况。
创建 SQL 映射文件 IUserinfoMapping.xml，代码如下：

```xml
<?xml version="1.0" encoding="UTF-8" ?>
<!DOCTYPE mapper PUBLIC "-//mybatis.org//DTD Mapper 3.0//EN" "mybatis-3-mapper.dtd">
<mapper namespace="com.ghy.www.sqlmapping.IUserinfoMapping">
    <select id="test1" resultType="int">
        select count(*) from
        userinfo
    </select>

    <select id="test2" resultType="com.ghy.www.entity.Userinfo">
        select * from userinfo
        where id=1
    </select>

    <select id="test31" resultType="map">
        select * from userinfo
        where id=2
    </select>

    <select id="test32" resultType="map">
        select sum(age) "sumAge" from
        userinfo
    </select>

    <select id="test33" resultType="map">
        select sum(age) from userinfo
    </select>

    <select id="test4" resultType="com.ghy.www.entity.Userinfo">
        select * from userinfo
        order by id
        asc
    </select>

    <select id="test5" resultType="map">
        select * from userinfo
        order by id
        asc
    </select>

    <select id="test6" resultType="long">
        select id from userinfo
        order by id
```

```
            asc
    </select>
</mapper>
```

创建 Mapper 接口，代码如下：

```java
package com.ghy.www.sqlmapping;

import com.ghy.www.entity.Userinfo;
import org.apache.ibatis.annotations.Mapper;

import java.util.List;
import java.util.Map;

@Mapper
public interface IUserinfoMapping {
    public int test1();

    public Userinfo test2();

    public Map test31();

    public Map test32();

    public Map test33();

    public List<Userinfo> test4();

    public List<Map> test5();

    public List<Long> test6();
}
```

ORM 框架 MyBatis 的介绍结束，通过这部分介绍，读者可以熟练地在 Spring Boot 框架中使用 MyBatis 进行数据库的 CURD 操作。

第 5 章 MyBatis 3 核心技术之实战技能

本章示例涉及的都是 MyBatis 的高频知识点，熟练掌握本章内容会进一步加深读者对 MyBatis 掌握的程度。

5.1 实现输出日志

在 MyBatis 中输出日志的最主要目的是监测 SQL 语句的执行情况，比如查看执行的 SQL 语句的字符串、参数名及参数值等信息。

创建测试项目 log。

添加 yml 配置代码：

```
logging:
  level:
    com.ghy.www.sqlmapping: DEBUG #属性是 SQL 映射接口所在的包名
```

SQL 映射文件中的代码如下：

```
<select id="test1" resultType="com.ghy.www.entity.Userinfo">
    select * from
    userinfo where id=#{param1} or id=#{param2}
</select>
```

程序运行结果如图 5-1 所示。

```
c.g.w.sqlmapping.IUserinfoMapping.test1  : ==>  Preparing: select * from userinfo where id=? or id=?
c.g.w.sqlmapping.IUserinfoMapping.test1  : ==> Parameters: 1(Integer), 2(Integer)
c.g.w.sqlmapping.IUserinfoMapping.test1  : <==      Total: 2
1 a 1 11 Mon Oct 12 00:00:00 CST 2020
2 b 2 22 Mon Oct 12 00:00:00 CST 2020
```

图 5-1 输出日志

输出的日志中打印出 SQL 语句，以及传给 SQL 语句的参数值。

5.2 SQL 语句中特殊符号的处理

如果 SQL 语句中有一些特殊符号，则必须使用如下格式进行 SQL 语句的设计：

```xml
<![CDATA[ 特殊符号 ]]>
```
创建测试项目 hasOtherChar。

SQL 映射文件的代码如下：

```xml
<mapper namespace="com.ghy.www.sqlmapping.IUserinfoMapping">
    <select id="test1" parameterType="int"
            resultType="com.ghy.www.entity.Userinfo">
        select * from userinfo
        where
        id <![CDATA[<]]>
        #{idParam} order by id desc
    </select>
</mapper>
```

小于号"<"在 XML 文件中是特殊的符号，它要放在<![CDATA[特殊符号]]>中。

5.3 使用别名

别名的作用就是使用短名称代替冗长的全名称，以简化配置。

别名分为系统预定义别名和自定义别名。

5.3.1 系统预定义别名

在 SQL 映射文件中使用了系统预定义别名，比如 SQL 映射代码如下：

```xml
<select id="getUserinfo" parameterType="long"
        resultType="entity.Userinfo">
    select * from userinfo
    where
    id <![CDATA[<]]>
    #{idParam} order by id asc
</select>
```

SQL 映射中的 long 值是 java.lang.Long 的别名，常见的数据类型别名都在 MyBatis 的源代码中进行了注册，源代码如下：

```java
public TypeAliasRegistry() {
    registerAlias("string", String.class);

    registerAlias("byte", Byte.class);
    registerAlias("long", Long.class);
    registerAlias("short", Short.class);
    registerAlias("int", Integer.class);
    registerAlias("integer", Integer.class);
    registerAlias("double", Double.class);
    registerAlias("float", Float.class);
    registerAlias("boolean", Boolean.class);

    registerAlias("byte[]", Byte[].class);
    registerAlias("long[]", Long[].class);
    registerAlias("short[]", Short[].class);
    registerAlias("int[]", Integer[].class);
    registerAlias("integer[]", Integer[].class);
```

```java
registerAlias("double[]", Double[].class);
registerAlias("float[]", Float[].class);
registerAlias("boolean[]", Boolean[].class);

registerAlias("_byte", byte.class);
registerAlias("_long", long.class);
registerAlias("_short", short.class);
registerAlias("_int", int.class);
registerAlias("_integer", int.class);
registerAlias("_double", double.class);
registerAlias("_float", float.class);
registerAlias("_boolean", boolean.class);

registerAlias("_byte[]", byte[].class);
registerAlias("_long[]", long[].class);
registerAlias("_short[]", short[].class);
registerAlias("_int[]", int[].class);
registerAlias("_integer[]", int[].class);
registerAlias("_double[]", double[].class);
registerAlias("_float[]", float[].class);
registerAlias("_boolean[]", boolean[].class);

registerAlias("date", Date.class);
registerAlias("decimal", BigDecimal.class);
registerAlias("bigdecimal", BigDecimal.class);
registerAlias("biginteger", BigInteger.class);
registerAlias("object", Object.class);

registerAlias("date[]", Date[].class);
registerAlias("decimal[]", BigDecimal[].class);
registerAlias("bigdecimal[]", BigDecimal[].class);
registerAlias("biginteger[]", BigInteger[].class);
registerAlias("object[]", Object[].class);

registerAlias("map", Map.class);
registerAlias("hashmap", HashMap.class);
registerAlias("list", List.class);
registerAlias("arraylist", ArrayList.class);
registerAlias("collection", Collection.class);
registerAlias("iterator", Iterator.class);

registerAlias("ResultSet", ResultSet.class);
}
```

5.3.2 使用 type-aliases-package 配置设置别名

在执行 SQL 语句 select 或 insert 时，要在 parameterType 或 resultType 属性中写明完整的实体类路径，路径中需要包含完整的包名，示例代码如下：

```xml
<insert id="insertUserinfo" parameterType="sqlmapping.Userinfo">
    <selectKey resultType="java.lang.Long" keyProperty="id"
            order="BEFORE">
        select idauto.nextval from dual
    </selectKey>
    insert into
    userinfo(id,username,password,age)
```

```xml
        values(#{id},#{username},#{password},#{age})
</insert>
<select id="getUserinfoById" parameterType="long" resultType="sqlmapping.Userinfo">
    select * from
    userinfo where id=#{id}
</select>
```

如果包名嵌套层级较多,会出现大量冗余的配置代码,这时可以使用 typeAliases 配置对数据类型进行别名简化处理。

创建测试项目 typeAliasTest。

添加 yml 配置代码:

```
mybatis:
    type-aliases-package: com.ghy.www.entity
```

配置的作用是扫描指定包下的类,这些类都被自动赋予了与类同名的别名,别名不区分大小写,别名中不包含包名。

SQL 映射文件的代码如下:

```xml
<?xml version="1.0" encoding="UTF-8" ?>
<!DOCTYPE mapper PUBLIC "-//mybatis.org//DTD Mapper 3.0//EN" "mybatis-3-mapper.dtd">
<mapper namespace="com.ghy.www.sqlmapping.IUserinfoMapping">
    <select id="test1" parameterType="int"
            resultType="USERINFO">
        select * from userinfo
        where
        id <![CDATA[<]]>
        #{idParam} order by id desc
    </select>

    <select id="test2" parameterType="int"
            resultType="userinfo">
        select * from userinfo
        where
        id <![CDATA[<]]>
        #{idParam} order by id desc
    </select>

    <select id="test3" parameterType="int"
            resultType="USERinfo">
        select * from userinfo
        where
        id <![CDATA[<]]>
        #{idParam} order by id desc
    </select>
</mapper>
```

5.3.3 别名重复的解决办法

创建测试项目 typeAliasTestSame。

在使用如下配置进行别名处理时:

```
mybatis:
  type-aliases-package: com.ghy.www.entity1,com.ghy.www.entity2
```

5.3 使用别名

在不同的包中出现相同实体类名的情况下,项目启动时会提示异常信息:

Factory method 'sqlSessionFactory' threw exception; nested exception is org.apache.ibatis.type.TypeException: The alias 'Userinfo' is already mapped to the value 'com.ghy.www.entity2.Userinfo'.

要处理这个异常,可以在实体类上方使用@Alias注解来自定义别名:

```
@Alias(value = "userinfo1")
public class Userinfo {

@Alias(value = "userinfo2")
public class Userinfo {
```

然后在 SQL 映射文件中引用这个别名即可,映射代码如下:

```xml
<select id="test1" parameterType="int"
        resultType="userinfo1">
    select * from userinfo
    where
    id <![CDATA[<]]>
    #{idParam} order by id desc
</select>
<select id="test2" parameterType="int"
        resultType="userinfo2">
    select * from userinfo
    where
    id <![CDATA[<]]>
    #{idParam} order by id desc
</select>
```

创建 Mapper 接口,代码如下:

```java
package com.ghy.www.sqlmapping;

import org.apache.ibatis.annotations.Mapper;

import java.util.List;

@Mapper
public interface IUserinfoMapping {
    public List<com.ghy.www.entity1.Userinfo> test1(int userId);

    public List<com.ghy.www.entity2.Userinfo> test2(int userId);
}
```

创建控制层,代码如下:

```java
package com.ghy.www.controller;

import com.ghy.www.sqlmapping.IUserinfoMapping;
import org.springframework.beans.factory.annotation.Autowired;
import org.springframework.stereotype.Controller;
import org.springframework.transaction.annotation.Transactional;
import org.springframework.web.bind.annotation.RequestMapping;

import javax.servlet.http.HttpServletRequest;
import javax.servlet.http.HttpServletResponse;
import java.util.List;
```

```java
@Controller
@Transactional
public class TestController {
    @Autowired
    private IUserinfoMapping userinfoMapping;

    @RequestMapping("test1")
    public void test1(HttpServletRequest request, HttpServletResponse response) {
        {
            List<com.ghy.www.entity1.Userinfo> listUserinfo = userinfoMapping.test1(100);
            for (int i = 0; i < listUserinfo.size(); i++) {
                com.ghy.www.entity1.Userinfo userinfo = listUserinfo.get(i);
                System.out.println(userinfo);
            }
        }
        System.out.println();
        {
            List<com.ghy.www.entity2.Userinfo> listUserinfo = userinfoMapping.test2(99);
            for (int i = 0; i < listUserinfo.size(); i++) {
                com.ghy.www.entity2.Userinfo userinfo = listUserinfo.get(i);
                System.out.println(userinfo);
            }
        }
    }
}
```

程序运行结果如下：

```
com.ghy.www.entity1.Userinfo@65fd7956
com.ghy.www.entity1.Userinfo@74aaa8f
com.ghy.www.entity1.Userinfo@5626f391

com.ghy.www.entity2.Userinfo@72c90cdd
com.ghy.www.entity2.Userinfo@43a37b2
com.ghy.www.entity2.Userinfo@147d01e1
```

5.4 对 yml 文件中的数据库密码进行加密

对 yml 文件中的内容进行加密，有助于增强数据安全性。

创建测试项目 passwordsafe。

添加 pom.xml 依赖配置：

```xml
<dependency>
    <groupId>com.github.ulisesbocchio</groupId>
    <artifactId>jasypt-spring-boot-starter</artifactId>
    <version>3.0.3</version>
</dependency>
```

执行如下代码生成加密后的密码：

```java
package com.ghy.www.jasypt.test;

import org.jasypt.encryption.pbe.StandardPBEStringEncryptor;
import org.jasypt.encryption.pbe.config.EnvironmentPBEConfig;
```

5.4 对 yml 文件中的数据库密码进行加密

```java
public class Test1 {
    public static void main(String[] args) {
        StandardPBEStringEncryptor standardPBEStringEncryptor = new StandardPBEStringEncryptor();
        EnvironmentPBEConfig config = new EnvironmentPBEConfig();

        config.setAlgorithm("PBEWithMD5AndDES");          // 加密的算法，这个算法是默认的
        config.setPassword("ghy");                         // 加密的密钥
        standardPBEStringEncryptor.setConfig(config);
        String plainText = "123123";//真实的数据库密码
        String encryptedText = standardPBEStringEncryptor.encrypt(plainText);
        System.out.println(encryptedText);
    }
}
```

输出如下信息：

```
1c7rqhQe63upX6dpE3bVUw==
```

将加密后的密码进行解密，代码如下：

```java
package com.ghy.www.jasypt.test;

import org.jasypt.encryption.pbe.StandardPBEStringEncryptor;
import org.jasypt.encryption.pbe.config.EnvironmentPBEConfig;

public class Test2 {
    public static void main(String[] args) {
        StandardPBEStringEncryptor standardPBEStringEncryptor = new StandardPBEStringEncryptor();
        EnvironmentPBEConfig config = new EnvironmentPBEConfig();

        config.setAlgorithm("PBEWithMD5AndDES");
        config.setPassword("ghy");
        standardPBEStringEncryptor.setConfig(config);
        String encryptedText = "1c7rqhQe63upX6dpE3bVUw==";
        String plainText = standardPBEStringEncryptor.decrypt(encryptedText);
        System.out.println(plainText);
    }
}
```

输出如下信息：

```
123123
```

由此可见，这里正确地对密码进行了加密和解密。

配置文件 application.yml 的代码如下：

```yml
spring:
  datasource:
    driver-class-name: com.mysql.cj.jdbc.Driver
    hikari:
      auto-commit: true
      connection-test-query: SELECT 1
      connection-timeout: 30000
      idle-timeout: 30000
      max-lifetime: 1800000
      maximum-pool-size: 15
      minimum-idle: 5
      pool-name: MyHikariCP
```

```yaml
    password: ENC(1c7rqhQe63upX6dpE3bVUw==)
    type: com.zaxxer.hikari.HikariDataSource
    url: jdbc:mysql://localhost:3306/y2?serverTimezone=Asia/Shanghai
    username: root
server:
  servlet:
    encoding:
      charset: utf-8
      enabled: true
      force: true
logging:
  level:
    com.ghy.www.sqlmapping: DEBUG  #属性是 SQL 映射接口所在的包名
jasypt:
  encryptor:
    password: ghy
    algorithm: PBEWithMD5AndDES
    iv-generator-classname: org.jasypt.iv.NoIvGenerator
```

5.5 不同数据库对执行不同 SQL 语句的支持

在设计软件系统时经常需要考虑多数据库的支持，也就是在 Java 代码与 SQL 映射代码都不变的情况下执行不同的 SQL 语句。

5.5.1 使用<databaseIdProvider type="DB_VENDOR">实现执行不同的 SQL 语句

创建名称为 databaseIdTest 的项目。

SQL 映射文件 UserinfoMapper.xml 的代码如下：

```xml
<mapper namespace="com.ghy.www.sqlmapping.IUserinfoMapping">
    <select id="test1" parameterType="int"
            resultType="com.ghy.www.entity.Userinfo" databaseId="xxxOracle">
        select * from userinfo
        where
        id <![CDATA[<]]>
        #{id} order by id desc
    </select>
    <select id="test1" parameterType="int"
            resultType="com.ghy.www.entity.Userinfo" databaseId="xxxMySQL">
        select * from userinfo
        where
        id <![CDATA[<]]>
        #{id} order by id asc
    </select>
</mapper>
```

创建配置类，代码如下：

```
package com.ghy.www.config;

import org.apache.ibatis.mapping.DatabaseIdProvider;
```

```
import org.apache.ibatis.mapping.VendorDatabaseIdProvider;
import org.springframework.context.annotation.Bean;
import org.springframework.context.annotation.Configuration;

import java.util.Properties;

@Configuration
public class SpringConfig {
    @Bean
    public DatabaseIdProvider getDatabaseIdProvider() {
        DatabaseIdProvider databaseIdProvider = new VendorDatabaseIdProvider();
        Properties properties = new Properties();
        properties.setProperty("Oracle", "xxxOracle");
        properties.setProperty("MySQL", "xxxMySQL");
        databaseIdProvider.setProperties(properties);
        return databaseIdProvider;
    }
}
```

程序代码如下：

```
properties.setProperty("Oracle", "xxxOracle");
properties.setProperty("MySQL", "xxxMySQL");
```

这里 key 的值"Oracle"和"MySQL"是通过如下代码获得的：

```
System.out.println("getDatabaseProductName()=" + factory.getConfiguration().getEnvironment().getDataSource()
    .getConnection().getMetaData().getDatabaseProductName());
```

由此可知，key 的值不能随意填写，并且还要区分大小写。key 的值代表数据库的产品名称，而 value 属性代表这个数据库产品名称的别名。value 属性可以随意命名，但命名应尽量有意义。然后在 SQL 映射文件中使用如下代码进行引用：

```
<select id="test1" parameterType="int"
        resultType="com.ghy.www.entity.Userinfo" databaseId="xxxOracle">
    select * from userinfo
    where
    id <![CDATA[<]]>
    #{id} order by id desc
</select>
<select id="test1" parameterType="int"
        resultType="com.ghy.www.entity.Userinfo" databaseId="xxxMySQL">
    select * from userinfo
    where
    id <![CDATA[<]]>
    #{id} order by id asc
</select>
```

在不同的 SQL 映射上引用不同的数据库别名 xxxOracle 和 xxxMySQL，就可以实现。虽然 SQL 映射的 id 值一样，但在不同的数据库中可以执行相应的 SQL 语句。

使用不同的数据库可以执行相应的 SQL 语句，Oracle 采用倒序排列 SQL 语句，而 MySQL 采用正序。

5.5.2　如果 SQL 映射的 id 值相同，有无 databaseId 的优先级

如果 SQL 映射的 id 值相同，而一个 SQL 映射有 databaseId，另一个 SQL 映射无 databaseId，

那么优先级会是什么样呢？

创建测试项目 databaseIdTest2。

SQL 映射文件的代码如下：

```
<select id="test1" parameterType="int"
    resultType="com.ghy.www.entity.Userinfo">
    select * from userinfo
    where
    id <![CDATA[<]]>
    #{id} order by id desc
</select>
<select id="test1" parameterType="int"
    resultType="com.ghy.www.entity.Userinfo" databaseId="xxxMySQL">
    select * from userinfo
    where
    id <![CDATA[<]]>
    #{id} order by id asc
</select>
```

程序运行结果是正序排列，说明具有 databaseId="xxxMySQL"属性的优先级高。

5.6 动态 SQL

因为 MyBatis 是基于 SQL 映射的，所以 SQL 映射文件在此框架中非常重要，而动态 SQL 是指 MyBatis 提供的根据指定的条件来执行指定的 SQL 语句，使 SQL 映射文件中的 SQL 语句在执行时具有动态性。SQL 映射文件与动态 SQL 被设计得非常简单，本节中将介绍 SQL 映射文件中常见示例的使用。

5.6.1 使用<resultMap>标签实现映射

如果数据表中字段的名称和 Java 实体类中属性的名称不一致，就要使用<resultMap>标签来实现一个映射。

创建测试项目 resultMapTest。

数据表 userinfo 中的列名如图 5-2 所示。实体类的类名和属性名如图 5-3 所示。

```
public class UserinfoABC {
    private Integer idABC;
    private String usernameABC;
    private String passwordABC;
    private Integer ageABC;
    private Date insertdateABC;
```

图 5-2 数据表 userinfo 中的列名　　　　图 5-3 实体类的类名和属性名

SQL 映射文件的代码如下：

```
<resultMap type="com.ghy.www.entity.UserinfoABC" id="userinfo">
    <result column="id" property="idABC"/>
    <result column="username" property="usernameABC"/>
    <result column="password" property="passwordABC"/>
```

```xml
        <result column="age" property="ageABC"/>
        <result column="insertdate" property="insertdateABC"/>
</resultMap>

<select id="test1" resultMap="userinfo">
    select * from userinfo order
    by
    id desc
</select>
```

在<select>标签中使用 resultMap 属性来引用<resultMap>的 id 属性值，形成映射关系。

5.6.2 <resultMap>标签与实体类有参构造方法

创建测试项目 resultMapConstructParam。

SQL 映射文件的代码如下：

```xml
<resultMap type="com.ghy.www.entity.UserinfoABC" id="userinfo">
    <constructor>
        <arg column="id" javaType="int"/>
        <arg column="username" javaType="String"/>
        <arg column="password" javaType="String"/>
        <arg column="age" javaType="int"/>
        <arg column="insertdate" javaType="java.util.Date"/>
    </constructor>
</resultMap>

<select id="test1" resultMap="userinfo">
    select * from userinfo order
    by
    id desc
</select>
```

实体类 UserinfoABC 的代码如图 5-4 所示。

```java
public class UserinfoABC {
    private Integer idABC;
    private String usernameABC;
    private String passwordABC;
    private Integer ageABC;
    private Date insertdateABC;

    public UserinfoABC(Integer idABC, String usernameABC, String passwordABC, Integer ageABC, Date insertdateABC) {
        this.idABC = idABC;
        this.usernameABC = usernameABC;
        this.passwordABC = passwordABC;
        this.ageABC = ageABC;
        this.insertdateABC = insertdateABC;
        System.out.println("执行了UserinfoABC有参构造方法");
    }
}
```

图 5-4　实体类有参构造方法

程序运行结果如下：

```
执行了 UserinfoABC 有参构造方法
执行了 UserinfoABC 有参构造方法
执行了 UserinfoABC 有参构造方法
执行了 UserinfoABC 有参构造方法
```

```
执行了 UserinfoABC 有参构造方法
执行了 UserinfoABC 有参构造方法
执行了 UserinfoABC 有参构造方法
执行了 UserinfoABC 有参构造方法
8
7
6
5
4
3
2
1
```

实例化 UserinfoABC 实体类时执行的是有参构造方法。

5.6.3　使用${}拼接 SQL 语句

#{}表示向 SQL 语句传入参数值,而${}表示拼接 SQL 语句。

创建测试项目 sqlStringVar。

SQL 映射文件的代码如下:

```xml
<select id="test1" parameterType="map"
        resultType="com.ghy.www.entity.Userinfo">
    select * from userinfo where id>#{userId} order by
    ${orderSQL}
</select>
```

控制层的代码如下:

```java
@RequestMapping("test1")
public void test1(HttpServletRequest request, HttpServletResponse response) {
    Map map = new HashMap();
    map.put("userId", 4);
    map.put("orderSQL", " id desc");

    List<Userinfo> listUserinfo = userinfoMapping.test1(map);
    for (int i = 0; i < listUserinfo.size(); i++) {
        Userinfo userinfo = listUserinfo.get(i);
        System.out.println(userinfo.getId());
    }
}
```

注意:使用${}拼接 SQL 语句时容易发生 SQL 注入,对系统安全有影响。

5.6.4　<sql>标签的使用

出现重复的 SQL 语句不可避免,可以使用<sql>标签复用重复的 SQL 语句。

1．静态传值

静态传值是指向<sql>标签传入常量值。

创建测试项目 sqlTest。

SQL 映射文件的代码如下:

```xml
<sql id="userinfo5Column1">id "id",username "username",password,age,insertdate</sql>
<select id="test1" resultType="com.ghy.www.entity.Userinfo">
```

```xml
    select
    <include refid="userinfo5Column1"></include>
    from userinfo order
    by id desc
</select>

<sql id="userinfo5Column2">${col1} ,${col2}</sql>
<select id="test2" resultType="com.ghy.www.entity.Userinfo">
    select
    <include refid="userinfo5Column2">
        <property name="col1" value="id"/>
        <property name="col2" value="password"/>
    </include>
    from userinfo order
    by id desc
</select>
```

上面代码中的 id、username、password、age 和 insertdate 这 5 个字段在 SQL 映射文件中多处出现，可以将这 5 个字段封装进<sql id="userinfo5Column1">标签中，以减少配置的代码量。

配置代码：

```xml
<include refid="userinfo5Column2">
    <property name="col1" value="id"/>
    <property name="col2" value="password"/>
</include>
```

它的作用是向<sql id="userinfo5Column2">传递参数。

如果<select>映射的 resultType 属性值是"map"，则表示通过从 map 里使用 map.get(字段名称)方法的形式取得字段对应的值，但字段名称在 Oracle 中是大写字母形成，所以要使用 map.get(大写字段名称)的写法获得列值。为了支持小写字母形式，可以在 SQL 映射文件中定义的 SQL 语句为字段另起一个别名：

```
select id "id",username "username",password "password",age "age" from userinfo
```

也可以使用<sql>标签来进行声明，示例代码如下：

```
<sql id="userinfo5Column1">id "id",username "username",password,age,insertdate</sql>
```

这样从 map 中就可以以小写字母的形式取得字段值了。

2．向<sql>标签动态传值

动态传值是指向<sql>标签传入变量值，变量值来自 parameterType 属性。

上面的静态传值中使用配置代码：

```xml
<include refid="userinfo5Column2">
    <property name="col1" value="id"/>
    <property name="col2" value="password"/>
</include>
```

以静态的方式向<sql>标签的参数传入常量参数值。在 MyBatis 中还支持动态地传入参数值。

创建测试项目 sqlDynamicParam。

SQL 映射接口的代码如下：

```java
@Mapper
public interface IUserinfoMapping {
```

```
    public List<Userinfo> test1(Map map);
}
```

SQL 映射文件的代码如下:

```xml
<sql id="sqlTemplate">${col1},${col2}</sql>
<select id="test1" parameterType="map" resultType="com.ghy.www.entity.Userinfo">
    select
    <include refid="sqlTemplate">
        <property name="col1" value="${idCol}"/>
        <property name="col2" value="${usernameCol}"/>
    </include>
    from userinfo order
    by id asc
</select>
```

控制层的代码如下:

```java
@RequestMapping("test1")
public void test1(HttpServletRequest request, HttpServletResponse response) {
    Map map = new HashMap();
    map.put("idCol", "id");
    map.put("usernameCol", "username");

    List<Userinfo> listUserinfo = userinfoMapping.test1(map);
    for (int i = 0; i < listUserinfo.size(); i++) {
        Userinfo userinfo = listUserinfo.get(i);
        System.out.println(userinfo.getId() + " " + userinfo.getUsername() + " " +
userinfo.getPassword() + " " + userinfo.getAge() + " " + userinfo.getInsertdate());
    }
}
```

程序运行结果如下:

```
1 a null null null
2 b null null null
3 c null null null
4 d null null null
5 e null null null
6 f null null null
7 g null null null
8 h null null null
```

5.6.5 \<if>标签的使用

\<if>标签具有判断的功能。

创建测试项目 iftest。

SQL 映射文件的代码如下:

```xml
<select id="test1" parameterType="map" resultType="com.ghy.www.entity.Userinfo">
    select *
    from userinfo
    order by id
    <if test="orderType=='abc'">
        asc
    </if>
    <if test="orderType=='xyz'">
        desc
```

```
        </if>
    </select>
```

控制层的代码如下：

```java
@RequestMapping("test1")
public void test1(HttpServletRequest request, HttpServletResponse response) {
    {
        Map map = new HashMap();
        map.put("orderType", "abc");
        List<Userinfo> listUserinfo = userinfoMapping.test1(map);
        for (int i = 0; i < listUserinfo.size(); i++) {
            Userinfo userinfo = listUserinfo.get(i);
            System.out.println(userinfo.getId() + " " + userinfo.getUsername() + " " +
userinfo.getPassword() + " " + userinfo.getAge() + " " + userinfo.getInsertdate());
        }
    }
    System.out.println();
    System.out.println();
    {
        Map map = new HashMap();
        map.put("orderType", "xyz");
        List<Userinfo> listUserinfo = userinfoMapping.test1(map);
        for (int i = 0; i < listUserinfo.size(); i++) {
            Userinfo userinfo = listUserinfo.get(i);
            System.out.println(userinfo.getId() + " " + userinfo.getUsername() + " " +
userinfo.getPassword() + " " + userinfo.getAge() + " " + userinfo.getInsertdate());
        }
    }
}
```

5.6.6 <where>标签的使用

标签<where>的主要作用就是生成 where 语句，可以用在 delete、update 以及 select 语句中。
创建测试项目 whereTAG。
SQL 映射文件的代码如下：

```xml
<select id="test1" parameterType="map"
        resultType="com.ghy.www.entity.Userinfo">
    select * from userinfo
    <where>
        <if test="username!=null">and username like #{username}</if>
        <if test="password!=null">and password like #{password}</if>
    </where>
</select>
```

运行类的代码如下：

```java
@RequestMapping("test1")
public void test1(HttpServletRequest request, HttpServletResponse response) {
    Map map = new HashMap();
    map.put("username", "%中国%");
    map.put("password", "%中国人%");

    List<Userinfo> listUserinfo = userinfoMapping.test1(map);
    for (int i = 0; i < listUserinfo.size(); i++) {
        Userinfo userinfo = listUserinfo.get(i);
```

```
        System.out.println(userinfo.getId() + " " + userinfo.getUsername() + " " +
userinfo.getPassword() + " " + userinfo.getAge() + " " + userinfo.getInsertdate());
    }
}
```

如果在代码中不对 username 传递值,则<where>标签也能自动去掉语句"and password like #{password}"中的"and"关键字而成功执行 SQL 查询语句。

5.6.7 针对 Oracle/MySQL 实现 like 模糊查询

创建测试项目 2dbLikeTest。

SQL 映射文件的代码如下:

```xml
<mapper namespace="com.ghy.www.sqlmapping.IUserinfoMapping">
    <!--Oracle 数据库的写法: -->
    <!-- 使用$符号拼接字符串,不规范 -->
    <select id="testOracle1" parameterType="String"
            resultType="com.ghy.www.entity.Userinfo">
        select * from userinfo
        where username like '%${value}%'
    </select>

    <!-- 使用#符号传 1 个参数,规范 -->
    <select id="testOracle2" parameterType="String"
            resultType="com.ghy.www.entity.Userinfo">
        select * from userinfo
        where username like
        '%'||#{value}||'%'
    </select>

    <!-- 使用#符号传 2 个参数,规范 -->
    <select id="testOracle3" resultType="com.ghy.www.entity.Userinfo">
        select * from userinfo
        where
        username like
        '%'||#{username}||'%'and password like '%'||#{password}||'%'
    </select>

    <!--MySQL 数据库的写法: -->
    <!-- 使用$符号拼接字符串,不规范 -->
    <select id="testMySQL1" parameterType="String"
            resultType="com.ghy.www.entity.Userinfo">
        select * from userinfo
        where username like '%${value}%'
    </select>

    <!-- 使用#符号传 1 个参数,规范 -->
    <select id="testMySQL2" parameterType="String"
            resultType="com.ghy.www.entity.Userinfo">
        select * from userinfo
        where username like
        concat("%",#{value},"%")
    </select>

    <!-- 使用#符号传 2 个参数,规范 -->
    <select id="testMySQL3" resultType="com.ghy.www.entity.Userinfo">
```

```
        select * from userinfo
        where
        username like
        concat("%",#{username},"%") and password like
        concat("%",#{password},"%")
    </select>
</mapper>
```

Mapper 映射接口中方法的声明如下：

```
@Mapper
public interface IUserinfoMapping {
    public List<Userinfo> testOracle1(String username);

    public List<Userinfo> testOracle2(String username);

    public List<Userinfo> testOracle3(String username, String password);

    public List<Userinfo> testMySQL1(String username);

    public List<Userinfo> testMySQL2(String username);

    public List<Userinfo> testMySQL3(String username, String password);
}
```

上面的写法针对 Oracle 和 MySQL 数据库。

5.6.8 <choose>标签的使用

<choose>标签的作用是在众多的条件中选择一个，有些类似于 Java 语言中的 switch+break 语句的作用。

创建测试项目 chooseTAG。

SQL 映射文件的代码如下：

```
<select id="test1" parameterType="map"
        resultType="com.ghy.www.entity.Userinfo">
    select * from userinfo where 1=1
    <choose>
        <when test="username!=null">and username like #{username}</when>
        <when test="password!=null">and password like #{password}</when>
        <otherwise>and age=100</otherwise>
    </choose>
</select>
```

控制层的代码如下：

```
@RequestMapping("test1")
public void test1(HttpServletRequest request, HttpServletResponse response) {
    Map map1 = new HashMap();
    map1.put("username", "%中国%");

    Map map2 = new HashMap();
    map2.put("password", "%中国人%");

    Map map3 = new HashMap();

    {
```

```
            List<Userinfo> listUserinfo = userinfoMapping.test1(map1);
            for (int i = 0; i < listUserinfo.size(); i++) {
                Userinfo userinfo = listUserinfo.get(i);
                System.out.println(userinfo.getId() + " " + userinfo.getUsername() + " " +
userinfo.getPassword() + " " + userinfo.getAge() + " " + userinfo.getInsertdate());
            }
        }
        System.out.println();
        System.out.println();
        {
            List<Userinfo> listUserinfo = userinfoMapping.test1(map2);
            for (int i = 0; i < listUserinfo.size(); i++) {
                Userinfo userinfo = listUserinfo.get(i);
                System.out.println(userinfo.getId() + " " + userinfo.getUsername() + " " +
userinfo.getPassword() + " " + userinfo.getAge() + " " + userinfo.getInsertdate());
            }
        }
        System.out.println();
        System.out.println();
        {
            List<Userinfo> listUserinfo = userinfoMapping.test1(map3);
            for (int i = 0; i < listUserinfo.size(); i++) {
                Userinfo userinfo = listUserinfo.get(i);
                System.out.println(userinfo.getId() + " " + userinfo.getUsername() + " " +
userinfo.getPassword() + " " + userinfo.getAge() + " " + userinfo.getInsertdate());
            }
        }
    }
```

5.6.9 <set>标签的使用

<set>标签可以用在 update 语句中,作用是动态指定要更新的列。

创建测试项目 setTAG。

SQL 映射文件的代码如下:

```
<update id="test1" parameterType="map">
    update userinfo
    <set>
        <if test="username!=null">username=#{username},</if>
        <if test="password!=null">password=#{password},</if>
        <if test="age!=null">age=#{age},</if>
        <if test="insertdate!=null">insertdate=#{insertdate},</if>
    </set>
    where id=#{id}
</update>
```

最后一个 if 条件中的逗号可以不去掉,<set>标签会自动删除。

控制层的代码如下:

```
@RequestMapping("test1")
public void test1(HttpServletRequest request, HttpServletResponse response) {
    Map map1 = new HashMap();
    map1.put("id", 1);
    map1.put("username", "xxxxx");
    map1.put("password", null);
    map1.put("age", 999);
```

```
        map1.put("insertdate", new Date());

        userinfoMapping.test1(map1);
}
```

程序运行结果是 password 列未更新,其他列都已更新。

5.6.10 <foreach>标签的使用

<foreach>标签有循环功能,可以用来生成有规律的 SQL 语句。

<foreach>标签的主要属性有 item、index、collection、open、separator 和 close。

item 表示集合中每一个元素迭代时的别名,index 指定一个名字,用于表示在迭代过程中每次迭代的位置,open 表示该语句从哪里开始,separator 表示在两次迭代之间以什么符号作为分隔符,close 表示该语句到哪里结束。

创建测试项目 foreachTest。

SQL 映射文件的代码如下:

```xml
<mapper namespace="com.ghy.www.sqlmapping.IUserinfoMapping">
    <select id="selectAll1" parameterType="list"
            resultType="com.ghy.www.entity.Userinfo">
        select *
        from userinfo where id in
        <foreach item="eachId" collection="list" open="("
                separator="," close=")">
            #{eachId}
        </foreach>
    </select>

    <select id="selectAll2" parameterType="com.ghy.www.queryentity.QueryEntity"
            resultType="com.ghy.www.entity.Userinfo">
        select *
        from userinfo where id in
        <foreach item="eachId" collection="xxxxxxxxx" open="("
                separator="," close=")">
            #{eachId}
        </foreach>
    </select>

    <select id="selectAll3" parameterType="map"
            resultType="com.ghy.www.entity.Userinfo">
        select *
        from userinfo where id in
        <foreach item="eachId" collection="yyyyyyyyyyyy" open="("
                separator="," close=")">
            #{eachId}
        </foreach>
    </select>
</mapper>
```

控制层的代码如下:

```java
@RequestMapping("test1")
public void test1(HttpServletRequest request, HttpServletResponse response) {

    List idList = new ArrayList();
```

```java
            idList.add(1);
            idList.add(2);
            
            List<Userinfo> listUserinfo = userinfoMapping.selectAll1(idList);
            for (int i = 0; i < listUserinfo.size(); i++) {
                Userinfo userinfo = listUserinfo.get(i);
                System.out.println(userinfo.getId() + " " + userinfo.getUsername() + " " + userinfo.getPassword() + " "
                        + userinfo.getAge() + " " + userinfo.getInsertdate());
            }
            
            System.out.println();
            System.out.println();
            
            QueryEntity entity = new QueryEntity();
            idList.add(3);
            entity.setXxxxxxxxx(idList);
            listUserinfo = userinfoMapping.selectAll2(entity);
            for (int i = 0; i < listUserinfo.size(); i++) {
                Userinfo userinfo = listUserinfo.get(i);
                System.out.println(userinfo.getId() + " " + userinfo.getUsername() + " " + userinfo.getPassword() + " "
                        + userinfo.getAge() + " " + userinfo.getInsertdate());
            }
            
            System.out.println();
            System.out.println();
            
            HashMap map = new HashMap();
            idList.add(4);
            map.put("yyyyyyyyyyyy", idList);
            listUserinfo = userinfoMapping.selectAll3(map);
            for (int i = 0; i < listUserinfo.size(); i++) {
                Userinfo userinfo = listUserinfo.get(i);
                System.out.println(userinfo.getId() + " " + userinfo.getUsername() + " " + userinfo.getPassword() + " "
                        + userinfo.getAge() + " " + userinfo.getInsertdate());
            }
        }
```

5.6.11 使用<foreach>执行批量插入

如何实现批量插入（insert）操作呢？Oracle 使用如下格式的 SQL 语句实现：

```
INSERT INTO userinfo (id,
                    username,
                    password,
                    age,
                    insertdate)
    SELECT idauto.NEXTVAL,
          username,
          password,
          age,
          insertdate
      FROM (SELECT 'a' username,
                  'aa' password,
                  1 age,
```

```
                TO_DATE ('2000-1-1', 'yyyy-MM-dd') insertdate
           FROM DUAL
        UNION ALL
        SELECT 'b' username,
               'bb' password,
               1 age,
               TO_DATE ('2000-1-1', 'yyyy-MM-dd') insertdate
          FROM DUAL)
```

MySQL 使用如下格式的 SQL 语句实现批量插入:

```
insert into userinfo(username,password,age,insertdate)
values('a','aa',1,'2000-1-1'),
('b','aa',1,'2000-1-1'),
('c','aa',1,'2000-1-1'),
('d','aa',1,'2000-1-1'),
('e','aa',1,'2000-1-1')
```

由于批量插入的 SQL 语句在每种数据库中不一样, 导致在 MyBatis 中 SQL 映射文件的写法也不一样。

创建测试项目 batchOracle。

下面我们来看一看针对 Oracle 数据库的 SQL 映射文件的代码:

```
<insert id="insertOracle" parameterType="list">
    INSERT INTO userinfo (id,
    username,
    password,
    age,
    insertdate)
    select
    idauto.nextval,username,password,age,insertdate from (
    <foreach collection="list" item="eachUserinfo" separator="union all">
        select
        #{eachUserinfo.username} username,#{eachUserinfo.password}
        password,#{eachUserinfo.age} age,#{eachUserinfo.insertdate}
        insertdate
        from dual
    </foreach>
    )
</insert>
```

控制层的代码如下:

```
@RequestMapping("test1")
public void test1(HttpServletRequest request, HttpServletResponse response) {
    List list = new ArrayList();

    for (int i = 0; i < 5; i++) {
        Userinfo userinfo = new Userinfo();
        userinfo.setUsername("中国" + (i + 1));
        userinfo.setPassword("中国人" + (i + 1));
        userinfo.setAge(100);
        userinfo.setInsertdate(new Date());
        list.add(userinfo);
    }
    userinfoMapping.insertOracle(list);
}
```

创建测试项目 batchMySQL。

下面我们来看一看针对 MySQL 数据库的 SQL 映射文件的代码：

```xml
<insert id="insertMySQL" parameterType="list">
    INSERT INTO userinfo (
    username,
    password,
    age,
    insertdate)
    values
    <foreach collection="list" item="eachUserinfo" separator=",">
       (
       #{eachUserinfo.username},#{eachUserinfo.password},
       #{eachUserinfo.age},#{eachUserinfo.insertdate}
       )
    </foreach>
</insert>
```

控制层的代码如下：

```java
@RequestMapping("test1")
public void test1(HttpServletRequest request, HttpServletResponse response) {
    List list = new ArrayList();

    for (int i = 0; i < 5; i++) {
        Userinfo userinfo = new Userinfo();
        userinfo.setUsername("中国" + (i + 1));
        userinfo.setPassword("中国人" + (i + 1));
        userinfo.setAge(100);
        userinfo.setInsertdate(new Date());
        list.add(userinfo);
    }
    userinfoMapping.insertMySQL(list);
}
```

5.6.12 使用<bind>标签对 like 语句进行适配

在 SQL 语句中使用 like 查询时，MySQL 在拼接字符串时使用 concat()方法，而 Oracle 使用 "||" 运算符。两者的 SQL 语句不相同，这就需要创建两个 SQL 映射语句，SQL 映射文件的示例代码如下：

```xml
<mapper namespace="mapping.UserinfoMapper">
    <select id="selectAllOracle" parameterType="map"
        resultType="entity.Userinfo">
        select * from userinfo
        where username like
        '%'||#{username}||'%'
    </select>
    <select id="selectAllMySQL" parameterType="map"
        resultType="entity.Userinfo">
        select * from userinfo
        where username like
        concat('%',#{username},'%')
    </select>
</mapper>
```

我们可以使用<bind>标签在多个数据库之间进行适配。

创建测试项目 bindTAGNew。

SQL 映射文件的代码如下：

```xml
<mapper namespace="com.ghy.www.sqlmapping.IUserinfoMapping">
    <select id="selectAll1" parameterType="string"
            resultType="com.ghy.www.entity.Userinfo">
        <bind name="querySQL" value="'%'+_parameter+'%'"/>
        select * from userinfo where username like #{querySQL}
    </select>

    <select id="selectAll2" parameterType="com.ghy.www.entity.Userinfo"
            resultType="com.ghy.www.entity.Userinfo">
        <bind name="querySQL" value="'%'+_parameter.getUsername()+'%'"/>
        select * from userinfo where username like #{querySQL}
    </select>

    <select id="selectAll3" parameterType="com.ghy.www.entity.Userinfo"
            resultType="com.ghy.www.entity.Userinfo">
        <bind name="querySQL" value="'%'+username+'%'"/>
        select * from userinfo where username like #{querySQL}
    </select>

    <select id="selectAll4" parameterType="com.ghy.www.entity.Userinfo"
            resultType="com.ghy.www.entity.Userinfo">
        <bind name="querySQL" value="'%'+_parameter.username+'%'"/>
        select * from userinfo where username like #{querySQL}
    </select>

    <select id="selectAll5" parameterType="com.ghy.www.entity.Userinfo"
            resultType="com.ghy.www.entity.Userinfo">
        <bind name="querySQL" value="'%'+#root._parameter.username+'%'"/>
        select * from userinfo where username like #{querySQL}
    </select>

    <select id="selectAll6" parameterType="map"
            resultType="com.ghy.www.entity.Userinfo">
        <bind name="querySQL" value="'%'+username+'%'"/>
        select * from userinfo where username like #{querySQL}
    </select>
</mapper>
```

id 为 selectAll6 的<select>标签的子标签<bind>中的代码 value="'%'+username+'%'"，其含义是从 Map 中根据 key 为 username 找到对应的值，然后拼接成%value%的形式。

<bind>标签的 value 属性值中的运算使用 OGNL 表达式，OGNL 表达式的底层使用 Java 语言，所以在 Java 语言中可以直接使用+号进行字符串的拼接，而不像在 Oracle 或 MySQL 中要使用"||"运算符或 concat()方法。

控制层的代码如下：

```java
@RequestMapping("test1")
public void test1(HttpServletRequest request, HttpServletResponse response) {
    List<Userinfo> listUserinfo = userinfoMapping.selectAll1("中国");
    for (int i = 0; i < listUserinfo.size(); i++) {
        Userinfo userinfo = listUserinfo.get(i);
        System.out.println(userinfo.getId() + " " + userinfo.getUsername() + " " +
```

```java
userinfo.getPassword() + " "
            + userinfo.getAge() + " " + userinfo.getInsertdate());
}
System.out.println();
System.out.println();
Userinfo userinfo = new Userinfo();
userinfo.setUsername("中国");
listUserinfo = userinfoMapping.selectAll2(userinfo);
for (int i = 0; i < listUserinfo.size(); i++) {
    Userinfo eachUserinfo = listUserinfo.get(i);
    System.out.println(eachUserinfo.getId() + " " + eachUserinfo.getUsername() + " "
            + eachUserinfo.getPassword() + " " + eachUserinfo.getAge() + " " + eachUserinfo.getInsertdate());
}
System.out.println();
System.out.println();
listUserinfo = userinfoMapping.selectAll3(userinfo);
for (int i = 0; i < listUserinfo.size(); i++) {
    Userinfo eachUserinfo = listUserinfo.get(i);
    System.out.println(eachUserinfo.getId() + " " + eachUserinfo.getUsername() + " "
            + eachUserinfo.getPassword() + " " + eachUserinfo.getAge() + " " + eachUserinfo.getInsertdate());
}
System.out.println();
System.out.println();
listUserinfo = userinfoMapping.selectAll4(userinfo);
for (int i = 0; i < listUserinfo.size(); i++) {
    Userinfo eachUserinfo = listUserinfo.get(i);
    System.out.println(eachUserinfo.getId() + " " + eachUserinfo.getUsername() + " "
            + eachUserinfo.getPassword() + " " + eachUserinfo.getAge() + " " + eachUserinfo.getInsertdate());
}
System.out.println();
System.out.println();
listUserinfo = userinfoMapping.selectAll5(userinfo);
for (int i = 0; i < listUserinfo.size(); i++) {
    Userinfo eachUserinfo = listUserinfo.get(i);
    System.out.println(eachUserinfo.getId() + " " + eachUserinfo.getUsername() + " "
            + eachUserinfo.getPassword() + " " + eachUserinfo.getAge() + " " + eachUserinfo.getInsertdate());
}
System.out.println();
System.out.println();
Map map = new HashMap();
map.put("username", "中国");
listUserinfo = userinfoMapping.selectAll6(map);
for (int i = 0; i < listUserinfo.size(); i++) {
    Userinfo eachUserinfo = listUserinfo.get(i);
    System.out.println(eachUserinfo.getId() + " " + eachUserinfo.getUsername() + " "
            + eachUserinfo.getPassword() + " " + eachUserinfo.getAge() + " " + eachUserinfo.getInsertdate());
}
}
```

5.6.13 使用<trim>标签规范 SQL 语句

创建测试项目 trimtest。

5.6 动态 SQL

SQL 映射文件的代码如下：

```xml
<select id="selectAll1" parameterType="map"
      resultType="com.ghy.www.entity.Userinfo">
    select * from userinfo where 1=1
    <if test="username!=null">and username like #{username}</if>
</select>
```

控制层的代码如下：

```java
@RequestMapping("test1")
public void test1(HttpServletRequest request, HttpServletResponse response) {
    Map map = new HashMap();
    map.put("username", "%中%");

    List<Userinfo> listUserinfo = userinfoMapping.selectAll1(map);
    for (int i = 0; i < listUserinfo.size(); i++) {
        Userinfo userinfo = listUserinfo.get(i);
        System.out.println(userinfo.getId() + " " + userinfo.getUsername() + " " + userinfo.getPassword() + " "
                + userinfo.getAge() + " " + userinfo.getInsertdate());
    }
    System.out.println();
    System.out.println();
    listUserinfo = userinfoMapping.selectAll1(new HashMap());
    for (int i = 0; i < listUserinfo.size(); i++) {
        Userinfo userinfo = listUserinfo.get(i);
        System.out.println(userinfo.getId() + " " + userinfo.getUsername() + " " + userinfo.getPassword() + " "
                + userinfo.getAge() + " " + userinfo.getInsertdate());
    }
}
```

程序运行后可以得到正确的结果，但在 SQL 语句中使用 where 1=1 有些不规范，如图 5-5 所示。

```
==> Preparing: select * from userinfo where 1=1 and username like ?
==> Parameters: %中%(String)
<==      Total: 1



==> Preparing: select * from userinfo where 1=1
==> Parameters: 
<==      Total: 9
```

图 5-5 不规范的 SQL 语句

为此，创建新的 SQL 映射文件，代码如下：

```xml
<select id="selectAll2" parameterType="map"
      resultType="com.ghy.www.entity.Userinfo">
    select * from userinfo where
    <if test="username!=null">and username like #{username}</if>
</select>
```

控制层的代码如下:

```
@RequestMapping("test2")
public void test2(HttpServletRequest request, HttpServletResponse response) {
    Map map = new HashMap();
    map.put("username", "%中%");

    List<Userinfo> listUserinfo = userinfoMapping.selectAll2(map);
    for (int i = 0; i < listUserinfo.size(); i++) {
        Userinfo userinfo = listUserinfo.get(i);
        System.out.println(userinfo.getId() + " " + userinfo.getUsername() + " " + userinfo.getPassword() + " "
                + userinfo.getAge() + " " + userinfo.getInsertdate());
    }
}
```

程序运行后出现异常的信息如下:

java.sql.SQLSyntaxErrorException: You have an error in your SQL syntax; check the manual that corresponds to your MySQL server version for the right syntax to use near 'and username like '%中%'' at line 2

异常信息提示 SQL 语句不正确,原因是 where 语句后面跟了一个 and,日志输出 SQL 语句:

select * from userinfo where and username like ?

要去掉"and",可以使用<trim>标签。因此,创建新的 SQL 映射文件,代码如下:

```
<select id="selectAll3" parameterType="map"
        resultType="com.ghy.www.entity.Userinfo">
    select * from userinfo
    <where>
        <trim prefixOverrides="and">
            <if test="username!=null">and username like #{username}</if>
        </trim>
    </where>
</select>
```

以上写法对于无 where 子句的情况,需要使用<where>标签生成 where 子句。

还可以使用另一种写法:

```
<select id="selectAll4" parameterType="map" resultType="com.ghy.www.entity.Userinfo">
    select * from userinfo where
    <trim prefixOverrides="and">
        <if test="username!=null">and username like #{username}</if>
    </trim>
</select>
```

以上写法对于有 where 子句的情况,不需要使用<where>标签。

控制层的代码如下:

```
@RequestMapping("test3")
public void test3(HttpServletRequest request, HttpServletResponse response) {
    Map map = new HashMap();
    map.put("username", "%中%");

    List<Userinfo> listUserinfo = userinfoMapping.selectAll3(map);
    for (int i = 0; i < listUserinfo.size(); i++) {
        Userinfo userinfo = listUserinfo.get(i);
        System.out.println(userinfo.getId() + " " + userinfo.getUsername() + " " +
```

```
            userinfo.getPassword() + " "
                    + userinfo.getAge() + " " + userinfo.getInsertdate());
        }
    }

    @RequestMapping("test4")
    public void test4(HttpServletRequest request, HttpServletResponse response) {
        Map map = new HashMap();
        map.put("username", "%中%");

        List<Userinfo> listUserinfo = userinfoMapping.selectAll4(map);
        for (int i = 0; i < listUserinfo.size(); i++) {
            Userinfo userinfo = listUserinfo.get(i);
            System.out.println(userinfo.getId() + " " + userinfo.getUsername() + " " +
userinfo.getPassword() + " "
                    + userinfo.getAge() + " " + userinfo.getInsertdate());
        }
    }
```

程序运行后不再出现异常。

如果向 parameterType 传入 String 类型，则 SQL 映射文件的代码可以设计为如下：

```
<select id="selectAll5" parameterType="String" resultType="com.ghy.www.entity.Userinfo">
    select * from userinfo
    <where>
        <trim prefixOverrides="and">
            <if test="_parameter!=null">and username like #{value}</if>
        </trim>
    </where>
</select>
```

或者使用如下映射也可以实现同样的功能：

```
<select id="selectAll6" parameterType="String" resultType="com.ghy.www.entity.Userinfo">
    select * from userinfo
    <where>
        <trim prefixOverrides="and">
            <if test="value!=null">and username like #{value}</if>
        </trim>
    </where>
</select>
```

5.7 读写大文本类型的数据

MyBatis 对数据库中大文本数据类型的支持也非常好，不需要特别的环境配置即可完成大文本数据类型字段的读写操作。

5.7.1 操作 Oracle 数据库

创建测试项目 moretext-oracle。
SQL 映射文件的代码如下：

```
<insert id="insertBigtext" parameterType="com.ghy.www.entity.BigText">
    <selectKey order="BEFORE" resultType="java.lang.Long"
```

```xml
            keyProperty="id">
    select idauto.nextval from dual
</selectKey>
insert into bigtext(id,bigtext) values(#{id},#{bigtext})
</insert>

<select id="selectById1" resultType="com.ghy.www.entity.BigText">
    select *
    from bigtext where
    id=123
</select>

<select id="selectById2" resultType="map">
    select *
    from bigtext where
    id=123
</select>
```

其中，列 bigtext 是大文本类型。

实体类的代码如下：

```java
public class BigText {
    private long id;
    private String bigtext;
```

数据表 bigtext 中 id 为 123 的 bigtext 列中存储了大文本数据。

执行查询操作，将数据封装进实体类，代码如下：

```java
@RequestMapping("test1")
public void test1(HttpServletRequest request, HttpServletResponse response) {
    BigText bigText = bigTextMapper.selectById1();
    System.out.println(bigText.getBigtext());
}
```

程序运行后，在控制台输出了全部信息。

执行查询操作，将数据封装进 Map，代码如下：

```java
@RequestMapping("test2")
public void test2(HttpServletRequest request, HttpServletResponse response) throws SQLException, IOException {
    Map map = bigTextMapper.selectById2();
    oracle.sql.CLOB clobRef = (oracle.sql.CLOB) map.get("BIGTEXT");
    Reader reader = clobRef.getCharacterStream();
    char[] charArray = new char[10000];
    int readLength = reader.read(charArray);
    System.out.println("readLength=" + readLength);
    while (readLength != -1) {
        String newString = new String(charArray, 0, readLength);
        System.out.println(newString);
        readLength = reader.read(charArray);
    }
    reader.close();
}
```

程序运行后，在控制台输出了全部信息。

插入新的大文本记录，代码如下：

```java
@RequestMapping("test3")
public void test3(HttpServletRequest request, HttpServletResponse response) throws SQLException, IOException {
```

```
    BigText bigText = bigTextMapper.selectById1();
    bigTextMapper.insertBigtext(bigText);
}
```
运行代码后，在数据表中插入了另一条大文本记录。

5.7.2 操作 MySQL 数据库

创建测试项目 moretext-mysql。

SQL 映射文件的代码如下：

```
<insert id="insertBigtext" parameterType="com.ghy.www.entity.BigText" useGeneratedKeys=
"true" keyProperty="id">
    insert into bigtext(bigtext) values(#{bigtext})
</insert>

<select id="selectById1" resultType="com.ghy.www.entity.BigText">
    select *
    from bigtext where
    id=123
</select>
```

其中，列 bigtext 是 longtext 类型。

创建实体类，代码如下：

```
public class BigText {
    private long id;
    private String bigtext;
```

数据表 bigtext 中 id 为 123 的 bigtext 列中存储了大文本数据。

执行查询操作，将数据封装进实体类，代码如下：

```
@RequestMapping("test1")
public void test1(HttpServletRequest request, HttpServletResponse response) {
    BigText bigText = bigTextMapper.selectById1();
    System.out.println(bigText.getBigtext());
}
```

程序运行后，在控制台输出了全部信息。

插入新的 longtext 记录，代码如下：

```
@RequestMapping("test2")
public void test2(HttpServletRequest request, HttpServletResponse response) throws
SQLException, IOException {
    BigText bigText = bigTextMapper.selectById1();
    bigTextMapper.insertBigtext(bigText);
}
```

运行代码后，在数据表中插入了另一条大文本记录。

5.8 实现数据分页

想要实现数据分页功能，就要先算出起始位置，起始位置的算法如下：

起始位置=（目标到达的页数−1）×一页显示多少条记录

示例代码如下：

```
public static void main(String[] args) throws IOException {
    String gotoPage = "4";
    int gotoPageInt = 1;
    try {
        gotoPageInt = Integer.parseInt(gotoPage);
        if (gotoPageInt <= 0) {
            gotoPageInt = 1;
        }
    } catch (NumberFormatException e) {
        gotoPageInt = 1;
    }
    System.out.println(gotoPageInt);
}
```

创建测试项目 pageRowBounds。

SQL 映射文件的配置代码如下：

```xml
<select id="getUserinfo" resultType="com.ghy.www.entity.Userinfo">
    select * from userinfo
    where
    username = #{username} order by id asc
</select>
```

Mapper 接口中方法的代码如下：

```java
public List<Userinfo> getUserinfo(String username, RowBounds rowBounds);
```

实现分页功能，代码如下：

```java
@RequestMapping("test1")
public void test1(HttpServletRequest request, HttpServletResponse response) {
    {
        List<Userinfo> listUserinfo = userinfoMapping.getUserinfo("中国", new RowBounds(0, 2));
        for (int i = 0; i < listUserinfo.size(); i++) {
            Userinfo userinfo = listUserinfo.get(i);
            System.out.println(userinfo.getId() + " " + userinfo.getUsername() + " " + userinfo.getPassword());
        }
    }
    System.out.println();
    {
        List<Userinfo> listUserinfo = userinfoMapping.getUserinfo("中国", new RowBounds(2, 2));
        for (int i = 0; i < listUserinfo.size(); i++) {
            Userinfo userinfo = listUserinfo.get(i);
            System.out.println(userinfo.getId() + " " + userinfo.getUsername() + " " + userinfo.getPassword());
        }
    }
}
```

MyBatis 提供的数据分页功能在执行效率上是比较低的，它的实现方法是先将数据表中符合条件的全部记录放入内存，然后在内存中进行分页，这会造成对内存占用率较高。我们推荐使用第三方的 MyBatis 分页插件 PageHelper 来优化分页的执行效率，或自己定制分页的 SQL 语句。

5.9　实现一对一级联

当 A 类中有 B 类的属性时，这种情况属于一对一级联，代码结构如图 5-6 所示。

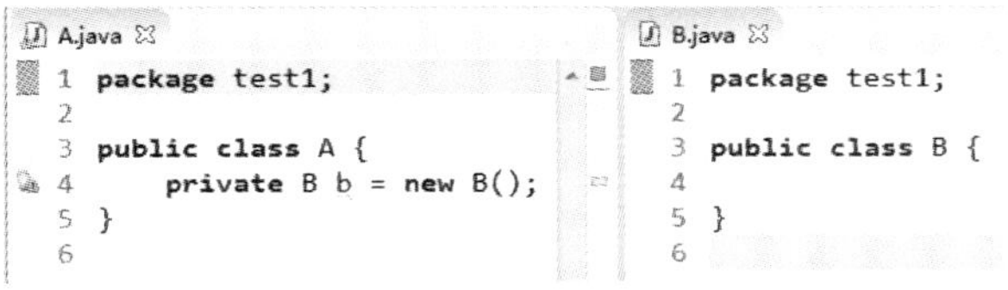

图 5-6　一对一级联

在一对一级联中，在取得 A 对象时就可以取得与 A 对象关联的 B 对象。

当 A 类中有 B 类的集合属性时，这种情况属于一对多级联，代码结构如图 5-7 所示。

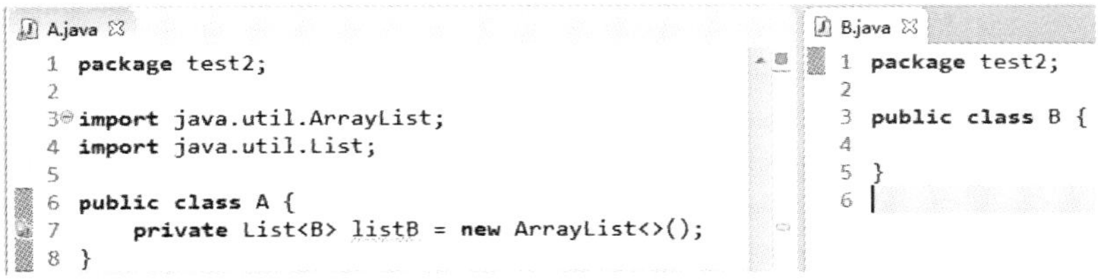

图 5-7　一对多级联

MyBatis 支持一对一和一对多的级联操作。

5.9.1　数据表结构和内容以及关系

我们先来测试一对一级联。

创建测试项目 one_one。

项目中使用的数据表及数据表内容的关联关系如图 5-8 所示。

每一个用户有一个身份证，用户和身份证之间属于一对一关联。

ID	USERN...	PASSWORD	AGE	INSERTDATE	CARDIDCOLUMN
1	中国1	中国人1	11	2000/1/1	101
2	中国2	中国人2	22	2000/1/1	102
3	中国3	中国人3	33	2000/1/1	103
4	中国4	中国人4	44	2000/1/1	104

ID	CARDNO
101	AA
102	BB
103	CC
104	DD

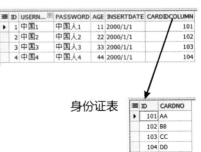

图 5-8　项目中使用的数据表

5.9.2　创建实体类

创建用户信息实体类，代码如下：

```
package com.ghy.www.entity;

import java.util.Date;

public class Userinfo {
    private long id;
    private String username;
    private String password;
    private long age;
    private Date insertdate;
    private IDCard idCard;

    public Userinfo() {
    }

    public long getId() {
```

```
            return id;
        }

        public void setId(long id) {
            this.id = id;
        }

        public String getUsername() {
            return username;
        }

        public void setUsername(String username) {
            this.username = username;
        }

        public String getPassword() {
            return password;
        }

        public void setPassword(String password) {
            this.password = password;
        }

        public long getAge() {
            return age;
        }

        public void setAge(long age) {
            this.age = age;
        }

        public Date getInsertdate() {
            return insertdate;
        }

        public void setInsertdate(Date insertdate) {
            this.insertdate = insertdate;
        }

        public IDCard getIdCard() {
            return idCard;
        }

        public void setIdCard(IDCard idCard) {
            this.idCard = idCard;
        }
}
```

创建身份证实体类，代码如下：

```
package com.ghy.www.entity;

public class IDCard {
    private long id;
    private String cardNo;

    public IDCard() {
    }
```

```java
    public long getId() {
        return id;
    }

    public void setId(long id) {
        this.id = id;
    }

    public String getCardNo() {
        return cardNo;
    }

    public void setCardNo(String cardNo) {
        this.cardNo = cardNo;
    }
}
```

5.9.3 创建 SQL 映射文件

创建身份证表对应的 SQL 映射文件，代码如下：

```xml
<mapper namespace="com.ghy.www.sqlmapping.IIDCard">
    <select id="selectIDCardById" parameterType="long"
            resultType="com.ghy.www.entity.IDCard">
        select * from idcard where id=#{id}
    </select>
</mapper>
```

Mapper 映射接口中方法的代码如下：

```java
@Mapper
public interface IIDCard {
    public IDCard selectIDCardById(long id);
}
```

创建用户信息表对应的 SQL 映射文件，代码如下：

```xml
<mapper namespace="com.ghy.www.sqlmapping.IUserinfoMapping">
    <resultMap type="com.ghy.www.entity.Userinfo" id="userinfoMap">
        <result column="username" property="username"/>
        <result column="password" property="password"/>
        <result column="age" property="age"/>
        <result column="insertdate" property="insertdate"/>
        <association property="idCard" column="CARDIDCOLUMN"
                select="com.ghy.www.sqlmapping.IIDCard.selectIDCardById"></association>
    </resultMap>

    <select id="getUserinfoById" parameterType="long"
            resultMap="userinfoMap">
        select * from
        userinfo where id=#{id}
    </select>

    <select id="getAllUserinfo" resultMap="userinfoMap">
        select * from userinfo
        order by id asc
    </select>
</mapper>
```

Mapper 映射接口中方法的代码如下：

```
@Mapper
public interface IUserinfoMapping {
    public Userinfo getUserinfoById(long userId);
    public List<Userinfo> getAllUserinfo();
}
```

5.9.4 级联解析

实现一对一级联的映射关系如图 5-9 所示。

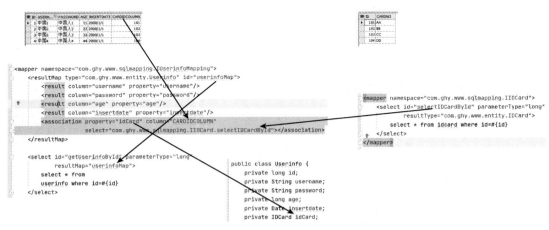

图 5-9　级联原理

5.9.5 根据 ID 查询记录

创建运行类，代码如下：

```
@RequestMapping("test1")
public void test1(HttpServletRequest request, HttpServletResponse response) {
    Userinfo userinfo = userinfoMapping.getUserinfoById(1);
    System.out.println(userinfo.getId() + " " + userinfo.getUsername() + " " + userinfo.getPassword() + " "
            + userinfo.getAge() + " " + userinfo.getInsertdate() + " " + userinfo.getIdCard().getId() + " "
            + userinfo.getIdCard().getCardNo());
}
```

程序运行后控制台输出的结果如图 5-10 所示。

```
com.zaxxer.hikari.HikariDataSource      : MyHikariCP - Start completed.
c.g.w.s.I.getUserinfoById               : ==>  Preparing: select * from userinfo where id=?
c.g.w.s.I.getUserinfoById               : ==> Parameters: 1(Long)
c.g.w.s.IIDCard.selectIDCardById        : ====>  Preparing: select * from idcard where id=?
c.g.w.s.IIDCard.selectIDCardById        : ====> Parameters: 101(Long)
c.g.w.s.IIDCard.selectIDCardById        : <====      Total: 1
c.g.w.s.I.getUserinfoById               : <==      Total: 1

1 中国1 中国人1 11 Sat Jan 01 00:00:00 CST 2000 101 AA
```

图 5-10　运行结果

从控制台输出的日志信息来看，使用的是 2 条 SQL 语句。如果查询更多的记录，执行的 SQL 语句次数是否会增加呢？下面我们继续测试。

5.9.6 查询所有记录

创建运行类，代码如下：

```
@RequestMapping("test2")
public void test2(HttpServletRequest request, HttpServletResponse response) {
    List<Userinfo> listUserinfo = userinfoMapping.getAllUserinfo();
    for (int i = 0; i < listUserinfo.size(); i++) {
        Userinfo userinfo = listUserinfo.get(i);
        System.out.println(userinfo.getId() + " " + userinfo.getUsername() + " " + userinfo.getPassword() + " "
                + userinfo.getAge() + " " + userinfo.getInsertdate() + " " + userinfo.getIdCard().getId() + " "
                + userinfo.getIdCard().getCardNo());
    }
}
```

程序运行后控制台输出的结果如图 5-11 所示。

```
com.zaxxer.hikari.HikariDataSource       : MyHikariCP - Start completed.
c.g.w.s.IUserinfoMapping.getAllUserinfo  : ==>  Preparing: select * from userinfo order by id asc
c.g.w.s.IUserinfoMapping.getAllUserinfo  : ==> Parameters:
c.g.w.s.IIDCard.selectIDCardById         : ====>  Preparing: select * from idcard where id=?
c.g.w.s.IIDCard.selectIDCardById         : ====> Parameters: 101(Long)
c.g.w.s.IIDCard.selectIDCardById         : <====      Total: 1
c.g.w.s.IIDCard.selectIDCardById         : ====>  Preparing: select * from idcard where id=?
c.g.w.s.IIDCard.selectIDCardById         : ====> Parameters: 102(Long)
c.g.w.s.IIDCard.selectIDCardById         : <====      Total: 1
c.g.w.s.IIDCard.selectIDCardById         : ====>  Preparing: select * from idcard where id=?
c.g.w.s.IIDCard.selectIDCardById         : ====> Parameters: 103(Long)
c.g.w.s.IIDCard.selectIDCardById         : <====      Total: 1
c.g.w.s.IIDCard.selectIDCardById         : ====>  Preparing: select * from idcard where id=?
c.g.w.s.IIDCard.selectIDCardById         : ====> Parameters: 104(Long)
c.g.w.s.IIDCard.selectIDCardById         : <====      Total: 1
c.g.w.s.IUserinfoMapping.getAllUserinfo  : <==      Total: 4

1 中国1 中国人1 11 Sat Jan 01 00:00:00 CST 2000 101 AA
2 中国2 中国人2 22 Sat Jan 01 00:00:00 CST 2000 102 BB
3 中国3 中国人3 33 Sat Jan 01 00:00:00 CST 2000 103 CC
4 中国4 中国人4 44 Sat Jan 01 00:00:00 CST 2000 104 DD
```

图 5-11 运行结果

果然，从控制台输出的日志信息来看，使用更多的 SQL 语句进行查询，会造成客户端与数据库通信次数过多，降低运行效率。下面我们来看一下优化的代码。

5.9.7 对 SQL 语句的执行次数进行优化

创建测试项目 one_one2。
创建用户信息表对应的 SQL 映射文件，代码如下：

```xml
<mapper namespace="com.ghy.www.sqlmapping.IUserinfoMapping">
    <resultMap type="com.ghy.www.entity.Userinfo" id="userinfoMap">
        <result column="id" property="id"/>
        <result column="username" property="username"/>
        <result column="password" property="password"/>
        <result column="age" property="age"/>
        <result column="insertdate" property="insertdate"/>
        <association property="idCard" javaType="com.ghy.www.entity.IDCard">
            <result column="cardid" property="id"/>
            <result column="cardNo" property="cardNo"/>
        </association>
    </resultMap>

    <select id="getAllUserinfo" resultMap="userinfoMap">
        select u.*,card.id
        cardid,card.cardNo
        from userinfo
        u,idcard card where
        u.cardidcolumn=card.id
        order by u.id
        asc
    </select>

    <select id="getUserinfoById" parameterType="long"
            resultMap="userinfoMap">
        select u.*,card.id cardid,card.cardNo
        from userinfo
        u,idcard
        card where
        u.cardidcolumn=card.id and u.id=#{id}
        order by u.id asc
    </select>
</mapper>
```

创建运行类，代码如下：

```
@RequestMapping("test1")
public void test1(HttpServletRequest request, HttpServletResponse response) {
    Userinfo userinfo = userinfoMapping.getUserinfoById(1);
    System.out.println(userinfo.getId() + " " + userinfo.getUsername() + " " + userinfo.getPassword() + " "
            + userinfo.getAge() + " " + userinfo.getInsertdate() + " " + userinfo.getIdCard().getId() + " "
            + userinfo.getIdCard().getCardNo());
}
```

程序运行后控制台输出的结果如图 5-12 所示。

图 5-12 使用一条 SQL 语句

从控制台输出的日志信息来看，使用了 1 条 SQL 语句进行查询，提高了运行效率。

如果查询所有记录也会使用 1 条 SQL 语句，运行类代码如下：

```
@RequestMapping("test2")
public void test2(HttpServletRequest request, HttpServletResponse response) {
    List<Userinfo> listUserinfo = userinfoMapping.getAllUserinfo();
    for (int i = 0; i < listUserinfo.size(); i++) {
       Userinfo userinfo = listUserinfo.get(i);
       System.out.println(userinfo.getId() + " " + userinfo.getUsername() + " " +
userinfo.getPassword() + " "
                 + userinfo.getAge() + " " + userinfo.getInsertdate() + " " + userinfo.
getIdCard().getId() + " "
                 + userinfo.getIdCard().getCardNo());
    }
}
```

程序运行后控制台输出的结果如图 5-13 所示。

```
DEBUG 10716 --- [nio-8080-exec-4] c.g.w.s.IUserinfoMapping.getAllUserinfo  : ==>  Preparing:
select u.*,card.id cardid,card.cardNo from userinfo u,idcard card where u.cardidcolumn=card.id order by u.id asc
DEBUG 10716 --- [nio-8080-exec-4] c.g.w.s.IUserinfoMapping.getAllUserinfo  : ==> Parameters:
DEBUG 10716 --- [nio-8080-exec-4] c.g.w.s.IUserinfoMapping.getAllUserinfo  : <==      Total: 4
1 中国1 中国人1 11 Sat Jan 01 00:00:00 CST 2000 101 AA
2 中国2 中国人2 22 Sat Jan 01 00:00:00 CST 2000 102 BB
3 中国3 中国人3 33 Sat Jan 01 00:00:00 CST 2000 103 CC
4 中国4 中国人4 44 Sat Jan 01 00:00:00 CST 2000 104 DD
```

图 5-13 使用一条 SQL 语句

从控制台输出的日志信息来看，使用了 1 条 SQL 语句进行查询，提高了运行效率。

5.10 实现一对多级联

MyBatis 支持一对多级联。

创建测试项目 one_more。

5.10.1 数据表结构和内容以及关系

项目中使用的数据表及数据表内容的关联关系如图 5-14 所示。

每一个省有多个市，省和市之间属于一对多关联。

图 5-14 项目中使用的数据表

5.10.2 创建实体类

创建实体类 Sheng，代码如下：

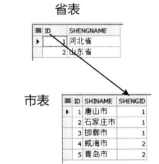

```
package com.ghy.www.entity;

import java.util.List;

public class Sheng {
    private long id;
    private String shengname;
    private List<Shi> shiList;
```

```java
    public Sheng() {
    }

    public long getId() {
        return id;
    }

    public void setId(long id) {
        this.id = id;
    }

    public String getShengname() {
        return shengname;
    }

    public void setShengname(String shengname) {
        this.shengname = shengname;
    }

    public List<Shi> getShiList() {
        return shiList;
    }

    public void setShiList(List<Shi> shiList) {
        this.shiList = shiList;
    }
}
```

创建实体类 Shi，代码如下：

```java
package com.ghy.www.entity;

public class Shi {
    private long id;
    private String shiname;
    private long shengid;

    public Shi() {
    }

    public long getId() {
        return id;
    }

    public void setId(long id) {
        this.id = id;
    }

    public String getShiname() {
        return shiname;
    }

    public void setShiname(String shiname) {
        this.shiname = shiname;
    }

    public long getShengid() {
```

```
            return shengid;
        }

        public void setShengid(long shengid) {
            this.shengid = shengid;
        }
}
```

5.10.3 创建 SQL 映射文件

创建省表对应的 SQL 映射文件,代码如下:

```
<mapper namespace="com.ghy.www.sqlmapping.IShengMapping">
    <resultMap type="com.ghy.www.entity.Sheng" id="shengMap">
        <result property="id" column="id"/>
        <result property="shengname" column="shengname"/>
        <collection property="shiList"
                select="com.ghy.www.sqlmapping.IShiMapping.getShiByShengId" column="id"></collection>
    </resultMap>

    <select id="getAllSheng" resultMap="shengMap">
        select * from sheng order by
        id asc
    </select>

    <select id="getShengById" parameterType="long"
            resultMap="shengMap">
        select * from
        sheng where id=#{id}
    </select>
</mapper>
```

Mapper 映射接口中方法的代码如下:

```
@Mapper
public interface IShengMapping {
    public Sheng getShengById(long id);

    public List<Sheng> getAllSheng();
}
```

创建市表对应的 SQL 映射文件,代码如下:

```
<mapper namespace="com.ghy.www.sqlmapping.IShiMapping">
    <select id="getShiByShengId" parameterType="long"
            resultType="com.ghy.www.entity.Shi">
        select * from
        shi where shengid=#{id}
    </select>
</mapper>
```

Mapper 映射接口中方法的代码如下:

```
@Mapper
public interface IShiMapping {
    public Shi getShiByShengId(long id);
}
```

5.10.4 级联解析

实现一对多级联的映射关系如图 5-15 所示。

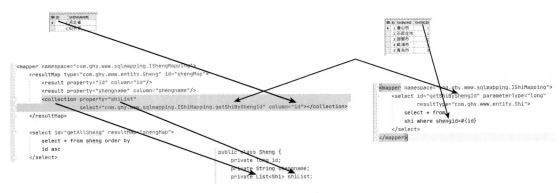

图 5-15 级联原理

5.10.5 根据 ID 查询记录

创建运行类，代码如下：

```
@RequestMapping("test1")
public void test1(HttpServletRequest request, HttpServletResponse response) {
    Sheng sheng = shengMapping.getShengById(1L);
    System.out.println(sheng.getId() + " " + sheng.getShengname());
    List<Shi> listShi = sheng.getShiList();
    for (int j = 0; j < listShi.size(); j++) {
        Shi shi = listShi.get(j);
        System.out.println("  " + shi.getId() + " " + shi.getShiname());
    }
}
```

程序运行后控制台输出的结果如图 5-16 所示。

```
com.zaxxer.hikari.HikariDataSource    : MyHikariCP - Start completed.
c.g.w.s.IShengMapping.getShengById    : ==>  Preparing: select * from sheng where id=?
c.g.w.s.IShengMapping.getShengById    : ==> Parameters: 1(Long)
c.g.w.s.IShiMapping.getShiByShengId   : ====>  Preparing: select * from shi where shengid=?
c.g.w.s.IShiMapping.getShiByShengId   : ====> Parameters: 1(Long)
c.g.w.s.IShiMapping.getShiByShengId   : <====      Total: 3
c.g.w.s.IShengMapping.getShengById    : <==      Total: 1

1 河北省
  1 唐山市
  2 石家庄市
  3 邯郸市
```

图 5-16 使用 2 条 SQL 语句

从控制台输出的日志信息来看，使用了 2 条 SQL 语句进行查询。

5.10.6 查询所有记录

创建运行类，代码如下：

```
@RequestMapping("test2")
public void test2(HttpServletRequest request, HttpServletResponse response) {
    List<Sheng> listSheng = shengMapping.getAllSheng();
    for (int i = 0; i < listSheng.size(); i++) {
        Sheng sheng = listSheng.get(i);
        System.out.println(sheng.getId() + " " + sheng.getShengname());
        List<Shi> listShi = sheng.getShiList();
        for (int j = 0; j < listShi.size(); j++) {
            Shi shi = listShi.get(j);
            System.out.println(" " + shi.getId() + " " + shi.getShiname());
        }
    }
}
```

程序运行后控制台输出的结果如图 5-17 所示。

```
c.g.w.s.IShengMapping.getAllSheng      : ==>  Preparing: select * from sheng order by id asc
c.g.w.s.IShengMapping.getAllSheng      : ==> Parameters:
c.g.w.s.IShiMapping.getShiByShengId    : ====>  Preparing: select * from shi where shengid=?
c.g.w.s.IShiMapping.getShiByShengId    : ====> Parameters: 1(Long)
c.g.w.s.IShiMapping.getShiByShengId    : <====      Total: 3
c.g.w.s.IShiMapping.getShiByShengId    : ====>  Preparing: select * from shi where shengid=?
c.g.w.s.IShiMapping.getShiByShengId    : ====> Parameters: 2(Long)
c.g.w.s.IShiMapping.getShiByShengId    : <====      Total: 2
c.g.w.s.IShengMapping.getAllSheng      : <==       Total: 2
1 河北省
 1 唐山市
 2 石家庄市
 3 邯郸市
2 山东省
 4 威海市
 5 青岛市
```

图 5-17 使用 3 条 SQL 语句

从控制台输出的日志信息来看，使用了 3 条 SQL 语句进行查询。

前两个测试分别使用 2 条和 3 条 SQL 语句进行查询。我们可以对 SQL 语句的执行次数进行优化，使 SQL 语句的执行次数变得更少。

5.10.7 对 SQL 语句的执行次数进行优化

创建测试项目 one_more2。

创建省表对应的 SQL 映射文件，代码如下：

```xml
<mapper namespace="com.ghy.www.sqlmapping.IShengMapping">
    <resultMap type="com.ghy.www.entity.Sheng" id="shengMap">
        <result property="id" column="id"/>
        <result property="shengname" column="shengname"/>
        <collection property="shiList" ofType="com.ghy.www.entity.Shi">
            <result property="id" column="shiId"/>
            <result property="shiname" column="shiname"/>
        </collection>
    </resultMap>

    <select id="getAllSheng" resultMap="shengMap">
        select sheng.*,shi.id
```

```
            shiId,shi.shiname from sheng,shi where sheng.id=shi.shengid
            order by
            sheng.id
            asc
        </select>

        <select id="getShengById" parameterType="long"
               resultMap="shengMap">
            select sheng.*,shi.id
            shiId,shi.shiname from sheng,shi where
            sheng.id=shi.shengid and sheng.id=#{id}
        </select>
</mapper>
```

配置代码如下：

```
<collection property="shiList" ofType="com.ghy.www.entity.Shi">
```

其中的 ofType 属性表示 List 中存储的是 Shi 对象。

创建运行类，代码如下：

```
@RequestMapping("test1")
public void test1(HttpServletRequest request, HttpServletResponse response) {
    Sheng sheng = shengMapping.getShengById(1L);
    System.out.println(sheng.getId() + " " + sheng.getShengname());
    List<Shi> listShi = sheng.getShiList();
    for (int j = 0; j < listShi.size(); j++) {
        Shi shi = listShi.get(j);
        System.out.println("    " + shi.getId() + " " + shi.getShiname());
    }
}
```

程序运行后控制台输出的结果如图 5-18 所示。

```
                    DEBUG 9924 --- [nio-8080-exec-1] c.g.w.s.IShengMapping.getShengById       : ==> Preparing:
select sheng.*,shi.id shiId,shi.shiname from sheng,shi where sheng.id=shi.shengid and sheng.id=?
                    DEBUG 9924 --- [nio-8080-exec-1] c.g.w.s.IShengMapping.getShengById       : ==> Parameters: 1(Long)
                    DEBUG 9924 --- [nio-8080-exec-1] c.g.w.s.IShengMapping.getShengById       : <==      Total: 3
1 河北省
    1 唐山市
    2 石家庄市
    3 邯郸市
```

图 5-18 使用 1 条 SQL 语句

从控制台输出的日志信息来看，使用了 1 条 SQL 语句进行查询，提高了运行效率。

如果查询所有记录，也会使用 1 条 SQL 语句，运行类代码如下：

```
@RequestMapping("test2")
public void test2(HttpServletRequest request, HttpServletResponse response) {
    List<Sheng> listSheng = shengMapping.getAllSheng();
    for (int i = 0; i < listSheng.size(); i++) {
        Sheng sheng = listSheng.get(i);
        System.out.println(sheng.getId() + " " + sheng.getShengname());
        List<Shi> listShi = sheng.getShiList();
        for (int j = 0; j < listShi.size(); j++) {
            Shi shi = listShi.get(j);
            System.out.println("    " + shi.getId() + " " + shi.getShiname());
```

            }
        }
    }

程序运行后控制台输出的结果如图 5-19 所示。

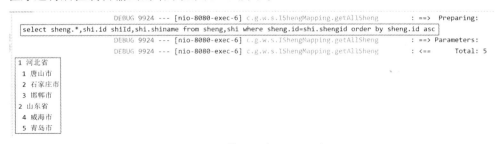

图 5-19 使用 1 条 SQL 语句

从控制台输出的日志信息来看，使用了 1 条 SQL 语句进行查询，提高了运行效率。

5.11 延迟加载

什么是延迟加载？延迟加载的反义词是立即加载，那么什么是立即加载呢？立即加载的含义是取得省对象时，省中所有的市对象都已经在内存中，所以立即加载非常耗费 CPU 和内存的资源。有的时候只需要取得省信息，而不需要取得市信息，但立即加载还是把市信息也一同取出来，在这种情况下就需要使用延迟加载了。延迟加载就是等到使用市对象时，再把市对象加载到内存，如果不使用市对象，就不加载到内存，这就是延迟加载。

5.11.1 默认采用立即加载策略

MyBatis 默认采用的加载策略是立即加载。

我们将 one_one 项目资源复制到新建的 lazyTest1 项目中，以完成立即加载的测试。

程序运行后在 for 语句处设置断点，调试结果如图 5-20 所示。

图 5-20 开始调试

执行了 getAllUserinfo()方法后，List 中的 Userinfo 对象的 idCard 属性值并不是 null，说明在取得 Userinfo 对象时，采用了立即加载策略，立即将 Userinfo 对象对应的 IDCard 对象加载到内存。

在一对多级联时，默认情况下也是采用立即加载，将 one_more 项目资源复制到新建的 lazyTest2 项目中，以完成立即加载的测试。

仍然在 for 语句处设置断点，调试效果如图 5-21 所示。从调试信息中可以发现，List 中的 Sheng 对象中的 shiList 集合中已经存在 Shi 对象，说明在取得 Sheng 对象时，采用了立即加载策略，将 Sheng 对象对应的 Shi 集合放入内存。

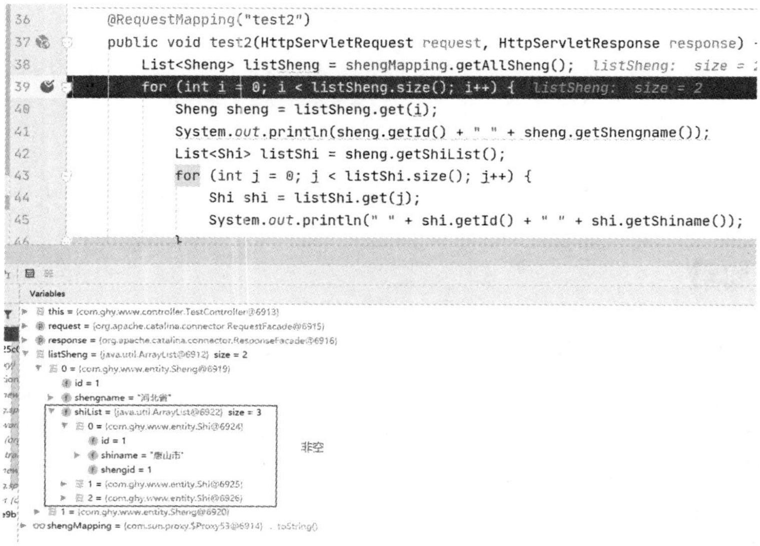

图 5-21　开始调试

如果不需要子表数据，采用立即加载策略会影响程序运行效率，可以使用延迟加载解决这个问题。

5.11.2　使用全局延迟加载策略与两种加载方式

在配置文件中使用配置代码

```
mybatis:
  configuration:
    lazy-loading-enabled: true
```

开启延迟加载。

但延迟加载的具体实现方式分为以下两种。

（1）取得子表时再加载子表，如调用 sheng.getShiList();代码时再向 List 中加载 Shi 对象。

（2）打印主表信息时加载子表，如调用如下代码：

```
System.out.println(sheng.getId() + " " + sheng.getShengname());
```

这两种延迟加载的方式可以使用配置

```
mybatis:
  configuration:
    aggressive-lazy-loading: true/false
```

5.11 延迟加载

进行切换,默认情况是第 1 种,默认值是 false。

在一对一或一对多级联中,这两种延迟加载的方式是一样的。

1. 测试一对一全局延迟加载

把 lazyTest1 项目资源复制到新建的 lazyTest3 项目中,以测试一对一延迟加载。

控制层的代码如下:

```
@RequestMapping("test2")
public void test2(HttpServletRequest request, HttpServletResponse response) {
    List<Userinfo> listUserinfo = userinfoMapping.getAllUserinfo();
    for (int i = 0; i < listUserinfo.size(); i++) {
        Userinfo userinfo = listUserinfo.get(i);
        System.out.print(userinfo.getId() + " " + userinfo.getUsername() + " " + userinfo.getPassword() + " "
                + userinfo.getAge() + " " + userinfo.getInsertdate());
        IDCard idCard = userinfo.getIdCard();
        System.out.println(idCard.getId() + " " + idCard.getCardNo());
    }
}
```

更改配置文件,代码如下:

```
logging:
  level:
    com.ghy.www.sqlmapping: DEBUG #属性是 SQL 映射接口所在的包名
mybatis:
  configuration:
    lazy-loading-enabled: true
```

执行 test2 控制层,结果如图 5-22 所示。

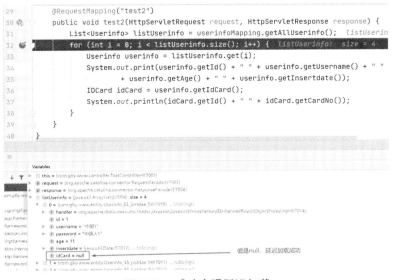

图 5-22 成功实现延迟加载

继续调试,结果如图 5-23 所示。

图 5-23　用到的时候才加载

从调试中可以发现，直到执行代码

```
IDCard idCard = userinfo.getIdCard();
```

时 IDCard 对象才被加载到内存。

这个特性和配置代码

```
logging:
  level:
    com.ghy.www.sqlmapping: DEBUG #属性是SQL映射接口所在的包名
mybatis:
  configuration:
    lazy-loading-enabled: true
    aggressive-lazy-loading: false
```

的作用是一样的，也就是说配置侵犯延迟加载（aggressive-lazy-loading）的默认值为 false，其含义是不侵犯延迟加载，保持当取得子表时再加载子表数据的特性。

如果使用配置

```
logging:
  level:
    com.ghy.www.sqlmapping: DEBUG #属性是SQL映射接口所在的包名
mybatis:
  configuration:
    lazy-loading-enabled: true
    aggressive-lazy-loading: true
```

则在调试的过程中，执行代码

```
System.out.print(userinfo.getId() + " " + userinfo.getUsername() + " " + userinfo.getPassword() + " "
        + userinfo.getAge() + " " + userinfo.getInsertdate());
```

时就将子表 IDCard 对象加载到内存。未加载子表数据的结果如图 5-24 所示。

5.11 延迟加载

图 5-24 未加载子表数据

加载子表数据的结果如图 5-25 所示。

图 5-25 加载子表数据

2. 测试一对多全局延迟加载

把 lazyTest2 项目资源复制到新建的 lazyTest4 项目中，以测试一对多延迟加载。控制层的代码如下：

```
@RequestMapping("test2")
public void test2(HttpServletRequest request, HttpServletResponse response) {
    List<Sheng> listSheng = shengMapping.getAllSheng();
    for (int i = 0; i < listSheng.size(); i++) {
        Sheng sheng = listSheng.get(i);
```

```
            System.out.println(sheng.getId() + " " + sheng.getShengname());
            List<Shi> listShi = sheng.getShiList();
            for (int j = 0; j < listShi.size(); j++) {
                Shi shi = listShi.get(j);
                System.out.println(" " + shi.getId() + " " + shi.getShiname());
            }
        }
    }
```

更改配置文件,代码如下:

```
logging:
  level:
    com.ghy.www.sqlmapping: DEBUG #属性是 SQL 映射接口所在的包名
mybatis:
  configuration:
    lazy-loading-enabled: true
```

执行 test2 控制层,结果如图 5-26 所示。

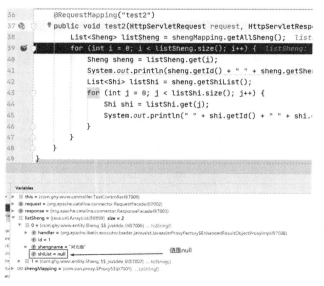

图 5-26 成功实现延迟加载

继续调试,结果如图 5-27 所示。

从调试中可以发现,直到执行代码

```
List<Shi> listShi = sheng.getShiList();
```

时 Shi 对象才被加载到内存。

这个特性和配置代码

```
logging:
  level:
    com.ghy.www.sqlmapping: DEBUG #属性是 SQL 映射接口所在的包名
mybatis:
  configuration:
    lazy-loading-enabled: true
    aggressive-lazy-loading: false
```

5.11 延迟加载

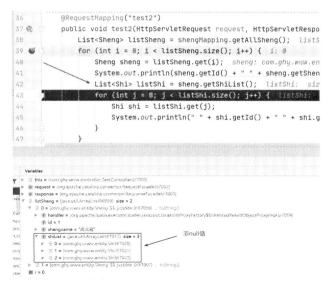

图 5-27 用到的时候才加载

的作用是一样的，也就是说配置 aggressive-lazy-loading 的默认值为 false，其含义是不侵犯延迟加载，保持当取得子表时再加载子表数据的特性。

如果使用配置

```
logging:
  level:
    com.ghy.www.sqlmapping: DEBUG  #属性是 SQL 映射接口所在的包名
mybatis:
  configuration:
    lazy-loading-enabled: true
    aggressive-lazy-loading: true
```

则在调试的过程中，执行代码

```
System.out.println(sheng.getId() + " " + sheng.getShengname());
```

时就将子表 Shi 对象加载到内存。未加载子表数据的结果如图 5-28 所示。

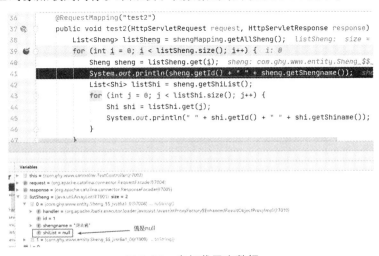

图 5-28 未加载子表数据

加载子表数据的结果如图 5-29 所示。

图 5-29 加载子表数据

5.11.3 使用 fetchType 属性设置局部加载策略

使用配置属性 lazy-loading-enabled 和 aggressive-lazy-loading 设置的是全局加载行为。如果想针对某一个映射来决定采用延迟加载还是立即加载，则可以在<collection>配置中使用 fetchType 属性设置，属性值 eager 表示立即加载，属性值 lazy 表示延迟加载。

1．测试一对一局部延迟加载

把 lazyTest3 项目资源复制到新建的 lazyTest5 项目中，以测试一对一延迟加载。
配置文件的代码如下：

```
logging:
  level:
    com.ghy.www.sqlmapping: DEBUG  #属性是SQL映射接口所在的包名
mybatis:
  configuration:
    lazy-loading-enabled: true
```

开启延迟加载。
省（sheng）表映射文件的配置代码如下：

```
<association property="idCard" column="CARDIDCOLUMN"
        select="com.ghy.www.sqlmapping.IIDCard.selectIDCardById" fetchType="eager">
</association>
```

取得 Sheng 对象时就把 Shi 对象也一同加载，即立即加载。
使用如下配置

```
<association property="idCard" column="CARDIDCOLUMN"
        select="com.ghy.www.sqlmapping.IIDCard.selectIDCardById" fetchType="lazy">
</association>
```

执行代码：

```
IDCard idCard = userinfo.getIdCard();
```

如果加载 IDCard 对象，则说明是延迟加载。

2．测试一对多局部延迟加载

把 lazyTest4 项目资源复制到新建的 lazyTest6 项目中，以测试一对多延迟加载。
配置代码

```
<collection property="shiList"
        select="com.ghy.www.sqlmapping.IShiMapping.getShiByShengId" column="id"
fetchType="eager"></collection>
```

表示立即加载。
配置代码

```
<collection property="shiList"
        select="com.ghy.www.sqlmapping.IShiMapping.getShiByShengId" column="id"
fetchType="lazy"></collection>
```

表示延迟加载，直到执行代码

```
List<Shi> listShi = sheng.getShiList();
```

时才加载 Shi 对象。

5.12 缓存的使用

我们可以将查询到的实体类放到缓存中，使后面的 select 查询语句可以复用该实体类，这会提高程序运行效率。

MyBatis 缓存分为一级缓存和二级缓存。

（1）一级缓存受 SqlSession 对象管理，每个 SqlSession 有自己所属的一级缓存。一级缓存中的数据是私有的。

（2）二级缓存受 SqlSessionFactory 对象管理，属于 Application 级。二级缓存中的数据是公有的。

5.12.1 一级缓存

一级缓存是 MyBatis 默认提供并开启的，每个 SqlSession 有自己所属的一级缓存。
创建测试项目 cacheTest1。
SQL 映射文件的代码如下：

```
<mapper namespace="com.ghy.www.sqlmapping.IShengMapping">
    <select id="getShengById" parameterType="long"
            resultType="com.ghy.www.entity.Sheng">
        select * from
        sheng where id=#{id}
    </select>
</mapper>
```

配置文件 application.yml 的代码如下：

```yaml
mybatis:
  mapper-locations: classpath*:com/ghy/www/sqlmapping/*.xml
```

运行类代码如下：

```java
@Controller
@Transactional
public class TestController {

    @Autowired
    private SqlSessionTemplate sqlSessionTemplate;

    @RequestMapping("test1")
    public void test1(HttpServletRequest request, HttpServletResponse response) {
        SqlSession sqlSession = sqlSessionTemplate.getSqlSessionFactory().openSession();
        Sheng sheng1 = sqlSession.selectOne("getShengById", 1L);
        Sheng sheng2 = sqlSession.selectOne("getShengById", 1L);
        Sheng sheng3 = sqlSession.selectOne("getShengById", 1L);

        System.out.println(sheng1.getId() + " " + sheng1.getShengname());
        System.out.println(sheng2.getId() + " " + sheng2.getShengname());
        System.out.println(sheng3.getId() + " " + sheng3.getShengname());

        sqlSession.commit();
        sqlSession.close();
    }

    @RequestMapping("test2")
    public void test2(HttpServletRequest request, HttpServletResponse response) {
    }
}
```

程序运行结果如图 5-30 所示。

```
com.zaxxer.hikari.HikariDataSource   : MyHikariCP - Start completed.
c.g.w.s.IShengMapping.getShengById   : ==>  Preparing: select * from sheng where id=?
c.g.w.s.IShengMapping.getShengById   : ==> Parameters: 1(Long)
c.g.w.s.IShengMapping.getShengById   : <==      Total: 1
1 河北省
1 河北省
1 河北省
```

图 5-30　缓存实体类成功

从程序运行结果来看，在一个 SqlSession 中复用了缓存中的 Sheng 对象。

不同 SqlSession 对象缓存中的数据不可以共享，创建示例代码如下：

```java
@RequestMapping("test2")
public void test2(HttpServletRequest request, HttpServletResponse response) {
    SqlSession sqlSession1 = sqlSessionTemplate.getSqlSessionFactory().openSession();
    SqlSession sqlSession2 = sqlSessionTemplate.getSqlSessionFactory().openSession();

    Sheng sheng1 = sqlSession1.selectOne("getShengById", 1L);
    Sheng sheng2 = sqlSession2.selectOne("getShengById", 1L);

    System.out.println(sheng1.getId() + " " + sheng1.getShengname());
```

```
        System.out.println(sheng2.getId() + " " + sheng2.getShengname());

        sqlSession1.commit();
        sqlSession2.commit();
        sqlSession1.close();
        sqlSession2.close();
    }
```

程序运行结果如图 5-31 所示。

```
com.zaxxer.hikari.HikariDataSource        : MyHikariCP - Start completed.
c.g.w.s.IShengMapping.getShengById        : ==>  Preparing: select * from sheng where id=?
c.g.w.s.IShengMapping.getShengById        : ==> Parameters: 1(Long)
c.g.w.s.IShengMapping.getShengById        : <==      Total: 1
c.g.w.s.IShengMapping.getShengById        : ==>  Preparing: select * from sheng where id=?
c.g.w.s.IShengMapping.getShengById        : ==> Parameters: 1(Long)
c.g.w.s.IShengMapping.getShengById        : <==      Total: 1
1 河北省
1 河北省
```

图 5-31 执行多次 SQL 语句

不同的 SqlSession 对象有不同的缓存。

如果我们想实现不同 SqlSession 缓存中的数据共享，则需要使用二级缓存。

5.12.2 二级缓存

创建二级缓存时只需要在 SQL 映射文件中添加如下配置代码：

`<cache></cache>`

在 SQL 映射文件中使用<cache></cache>配置后具有如下两个特性。

（1）当为 select 语句时：flushCache 属性默认为 false，表示语句被调用的任何时候都不会清空本地缓存和二级缓存，因为查询操作没有将数据更改，不需要刷新缓存中的数据。useCache 属性默认为 true，表示会将当前 select 语句的查询结果放入二级缓存。

（2）当为 insert、update、delete 语句时：flushCache 属性默认为 true，表示语句被调用的任何时候都会导致本地缓存和二级缓存被清空，因为数据被更改了。

我们先来看一看二级缓存为 Application 级的特性。

创建测试项目 cacheTest2。

SQL 映射文件的代码如下：

```
<mapper namespace="com.ghy.www.sqlmapping.IShengMapping">
    <cache></cache>
    <select id="getShengById" parameterType="long"
            resultType="com.ghy.www.entity.Sheng">
        select * from
        sheng where id=#{id}
    </select>

    <update id="updateUserinfoById">
        update sheng set
        shengname='xxx' where id=1
```

```
        </update>
    </mapper>
```

在 SQL 映射文件中使用<cache></cache>配置，以开启二级缓存。

实体类 Sheng 需要实现序列化接口（Serializable），部分代码如下：

```
public class Sheng implements Serializable {
```

控制层的代码如下：

```java
@RequestMapping("test2")
public void test2(HttpServletRequest request, HttpServletResponse response) {
    SqlSession sqlSession1 = sqlSessionTemplate.getSqlSessionFactory().openSession();
    SqlSession sqlSession2 = sqlSessionTemplate.getSqlSessionFactory().openSession();

    Sheng sheng1 = sqlSession1.selectOne("getShengById", 1L);
    sqlSession1.commit();
    sqlSession1.close();
    //注意：在查询完 sheng1 之后要将 sqlSession1 进行提交和关闭
    //目的是将 sheng1 对象放入二级缓存，以准备让新的 sqlSession2 对象从二级缓存中复用该对象

    Sheng sheng2 = sqlSession2.selectOne("getShengById", 1L);
    System.out.println(sheng1.getId() + " " + sheng1.getShengname());
    System.out.println(sheng2.getId() + " " + sheng2.getShengname());
    sqlSession2.commit();
    sqlSession2.close();
}
```

程序运行结果如图 5-32 所示。

```
com.zaxxer.hikari.HikariDataSource      : MyHikariCP - Start completed.
com.ghy.www.sqlmapping.IShengMapping    : Cache Hit Ratio [com.ghy.www.sqlmapping.IShengMapping]: 0.0
c.g.w.s.IShengMapping.getShengById      : ==>  Preparing: select * from sheng where id=?
c.g.w.s.IShengMapping.getShengById      : ==> Parameters: 1(Long)
c.g.w.s.IShengMapping.getShengById      : <==      Total: 1
com.ghy.www.sqlmapping.IShengMapping    : Cache Hit Ratio [com.ghy.www.sqlmapping.IShengMapping]: 0.5
1 河北省
1 河北省
```

图 5-32　缓存中的 Sheng 对象被复用

程序运行结果体现出不同的 SqlSession 可以共享二级缓存中实体类的数据。

此示例中的每个 selectOne()方法会返回不同的 Sheng 对象，但不同的 Sheng 对象中的属性值是相同的。

如果返回相同的 Sheng 对象，则在不同的线程中使用不同的 SqlSession 操作同一个 Sheng 对象时有可能会发生值覆盖的情况，从而出现非线程安全。

5.12.3　验证 update 语句具有清除二级缓存的特性

使用<cache></cache>配置后，update 语句具有清除二级缓存的特性。

SQL 映射文件的代码如下：

```xml
<mapper namespace="com.ghy.www.sqlmapping.IShengMapping">
    <cache></cache>
```

```xml
<select id="getShengById" parameterType="long"
        resultType="com.ghy.www.entity.Sheng">
    select * from
    sheng where id=#{id}
</select>

<update id="updateUserinfoById">
    update sheng set
    shengname='xxx' where id=1
</update>
</mapper>
```

Mapper 映射接口的代码如下：

```java
@Mapper
public interface IShengMapping {
    public Sheng getShengById(long id);
    public void updateUserinfoById();
}
```

运行类代码如下：

```java
@RequestMapping("test1")
public void test1(HttpServletRequest request, HttpServletResponse response) {
    {
        Sheng sheng = shengMapping.getShengById(1L);
        System.out.println(sheng.getId() + " " + sheng.getShengname());
    }

    shengMapping.updateUserinfoById();

    {
        Sheng sheng = shengMapping.getShengById(1L);
        System.out.println(sheng.getId() + " " + sheng.getShengname());
    }
}
```

程序运行结果如图 5-33 所示。

```
com.zaxxer.hikari.HikariDataSource      : MyHikariCP - Start completed.
com.ghy.www.sqlmapping.IShengMapping    : Cache Hit Ratio [com.ghy.www.sqlmapping.IShengMapping]: 0.0
c.g.w.s.IShengMapping.getShengById      : ==>  Preparing: select * from sheng where id=?
c.g.w.s.IShengMapping.getShengById      : ==> Parameters: 1(Long)
c.g.w.s.IShengMapping.getShengById      : <==      Total: 1
1 河北省
c.g.w.s.I.updateUserinfoById            : ==>  Preparing: update sheng set shengname='xxx' where id=1
c.g.w.s.I.updateUserinfoById            : ==> Parameters:
c.g.w.s.I.updateUserinfoById            : <==    Updates: 1
com.ghy.www.sqlmapping.IShengMapping    : Cache Hit Ratio [com.ghy.www.sqlmapping.IShengMapping]: 0.0
c.g.w.s.IShengMapping.getShengById      : ==>  Preparing: select * from sheng where id=?
c.g.w.s.IShengMapping.getShengById      : ==> Parameters: 1(Long)
c.g.w.s.IShengMapping.getShengById      : <==      Total: 1
1 xxx
```

图 5-33 运行结果

执行 update 语句后，二级缓存中 id 为 1 的 Userinfo 对象被清除，再执行第 2 个 select 语句时重新发起了 SQL 语句到数据库。

5.13　Spring 事务传播特性

不同业务层有不同的事务处理方式。当不同业务层之间嵌套调用时，不同的事务类型会互相影响，以进行不同的事务处理，这种机制称为事务传播特性。

创建测试项目 PropagationTest。

5.13.1　事务传播特性 REQUIRED

REQUIRED 表示如果已经存在一个事务，就加入这个事务中；如果当前没有事务，就新建一个事务。这是默认的事务传播设置。

测试代码如下：

```
@Service
public class UserinfoService1 {
    @Autowired
    private IUserinfoMapping userinfoMapping;

    @Transactional(propagation = Propagation.REQUIRED) //新建一个事务 A
    public void saveServiceMethod1() {
        Userinfo userinfo1 = new Userinfo();
        userinfo1.setUsername("中国1");
        userinfoMapping.insert(userinfo1);
        saveServiceMethod2();
    }

    @Transactional(propagation = Propagation.REQUIRED) //加入事务 A 中
    public void saveServiceMethod2() {
        Userinfo userinfo2 = new Userinfo();
        userinfo2.setUsername("中国2");
        userinfoMapping.insert(userinfo2);

        Userinfo userinfo3 = new Userinfo();
        userinfo3.setUsername(
            "中国2中国2中国2中国2中国2中国2中国2中国2中国2中国2中国2中国2中国2中国2中国2中国2中国2中国2中国2中国2中国2中国2中国2中国2中国2中国2");
        userinfoMapping.insert(userinfo3);
    }
}
```

控制层的代码如下：

```
@Controller
public class TestController1 {
    @Autowired
    private UserinfoService1 userinfoService;

    @RequestMapping("test1")
    public void test1(HttpServletRequest request, HttpServletResponse response) {
        userinfoService.saveServiceMethod1();
    }
}
```

程序运行后数据表中没有添加任何一条记录，两个业务使用相同的事务 A，事务 A 整体回滚。

5.13.2 事务传播特性 SUPPORTS

SUPPORTS 表示如果已经存在一个事务，就加入这个事务中；如果当前没有事务，就以非事务方式执行。

测试代码如下：

```
@Service
public class UserinfoService2 {
    @Autowired
    private IUserinfoMapping userinfoMapping;

    public void saveServiceMethod1() { //此方法无事务
        Userinfo userinfo1 = new Userinfo();
        userinfo1.setUsername("中国1");
        userinfoMapping.insert(userinfo1);
        saveServiceMethod2();
    }

    @Transactional(propagation = Propagation.SUPPORTS) //以非事务方式执行
    public void saveServiceMethod2() {
        Userinfo userinfo2 = new Userinfo();
        userinfo2.setUsername("中国2");
        userinfoMapping.insert(userinfo2);

        Userinfo userinfo3 = new Userinfo();
        userinfo3.setUsername(
                "中国2中国2中国2中国2中国2中国2中国2中国2中国2中国2中国2中国2中国2中国2中国2中国2中国2中国2中国2中国2中国2中国2中国2中国2中国2中国2中国2");
        userinfoMapping.insert(userinfo3);
    }
}
```

控制层的代码如下：

```
@Controller
public class TestController2 {
    @Autowired
    private UserinfoService2 userinfoService;

    @RequestMapping("test2")
    public void test1(HttpServletRequest request, HttpServletResponse response) {
        userinfoService.saveServiceMethod1();
    }
}
```

程序运行后数据表中添加了两条记录，分别是"中国1"和"中国2"。

5.13.3 事务传播特性 MANDATORY

MANDATORY 表示如果已经存在一个事务，就加入这个事务中；如果当前没有事务，就抛出异常。

测试代码如下：

```java
@Service
public class UserinfoService3 {
    @Autowired
    private IUserinfoMapping userinfoMapping;

    @Transactional(propagation = Propagation.MANDATORY) //此方法无事务
    public void saveServiceMethod1() {
        Userinfo userinfo1 = new Userinfo();
        userinfo1.setUsername("中国1");
        userinfoMapping.insert(userinfo1);
    }
}
```

控制层的代码如下：

```java
@Controller
public class TestController3 {
    @Autowired
    private UserinfoService3 userinfoService;

    @RequestMapping("test3")
    public void test1(HttpServletRequest request, HttpServletResponse response) {
        userinfoService.saveServiceMethod1();
    }
}
```

程序运行后控制台输出异常的信息如下：

```
No existing transaction found for transaction marked with propagation 'mandatory'
```

5.13.4 事务传播特性 REQUIRES_NEW

REQUIRES_NEW 表示新建事务，如果已经存在一个事务，就把这个事务挂起并创建新的事务。

1. 测试 1

测试代码 1 如下：

```java
@Service
public class UserinfoService4_1 {
    @Autowired
    private IUserinfoMapping userinfoMapping;
    @Autowired
    private UserinfoService4_2 userinfoService4_2;

    @Transactional(propagation = Propagation.REQUIRED) //新建事务 A
    public void saveServiceMethod1() {
        Userinfo userinfo1 = new Userinfo();
        userinfo1.setUsername("中国1");
        userinfoMapping.insert(userinfo1);
        try {
            userinfoService4_2.saveServiceMethod2();
        } catch (Exception e) {
            e.printStackTrace();
        }
```

        }
    }

测试代码 2 如下：

```
@Service
public class UserinfoService4_2 {
    @Autowired
    private IUserinfoMapping userinfoMapping;

    @Transactional(propagation = Propagation.REQUIRES_NEW) //将事务 A 挂起，再新建事务 B
    public void saveServiceMethod2() {
        Userinfo userinfo2 = new Userinfo();
        userinfo2.setUsername("中国 2");
        userinfoMapping.insert(userinfo2);

        Userinfo userinfo3 = new Userinfo();
        userinfo3.setUsername(
            "中国 3 中国 3 中国 3 中国 3 中国 3 中国 3 中国 3 中国 3 中国 3 中国 3 中国 3 中国 3 中国 3 中国 3 中国 3 中国 3 中国 3 中国 3 中国 3 中国 3 中国 3 中国 3 中国 3 中国 3 中国 3 中国 3 中国 3 中国 3 中国 3 中国 3 中国 3 中国 3 中国 3 中国 3 中国 3 中国 3 中国 3 中国 3");
        userinfoMapping.insert(userinfo3);
    }
}
```

控制层的代码如下：

```
@Controller
public class TestController4 {
    @Autowired
    private UserinfoService4_1 userinfoService;

    @RequestMapping("test4")
    public void test1(HttpServletRequest request, HttpServletResponse response) {
        userinfoService.saveServiceMethod1();
    }
}
```

程序运行后，在数据表 userinfo 中只添加了一条 "中国 1" 记录，其他两条记录由于在事务 B 中被回滚了，因此没有被插入。

2. 测试 2

测试代码 1 如下：

```
@Service
public class UserinfoService5_1 {
    @Autowired
    private IUserinfoMapping userinfoMapping;

    @Autowired
    private UserinfoService5_2 userinfoService5_2;

    @Transactional(propagation = Propagation.REQUIRED) //新建事务 A
    public void saveServiceMethod1() {
        Userinfo userinfo1 = new Userinfo();
        userinfo1.setUsername("中国 1");
        userinfoMapping.insert(userinfo1);
```

```
            userinfoService5_2.saveServiceMethod2();

            Userinfo userinfo3 = new Userinfo();
            userinfo3.setUsername(
                "中国3中国3中国3中国3中国3中国3中国3中国3中国3中国3中国3中国3中国3
3中国3中国3中国3中国3中国3中国3中国3中国3中国3中国3中国3中国3中国3中
国3中国3中国3中国3中国3中国3中国3中国3中国3中国3中国3中国3中国3");
            userinfoMapping.insert(userinfo3);
        }
    }
```

测试代码 2 如下：

```
@Service
public class UserinfoService5_2 {
    @Autowired
    private IUserinfoMapping userinfoMapping;

    @Transactional(propagation = Propagation.REQUIRES_NEW) //将事务A挂起，再新建事务B
    public void saveServiceMethod2() {
        Userinfo userinfo2 = new Userinfo();
        userinfo2.setUsername("中国2");
        userinfoMapping.insert(userinfo2);
    }
}
```

控制层的代码如下：

```
@Controller
public class TestController5 {
    @Autowired
    private UserinfoService5_1 userinfoService;

    @RequestMapping("test5")
    public void test1(HttpServletRequest request, HttpServletResponse response) {
        userinfoService.saveServiceMethod1();
    }
}
```

程序运行后，在数据表 userinfo 中只添加了一条 "中国 2" 记录，其他两条记录由于在事务 A 中被回滚了，因此没有被插入。

5.13.5　事务传播特性 NOT_SUPPORTED

如 5.13.2 节所述，SUPPORTS 表示如果已经存在一个事务，就加入这个事务中；如果当前没有事务，就以非事务方式执行。

NOT_SUPPORTED 表示如果已经存在一个事务，就把这个事务挂起，并以非事务方式执行操作。

测试代码 1 如下：

```
@Service
public class UserinfoService6_1 {
    @Autowired
    private IUserinfoMapping userinfoMapping;
```

```
    @Autowired
    private UserinfoService6_2 userinfoService6_2;

    @Transactional(propagation = Propagation.REQUIRED)  //新建事务A
    public void saveServiceMethod1() {
        Userinfo userinfo1 = new Userinfo();
        userinfo1.setUsername("中国1");
        userinfoMapping.insert(userinfo1);
        try {
            userinfoService6_2.saveServiceMethod2();
        } catch (Exception e) {
            e.printStackTrace();
        }
    }
}
```

测试代码2如下：

```
@Service
public class UserinfoService6_2 {
    @Autowired
    private IUserinfoMapping userinfoMapping;

    @Transactional(propagation = Propagation.NOT_SUPPORTED)  //将事务A挂起，再以非事务方式执行
    public void saveServiceMethod2() {
        Userinfo userinfo2 = new Userinfo();
        userinfo2.setUsername("中国2");
        userinfoMapping.insert(userinfo2);

        Userinfo userinfo3 = new Userinfo();
        userinfo3.setUsername(
                "中国3中国3中国3中国3中国3中国3中国3中国3中国3中国3中国3中国3中国3中国3中国3中国3中国3中国3中国3中国3中国3中国3中国3中国3中国3中国3中国3中国3中国3中国3中国3中国3中国3中国3中国3中国3中国3中国3中国3中国3中国3中国3中国3中国3中国3中国3中国3中国3中国3");
        userinfoMapping.insert(userinfo3);
    }
}
```

控制层的代码如下：

```
@Controller
public class TestController6 {
    @Autowired
    private UserinfoService6_1 userinfoService;

    @RequestMapping("test6")
    public void test1(HttpServletRequest request, HttpServletResponse response) {
        userinfoService.saveServiceMethod1();
    }
}
```

程序运行后，在数据表userinfo中只添加了两条记录，分别是"中国1"和"中国2"记录，剩余一条记录由于出现错误，因此没有被插入。

5.13.6 事务传播特性NEVER

NEVER表示如果已经存在一个事务，就抛出异常；如果当前没有事务，就以非事务方式执行。

1. 以非事务方式执行

测试代码如下：

```java
@Service
public class UserinfoService7 {
    @Autowired
    private IUserinfoMapping userinfoMapping;

    @Transactional(propagation = Propagation.NEVER) //以非事务方式执行
    public void saveServiceMethod1() {
        Userinfo userinfo1 = new Userinfo();
        userinfo1.setUsername("中国1");
        userinfoMapping.insert(userinfo1);

        Userinfo userinfo2 = new Userinfo();
        userinfo2.setUsername("中国2");
        userinfoMapping.insert(userinfo2);

        Userinfo userinfo3 = new Userinfo();
        userinfo3.setUsername(
            "中国3 中国3 中国3 中国3 中国3 中国3 中国3 中国3 中国3 中国3 中国3 中国3 中国3 中国3 中国3 中国3 中国3 中国3 中国3 中国3 中国3 中国3 中国3 中国3 中国3 中国3 中国3 中国3 中国3 中国3 中国3 中国3 中国3 中国3 中国3 中国3 中国3 中国3");
        userinfoMapping.insert(userinfo3);
    }
}
```

控制层的代码如下：

```java
@Controller
public class TestController7 {
    @Autowired
    private UserinfoService7 userinfoService;

    @RequestMapping("test7")
    public void test1(HttpServletRequest request, HttpServletResponse response) {
        userinfoService.saveServiceMethod1();
    }
}
```

程序运行后，在数据表 userinfo 中只添加了两条记录，分别是"中国1"和"中国2"记录，剩余一条记录由于出现错误，因此没有被插入。

2. 存在事务则抛出异常

业务类代码1如下：

```java
@Service
public class UserinfoService8_1 {
    @Autowired
    private IUserinfoMapping userinfoMapping;

    @Autowired
    private UserinfoService8_2 userinfoService8_2;

    @Transactional(propagation = Propagation.REQUIRED)
```

```
    public void insertServiceMethod1() {
        Userinfo userinfo1 = new Userinfo();
        userinfo1.setUsername("1");
        userinfo1.setPassword("11");
        userinfoMapping.insert(userinfo1);

        userinfoService8_2.insertServiceMethod2();
    }
}
```

业务类代码 2 如下：

```
@Service
public class UserinfoService8_2 {
    @Autowired
    private IUserinfoMapping userinfoMapping;

    @Transactional(propagation = Propagation.NEVER)
    public void insertServiceMethod2() {
        Userinfo userinfo2 = new Userinfo();
        userinfo2.setUsername("2");
        userinfo2.setPassword("22");
        userinfoMapping.insert(userinfo2);
    }
}
```

控制层的代码如下：

```
@Controller
public class TestController8 {
    @Autowired
    private UserinfoService8_1 userinfoService;

    @RequestMapping("test8")
    public void test1(HttpServletRequest request, HttpServletResponse response) {
        userinfoService.insertServiceMethod1();
    }
}
```

程序运行后出现异常的信息如下：

```
Exception in thread "main" org.springframework.transaction.IllegalTransactionStateException: Existing transaction found for transaction marked with propagation 'never'
```

数据表中没有添加任何一条记录。

5.13.7 事务传播特性 NESTED

NESTED 表示创建当前事务的子事务。

测试代码如下：

```
@Service
public class UserinfoService9_1 {
    @Autowired
    private IUserinfoMapping userinfoMapping;

    @Autowired
    private UserinfoService9_2 userinfoService9_2;
```

```java
@Transactional(propagation = Propagation.REQUIRED) //创建事务 A
public void saveServiceMethod1() {
    Userinfo userinfo1 = new Userinfo();
    userinfo1.setUsername("中国 1");
    userinfoMapping.insert(userinfo1);

    userinfoService9_2.saveServiceMethod2();

    Userinfo userinfo3 = new Userinfo();
    userinfo3.setUsername(
        "中国 3 中国 3 中国 3 中国 3 中国 3 中国 3 中国 3 中国 3 中国 3 中国 3 中国 3 中国 3 中国 3 中国 3 中国 3 中国 3 中国 3 中国 3 中国 3 中国 3 中国 3 中国 3 中国 3 中国 3 中国 3 中国 3 中国 3 中国 3 中国 3 中国 3 中国 3 中国 3 中国 3 中国 3");
    userinfoMapping.insert(userinfo3);
}
```

测试代码如下：

```java
@Service
public class UserinfoService9_2 {
    @Autowired
    private IUserinfoMapping userinfoMapping;

    @Transactional(propagation = Propagation.NESTED) //创建事务 A 的子事务 AA
    public void saveServiceMethod2() {
        Userinfo userinfo2 = new Userinfo();
        userinfo2.setUsername("中国 2");
        userinfoMapping.insert(userinfo2);
    }
}
```

控制层的代码如下：

```java
@Controller
public class TestController9 {
    @Autowired
    private UserinfoService9_1 userinfoService;

    @RequestMapping("test9")
    public void test1(HttpServletRequest request, HttpServletResponse response) {
        userinfoService.saveServiceMethod1();
    }
}
```

程序运行后，没有在数据表中添加任何一条新的记录，而是整体回滚。

Propagation.REQUIRES_NEW 和 Propagation.NESTED 的最大区别在于，Propagation.REQUIRES_NEW 完全是一个新的事务，而 Propagation.NESTED 则是外部事务的子事务。如果外部事务被提交，子事务也会被提交，这个规则同样适用于回滚。

5.13.8　事务传播特性总结

（1）REQUIRED 表示如果已经存在一个事务，就加入这个事务中；如果当前没有事务，就新建一个事务。这是默认的事务传播设置。

解释：如果马路上有车就搭车，如果马路上没有车就自己造车。

（2）SUPPORTS 表示如果已经存在一个事务，就加入这个事务中；如果当前没有事务，就以非事务方式执行。

解释：如果马路上有车就搭车，如果马路上没有车就步行。

（3）MANDATORY 表示如果已经存在一个事务，就加入这个事务中；如果当前没有事务，就抛出异常。

解释：如果马路上有车就搭车，如果马路上没有车就愤怒地爆炸。

（4）REQUIRES_NEW 表示新建事务，如果已经存在一个事务，就把这个事务挂起并创建新的事务。

解释：即使马路上有车也不搭车，仍然自己造车。

（5）NOT_SUPPORTED 表示如果已经存在一个事务，就把这个事务挂起，并以非事务方式执行。

解释：即使马路上有车也不搭车，而选择步行。

（6）NEVER 表示如果已经存在一个事务，就抛出异常；如果当前没有事务，以非事务方式执行。

解释：如果马路上有车就愤怒地爆炸；如果马路上没有车就步行。

（7）NESTED 表示创建当前事务的子事务。

解释：水和鱼的关系，鱼（子事务）没有了但不影响水（父事务），但水（父事务）没有了则影响鱼（子事务）。

第 6 章　模板引擎 FreeMarker 和 Thymeleaf 的使用

JSP 是 Java Web 开发中比较古老的视图层（View）技术，也是 Java Web 程序员比较熟悉的技术，它的功能丰富，包括以下几点。

（1）支持在 JSP 文件中编写 Java 代码，虽然这不符合 MVC 标准。

（2）JSP 文件中可以使用 JSTL 标签来实现逻辑处理。

（3）支持丰富的第三方 JSP 标签库。

（4）支持 EL 表达式语言。

（5）JSP 是官方的标准，用户群比较广泛。

（6）性能良好。

JSP 文件会被 Web 容器编译成 Servlet 的 Class 文件并执行，这是 JSP 技术的原理。

既然 JSP 技术有这么多的优点，那么为什么现在新的软件项目不再使用 JSP 作为视图层了呢？理由有很多，这里总结为以下两点。

（1）绑定 Web 环境，运行环境受限。

（2）可以在<%%>中写 Java 代码，破坏了 MVC 分层架构。

这两个缺点是最为致命的。我们先来看一看第 1 点，在某些情况下，生成的视图层并不在 Web 环境中进行展示，这时 JSP 就无法满足这样的需求，因为它的运行环境必须在 Web 中。我们再来看一看第 2 点，MVC 开发方式的主要作用就是实现代码的分层，便于软件的后期维护与升级，但在 JSP 页面中可以随意在<%%>中编写 Java 代码，这就破坏了 MVC 分层结构，项目整体清晰度遭到破坏，严重时会影响工作效率。现在大多数新的软件项目不再使用 JSP 作为视图层了。

那么，新项目的视图层采用什么技术呢？模板引擎。大多数模板引擎支持在 Web 或非 Web 环境中运行，不绑定运行环境，并且在生成视图层时，不允许在视图层中使用 Java 代码，这就实现了严格的 MVC 分层。模板引擎性能优良，支持缓存，可以使用内置函数以及自定义函数，有些功能是 JSP 所不具有的。

业界比较著名的模板引擎有 FreeMarker 和 Thymeleaf，本章的目的就是学习这两种模板引擎的使用。

模板引擎也被称为标签库，因为模板引擎中提供很多标签来生成最终的结果。

6.1 使用 FreeMarker 模板引擎

FreeMarker 是一个基于 Java 语言的开源的模版引擎，它是基于文本的模板输出工具。使用 FreeMarker 可以生成任意字符信息，包括 HTML、XML 和 Java 文件等文本资源。

FreeMarker 在软件项目中常常用来生成 HTML Web 页面。

6.1.1 FreeMarker 的优势

FreeMarker 的优势可以总结为以下 5 点。

（1）分离表现层和业务逻辑。使用 JSP 开发项目的过程中，可以在页面中使用<%%>来创建大量业务逻辑代码，包括 Java 代码、JavaScript 脚本、CSS 样式以及 HTML 代码，使页面内容混乱，使后期的代码维护变得非常困难。而 FreeMarker 不支持 Java 脚本代码。FreeMarker 的原理是"模板+数据模型=输出"。模板只负责数据在页面中的展现，不涉及任何逻辑，所有的逻辑都是由数据模型来处理的，用户看到的最终输出是模板和数据模型合并后的结果。

（2）减少 CPU 占用。在以往的开发中，使用的都是 JSP 页面来展现数据，即所谓的表现层。JSP 在第一次执行的时候需要转换成 Servlet 类，开发阶段进行代码调试时，需要频繁地修改 JSP，每次修改后都要编译和转换，这个过程大量耗时，会影响运行效率。而相对于 JSP 来说，FreeMarker 模板技术不存在编译和转换的问题，所以就不会在编译和转换上占用 CPU 资源。

（3）分工更加明确。以往用 JSP 展现数据时，程序员不熟悉界面设计技术，反之界面开发人员也不熟悉程序语言，如果他们一起编辑某一个文件，稍有不慎就可能会将页面元素删除或去掉某段程序代码，使页面变形或程序出错，这时就需要双方相互沟通协作，以解决出现的问题，从而减慢了项目开发进度，增加了额外的沟通成本。而使用 FreeMarker 后，作为界面开发人员，只需要专心创建 HTML 文件、图像以及 Web 页面的可视化内容，不需要考虑数据的来源，而程序开发人员则只需要专注于系统实现，负责为页面准备要展现的数据。

（4）不依赖于 Web 环境。FreeMarker 与 JSP 不同，JSP 需要依赖 Web 容器，而 FreeMarker 与 Web 容器无关，可以运行在非 Web 环境中。FreeMarker 不仅可以用作表现层的实现技术，还可以用于生成 XML、JSP 等文本资源。

（5）执行效率高。FreeMarker 的执行效率比 JSP 快一些，因为它支持缓存的处理。

6.1.2 FreeMarker 的输出

FreeMarker 执行流程如图 6-1 所示。

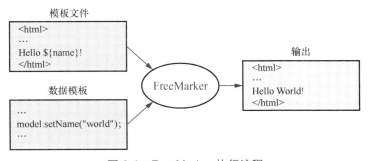

图 6-1 FreeMarker 执行流程

想要显示最终的用户界面（UI），可以使用 FreeMarker 来生成，但 FreeMarker 依赖两个必要元素。

（1）模板文件（Template）。

（2）数据模型（Java object）。

也就是：

模板 + 数据模型 = 输出

6.1.3　整合 Spring Boot 与输出常见数据类型

创建测试项目 simpleUse。最终的项目结构如图 6-2 所示。

图 6-2　项目结构

创建控制层，代码如下：

```
package com.ghy.www.controller;

import org.springframework.stereotype.Controller;
import org.springframework.ui.Model;
import org.springframework.web.bind.annotation.RequestMapping;

import javax.servlet.http.HttpServletRequest;
import javax.servlet.http.HttpServletResponse;
import java.math.BigDecimal;

@Controller
public class TestController {
    @RequestMapping("test")
```

```java
        public String test1(Model model, HttpServletRequest request, HttpServletResponse response) {
            model.addAttribute("charKey", 'z');
            model.addAttribute("byteKey", (byte) 123);
            model.addAttribute("intKey1", 124);
            model.addAttribute("intKey2", 1234567890);

            // float 是以科学计数法进行保存的,精度 7 位
            // 科学计算法值:1.23456789123456789Ex 精度前 7 位是指 1.234567
            float float1 = 0.000000012345678912345678F;
            // 科学计算法值:1.23456789123456789Ex
            float float2 = 0.000123456789123456789F;
            // 科学计算法值:1.23456789012345678912345678912345678Ex
            float float3 = 1234567890123456789.123456789123456789F;
            // 科学计算法值:1.2345126456789123456789Ex
            float float4 = 12345.126456789123456789F;
            // 科学计算法值:1.23456789123456789Ex
            float float5 = 123456789123456789F;

            model.addAttribute("floatKey1", float1);
            model.addAttribute("floatKey2", float2);
            model.addAttribute("floatKey3", float3);
            model.addAttribute("floatKey4", float4);
            model.addAttribute("floatKey5", float5);

            // double 是以科学计数法进行保存的,精度 16 位

            // 科学计算法值:1.234567890123456189Ex
            double double1 = 0.0000000123456789012345618D;
            // 科学计算法值:1.234567890123456189Ex
            double double2 = 0.000123456789012345618D;
            // 科学计算法值:1.23456789123456263123456789123456789Ex
            double double3 = 123456789123456263.123456789123456789D;
            // 科学计算法值:1.2345678901234568891 23456789Ex
            double double4 = 12345.678901234568891 23456789D;
            // 科学计算法值:1.2345678901234568891 23456789Ex
            double double5 = 12345678901234568891 23456789D;

            model.addAttribute("doubleKey1", double1);
            model.addAttribute("doubleKey2", double2);
            model.addAttribute("doubleKey3", double3);
            model.addAttribute("doubleKey4", double4);
            model.addAttribute("doubleKey5", double5);

            model.addAttribute("BigDecimalKey1", new BigDecimal("123456789123456263.123456789123456789"));
            model.addAttribute("BigDecimalKey2", new BigDecimal("123456789123456263123456789123456789"));

            model.addAttribute("StringKey", "我是中国人");

            return "test";
        }
    }
```

创建运行类,代码如下:

```
package com.ghy.www;

import org.springframework.boot.SpringApplication;
```

```
import org.springframework.boot.autoconfigure.SpringBootApplication;

@SpringBootApplication
public class Application {
    public static void main(String[] args) {
        SpringApplication.run(Application.class, args);
    }
}
```

标签文件 test.ftlh 的代码如下：

```
<!DOCTYPE html>
<html>
    <head>
        <meta charset="UTF-8">
        <title>Insert title here</title>
        <script src="js/myjs.js"></script>
        <link href="css/mycss.css" rel="stylesheet"/>
    </head>
    <body>
        <span class="myColorSize">测试样式的文字</span>
        <br/>
        <br/>
        <#-- 我是注释，我不显示，我不在 response 中返回-->
        charKey:${charKey}<br/>
        byteKey:${byteKey}<br/>
        intKey1:${intKey1}<br/>
        intKey2:${intKey2}<br/>
        <br/>
        floatKey1:${floatKey1}<br/>
        floatKey2:${floatKey2}<br/>
        floatKey3:${floatKey3}<br/>
        floatKey4:${floatKey4}<br/>
        floatKey5:${floatKey5}<br/>
        <br/>
        doubleKey1:${doubleKey1}<br/>
        doubleKey2:${doubleKey2}<br/>
        doubleKey3:${doubleKey3}<br/>
        doubleKey4:${doubleKey4}<br/>
        doubleKey5:${doubleKey5}<br/>
        <br/>
        BigDecimalKey1:${BigDecimalKey1}<br/>
        BigDecimalKey2:${BigDecimalKey2}<br/>
        StringKey:${StringKey}<br/>
    </body>
</html>
```

FreeMarker 标签文件扩展名是*.ftl 或*.ftlh，FTL 全称是"FreeMarker Template Language"。ftl 文件中的动态标签是 FreeMarker 的指令，它们是不会在输出中打印的，这些标签以#开头。

FreeMarker 模板文件主要由如下 4 个部分组成。

（1）文本：直接输出的部分。
（2）注释：<#-- ... -->格式部分，不会输出。
（3）插值：${...}格式部分，将使用数据模型中的部分替代输出。
（4）FTL 指令：具有动态处理能力。

FreeMarker 和 EL 表达式一样，使用${...}输出真实的值来替换大括号内的表达式。

6.1 使用 FreeMarker 模板引擎

注释和 HTML 的注释也很相似，使用<#-- text -->来标识，FTL 中的注释不会出现在输出中，因为 FreeMarker 会跳过它们。

配置文件 application.yml 的代码如下：

```yaml
spring:
  freemarker:
    charset: UTF-8
    content-type: text/html; charset=utf-8
    template-loader-path: classpath:/templates
    settings:
      template_exception_handler: html_debug

server:
  servlet:
    encoding:
      charset: utf-8
      enabled: true
      force: true
```

文件 mycss.css 的代码如下：

```css
.myColorSize {
    color: red;
    font-size: 30px;
}
```

文件 myjs.js 的代码如下：

```js
setTimeout(function () {
    alert("自动弹出");
}, 3000)
```

配置文件 pom.xml 的核心代码如下：

```xml
<parent>
    <groupId>org.springframework.boot</groupId>
    <artifactId>spring-boot-starter-parent</artifactId>
    <version>2.3.4.RELEASE</version>
</parent>

<properties>
    <project.build.sourceEncoding>UTF-8</project.build.sourceEncoding>
    <maven.compiler.source>1.8</maven.compiler.source>
    <maven.compiler.target>1.8</maven.compiler.target>
</properties>

<dependencies>
    <dependency>
        <groupId>junit</groupId>
        <artifactId>junit</artifactId>
        <version>4.11</version>
        <scope>test</scope>
    </dependency>

    <dependency>
        <groupId>org.springframework.boot</groupId>
        <artifactId>spring-boot-starter-web</artifactId>
    </dependency>
```

```xml
    <dependency>
        <groupId>org.springframework.boot</groupId>
        <artifactId>spring-boot-starter-freemarker</artifactId>
    </dependency>
</dependencies>
```

启动项目并运行 test 控制层，运行结果如图 6-3 所示。

图 6-3　运行结果

生成的 HTML 源代码如下：

```
<!DOCTYPE html>
<html>
    <head>
        <meta charset="UTF-8">
        <title>Insert title here</title>
        <script src="js/myjs.js"></script>
        <link href="css/mycss.css" rel="stylesheet"/>
    </head>
    <body>
        <span class="myColorSize">测试样式的文字</span>
        <br/>
        <br/>
        charKey:z<br/>
        byteKey:123<br/>
        intKey1:124<br/>
        intKey2:1,234,567,890<br/>
        <br/>
        floatKey1:0<br/>
        floatKey2:0<br/>
        floatKey3:1,234,567,939,550,609,410<br/>
        floatKey4:12,345.126<br/>
        floatKey5:123,456,790,519,087,104<br/>
        <br/>
        doubleKey1:0<br/>
        doubleKey2:0<br/>
```

```
        doubleKey3:123,456,789,123,456,256<br/>
        doubleKey4:12,345.679<br/>
        doubleKey5:1,234,567,890,123,456,900,000,000,000<br/>
        <br/>
        BigDecimalKey1:123,456,789,123,456,263.123<br/>
        BigDecimalKey2:123,456,789,123,456,263,123,456,789,123,456,789<br/>
        StringKey:我是中国人<br/>
    </body>
</html>
```

在源代码中没有生成注释。

6.1.4 输出布尔值

FreeMarker 输出布尔值时需要一些处理，在默认情况下不支持直接输出布尔值。

控制层的代码如下：

```
@RequestMapping("test2")
public String test2(Model model, HttpServletRequest request, HttpServletResponse response) {
    model.addAttribute("booleanValue1", true);
    model.addAttribute("booleanValue2", false);
    return "test2";
}
```

模板文件的代码如下：

```
<body>
    booleanValue1:${booleanValue1}<br/>
    booleanValue2:${booleanValue2}<br/>
</body>
```

程序运行结果出现异常，信息如下：

```
Can't convert boolean to string automatically, because the "boolean_format" setting was
"true,false", which is the legacy deprecated default, and we treat it as if no format was set.
This is the default configuration; you should provide the format explicitly for each place where
you print a boolean.

    ----
    Tip: Write something like myBool?string('yes', 'no') to specify boolean formatting in place.
    ----
    Tip: If you want "true"/"false" result as you are generating computer-language output (not
for direct human consumption), then use "?c", like ${myBool?c}. (If you always generate
computer-language output, then it's might be reasonable to set the "boolean_format" setting
to "c" instead.)
    ----
    Tip: If you need the same two values on most places, the programmers can set the
"boolean_format" setting to something like "yes,no". However, then it will be easy to unwillingly
format booleans like that.
```

要处理此异常，可以使用以下两种方式。

（1）自动转换。

（2）手动转换。

如果我们进行自动转换，需要使用如下 yml 配置代码：

```
spring:
  freemarker:
    charset: UTF-8
```

```
    content-type: text/html; charset=utf-8
    template-loader-path: classpath:/templates
    settings:
      template_exception_handler: html_debug
      boolean_format: 'c'
```

值'c'表示将布尔值转换成字符值。布尔值 true 会输出为 true，布尔值 false 会输出为 false。运行结果如图 6-4 所示。

如果我们进行手动转换，需要使用如下代码：

```
<body>
    booleanValue1:${booleanValue1?c}<br/>
    booleanValue2:${booleanValue2?c}<br/>
    <br/>
    <br/>
    booleanValue1:${booleanValue1?string('值是true','值是false')}<br/>
    booleanValue2:${booleanValue2?string('值是true','值是false')}<br/>
</body>
```

程序运行结果如图 6-5 所示。

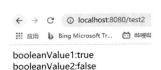

图 6-4　自动转换运行结果　　　　图 6-5　手动转换运行结果

6.1.5　输出 Date 数据类型

控制层的代码如下：

```
@RequestMapping("test4")
public String test4(Model model, HttpServletRequest request, HttpServletResponse response) {
    model.addAttribute("nowDate", new Date());
    return "test4";
}
```

模板文件的代码如下：

```
<body>
    nowDate=${nowDate}<br/>
</body>
```

程序运行后出现异常，信息如下：

```
Can't convert the date-like value to string because it isn't known if it's a date (no time part), time or date-time value.
The blamed expression:
==> nowDate  [in template "test4.ftlh" at line 8, column 19]


----
Tip: Use ?date, ?time, or ?datetime to tell FreeMarker the exact type.
----
Tip: If you need a particular format only once, use ?string(pattern), like ?string('dd.MM.yyyy HH:mm:ss'), to specify which fields to display.
```

异常信息提示不能直接将 Date 数据直接输出，而需要进行格式化处理。创建模板文件，代码如下：

```
<body>
    ${nowDate?date}<br/>
    ${nowDate?time}<br/>
    ${nowDate?datetime}<br/>
    ${nowDate?string('yyyy-MM-dd hh:mm:ss')}<br/>
    ${nowDate?string('yyyy年MM月dd日 hh时mm分ss秒')}<br/>
    ${nowDate?string('yyyy-MM-dd')}<br/>
    ${nowDate?string('hh:mm:ss')}<br/>
</body>
```

程序运行结果如图 6-6 所示。

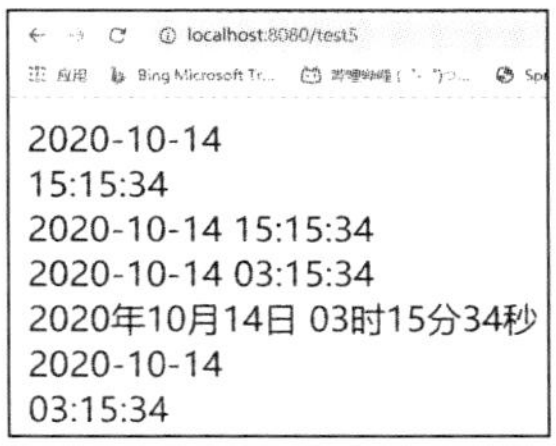

图 6-6 运行结果

6.1.6 循环集合中的数据

创建实体类，代码如下：

```
package com.ghy.www.entity;

import java.util.Date;

public class Userinfo {
    private int id;
    private String username;
    private String password;
    private int age;
    private Date insertDate;

    public Userinfo() {
    }

    public Userinfo(int id, String username, String password, int age, Date insertDate) {
        super();
        this.id = id;
        this.username = username;
        this.password = password;
        this.age = age;
        this.insertDate = insertDate;
    }

    public int getId() {
        return id;
    }

    public void setId(int id) {
        this.id = id;
    }

    public String getUsername() {
        return username;
    }

    public void setUsername(String username) {
        this.username = username;
    }
```

```java
    public String getPassword() {
        return password;
    }

    public void setPassword(String password) {
        this.password = password;
    }

    public int getAge() {
        return age;
    }

    public void setAge(int age) {
        this.age = age;
    }

    public Date getInsertDate() {
        return insertDate;
    }

    public void setInsertDate(Date insertDate) {
        this.insertDate = insertDate;
    }
}
```

控制层的代码如下：

```java
@RequestMapping("test6")
public String test6(Model model, HttpServletRequest request, HttpServletResponse response) {
    Userinfo userinfo1 = new Userinfo(1, "中国1", "中国人1", 11, new Date());
    Userinfo userinfo2 = new Userinfo(2, "中国2", "中国人2", 22, new Date());
    Userinfo userinfo3 = new Userinfo(3, "中国3", "中国人3", 33, new Date());

    String[] stringArray = new String[]{"我", "是", "美", "国", "人"};
    model.addAttribute("stringArray", stringArray);
    Userinfo[] userinfoArray = new Userinfo[]{userinfo1, userinfo2, userinfo3};
    model.addAttribute("userinfoArray", userinfoArray);

    List listString = new ArrayList();
    listString.add("中国人1");
    listString.add("中国人2");
    listString.add("中国人3");
    model.addAttribute("listString", listString);

    List listBean = new ArrayList();
    listBean.add(userinfo1);
    listBean.add(userinfo2);
    listBean.add(userinfo3);
    model.addAttribute("listBean", listBean);

    Set setString = new LinkedHashSet();
    setString.add("中国人1");
    setString.add("中国人2");
    setString.add("中国人3");
    model.addAttribute("setString", setString);

    Set setBean = new LinkedHashSet();
```

```
        setBean.add(userinfo1);
        setBean.add(userinfo2);
        setBean.add(userinfo3);
        model.addAttribute("setBean", setBean);

        Map mapString = new LinkedHashMap();
        mapString.put("key1", "中国人1");
        mapString.put("key2", "中国人2");
        mapString.put("key3", "中国人3");
        model.addAttribute("mapString", mapString);

        Map mapBean = new LinkedHashMap();
        mapBean.put("key1", userinfo1);
        mapBean.put("key2", userinfo2);
        mapBean.put("key3", userinfo3);
        model.addAttribute("mapBean", mapBean);

        return "test6";
}
```

模板文件的代码如下:

```
<body>
    stringArray:<br/>
    <#list stringArray as eachString>
        ${eachString}<br/>
    </#list>
    <br/>
    userinfoArray:<br/>
    <#list userinfoArray as userinfo>
        ${userinfo.id}__
        ${userinfo.username}__
        ${userinfo.password}__
        ${userinfo.age}__
        ${userinfo.insertDate?datetime}__
        ${userinfo.insertDate?string("yyyy-MM-dd")}__
        ${userinfo.insertDate?string("HH:mm:ss")}
        <br/>
    </#list>
    <br/>
    listString:<br/>
    <#list listString as eachString>
        ${eachString}<br/>
    </#list>
    <br/>
    listBean:<br/>
    <#list listBean as userinfo>
        ${userinfo.id}__
        ${userinfo.username}__
        ${userinfo.password}__
        ${userinfo.age}__
        ${userinfo.insertDate?datetime}__
        ${userinfo.insertDate?string("yyyy-MM-dd")}__
        ${userinfo.insertDate?string("HH:mm:ss")}
        <br/>
    </#list>
    <br/>
    list_use_#items:<br/>
```

```
        <#list listBean>
            <#items as userinfo>
                ${userinfo.id}__
                ${userinfo.username}__
                ${userinfo.password}__
                ${userinfo.age}__
                ${userinfo.insertDate?datetime}__
                ${userinfo.insertDate?string("yyyy-MM-dd")}__
                ${userinfo.insertDate?string("HH:mm:ss")}
                <br/>
            </#items>
        </#list>
        <br/>
        setString:<br/>
        <#list setString as eachString>
            ${eachString}<br/>
        </#list>
        <br/>
        setBean:<br/>
        <#list setBean as userinfo>
            ${userinfo.id}__
            ${userinfo.username}__
            ${userinfo.password}__
            ${userinfo.age}__
            ${userinfo.insertDate?datetime}__
            ${userinfo.insertDate?string("yyyy-MM-dd")}__
            ${userinfo.insertDate?string("HH:mm:ss")}
            <br/>
        </#list>
        <br/>
        mapString:<br/>
        <#list mapString?keys as key>
            ${key}__${mapString[key]}<br/>
        </#list>
        <br/>
        mapBean:<br/>
        <#list mapBean?keys as key>
            ${key}
            ${mapBean[key].id}__
            ${mapBean[key].username}__
            ${mapBean[key].password}__
            ${mapBean[key].age}__
            ${mapBean[key].insertDate?datetime}__
            ${mapBean[key].insertDate?string("yyyy-MM-dd")}__
            ${mapBean[key].insertDate?string("HH:mm:ss")}
            <br/>
        </#list>
    </body>
```

在循环 List 数据类型时，可以使用如下两种写法。

（1）`<#list listBean as userinfo></#list>`。

（2）`<#list listBean><#items as userinfo></#items></#list>`。

我们先来看一看第 1 种写法：

```
<table border="1">
    <#list listBean as eachUserinfo>
        <tr>
```

```
            <td>${eachUserinfo.id}</td>
            <td>${eachUserinfo.username}</td>
            <td>${eachUserinfo.password}</td>
            <td>${eachUserinfo.age}</td>
            <td>${eachUserinfo.insertdate?string('yyyy年MM月dd日 HH时mm分ss秒')}</td>
        </tr>
    </#list>
</table>
```

上面的代码可以生成 table、tr 和 td。<table>标签中有<#list></#list>写法会使代码整体不具有紧凑性，比如 tr 的父标签有多级时，在复制这段代码时还要额外注意不要遗漏父标签，并且还要注意父标签的双标签的<></>的匹配问题，这样会降低开发效率。因此，当使用 FreeMarker 循环生成具有父子关系的标签时，建议使用第 2 种写法，代码如下：

```
<#list listBean>
    <table border="1">
        <#items as eachUserinfo>
            <tr>
                <td>${eachUserinfo.id}</td>
                <td>${eachUserinfo.username}</td>
                <td>${eachUserinfo.password}</td>
                <td>${eachUserinfo.age}</td>
                <td>${eachUserinfo.insertdate?string('yyyy年MM月dd日 HH时mm分ss秒')}</td>
            </tr>
        </#items>
    </table>
</#list>
```

第 2 种写法将父标签和子标签全部存放在<#list></#list>中，因此，直接复制<#list></#list>代码段就会包含父标签和子标签，提高了开发效率。

6.1.7 使用 if 命令实现判断

控制层的代码如下：

```
@RequestMapping("test7")
public String test7(Model model, HttpServletRequest request, HttpServletResponse response) {
    model.addAttribute("username1", "高洪岩");
    model.addAttribute("username2", "岩洪高");
    model.addAttribute("age", 100);
    return "test7";
}
```

模板文件的代码如下：

```
<body>
    username:<#if username1=="高洪岩">是高洪岩</#if><br/>
    username:<#if username2=="高洪岩">是高洪岩<#else>不是高洪岩</#if><br/>
    age:<#if age==97>==97
<#elseif age==98>==987
<#elseif age==99>==99
<#else>==100
    </#if>
    <br/>
</body>
```

程序运行结果如图 6-7 所示。

username:是高洪岩
username:不是高洪岩
age:==100

图 6-7 运行结果

6.1.8 判断 List 的 size 值是否为 0

控制层的代码如下：

```
@RequestMapping("test8")
public String test8(Model model, HttpServletRequest request, HttpServletResponse response) {
    List listString = new ArrayList();
    model.addAttribute("listString", listString);
    return "test8";
}
```

模板文件的代码如下：

```
<body>
    <#if listString?size==0>
        listString 的 size 值是 0
    <#else>
        listString 的 size 值不是 0
    </#if>
</body>
```

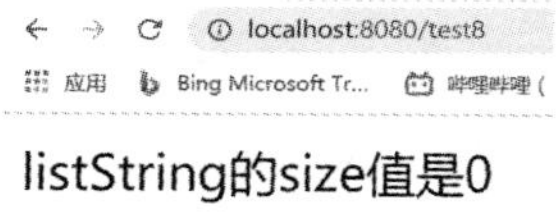

程序运行结果如图 6-8 所示。

程序运行结果是正确的，size 值是 0。

listString的size值是0

图 6-8 运行结果

代码<#if listString?size==0>中的?size 是 FreeMarker 的写法，表示取得 listString 这个对象的 size 属性值，如果使用<#if listString.size==0>这样的写法，那么 listString 必须要有针对 size 属性的 public int getSize()方法，这样的写法通过反射技术调用 public int getSize()方法，但 listString 的数据类型是 ArrayList，而 ArrayList 中没有 public int getSize() 方法，只有 public int size()方法，所以使用<#if listString.size==0>这样的写法会在页面上出现异常，示例代码如下：

```
<body>
    <#if size0.size==0>==0 <#else>!=0 </#if>
</body>
```

程序运行后出现如下异常：

```
The following has evaluated to null or missing:
==> size0  [in template "test9.ftlh" at line 8, column 14]

----
Tip: If the failing expression is known to legally refer to something that's sometimes
null or missing, either specify a default value like myOptionalVar!myDefault, or use <#if
myOptionalVar??>when-present<#else>when-missing</#if>. (These only cover the last step of the
expression; to cover the whole expression, use parenthesis: (myOptionalVar.foo)!myDefault,
(myOptionalVar.foo)??
```

如果在标签中使用点号访问属性值，则对象中必须要有属性的 get 方法。
创建实体类，代码如下：

```
package com.ghy.www.entity;

public class MyArrayList {
    private int sizesize;

    public MyArrayList() {
    }

    public MyArrayList(int sizesize) {
        super();
        this.sizesize = sizesize;
    }

    public int getSizesize() {
        return sizesize;
    }

    public void setSizesize(int sizesize) {
        this.sizesize = sizesize;
    }
}
```

控制层的代码如下：

```
@RequestMapping("test10")
public String test10(Model model, HttpServletRequest request, HttpServletResponse response) {
    model.addAttribute("size0", new MyArrayList());
    return "test10";
}
```

创建标签文件，代码如下：

```
<body>
    <#if size0.sizesize==0>==0 <#else>!=0 </#if>
</body>
```

程序运行结果如图 6-9 所示。

程序未出现异常的原因是 MyArrayList 类中有针对属性 private int sizesize 的 public int getSizesize()方法。使用<#if size0.sizesize==0>的写法会调用 MyArrayList 对象中的 public int getSizesize()方法，返回 size 属性值。

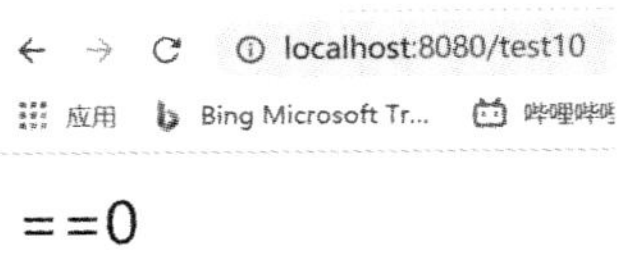

图 6-9 运行结果

6.1.9 处理 null 值

控制层的代码如下：

```
@RequestMapping("test11")
public String test11(Model model, HttpServletRequest request, HttpServletResponse response) {
    model.addAttribute("nullString", null);
    model.addAttribute("notNullString", "中国人");

    model.addAttribute("showString", "我是显示的字符串");
```

```
        Userinfo userinfo = new Userinfo();
        userinfo.setUsername(null);
        userinfo.setPassword("我是密码的值");
        model.addAttribute("userinfo", userinfo);

        return "test11";
    }
```

模板文件的代码如下：

```
<body>
    ${nullString}
</body>
```

程序运行后出现异常的信息如下：

```
The following has evaluated to null or missing:
==> nullString  [in template "test11.ftlh" at line 8, column 11]

----
Tip: If the failing expression is known to legally refer to something that's sometimes
null or missing, either specify a default value like myOptionalVar!myDefault, or use <#if
myOptionalVar??>when-present<#else>when-missing</#if>. (These only cover the last step of
the expression; to cover the whole expression, use parenthesis: (myOptionalVar.foo)!myDefault,
(myOptionalVar.foo)??
```

需要注意的是，FreeMarker 框架不能直接输出 null 值，需要进行一些处理。
更改模板文件，内容如下：

```
<body>
    |${nullString!}|<br/>
    |${nullString!""}|<br/>
    |${nullString!"nullString 是 null"}|<br/>
    |${nullString???string("nullString 不是 null","nullString 是 null")}|<br/>
    <br/>
    |${userinfo.username!}|<br/>
    |${userinfo.username!""}|<br/>
    |${userinfo.username!"userinfo.username 是 null"}|<br/>
    |${userinfo.username???string("userinfo.username 不是 null","userinfo.username 是 null")}
|<br/>
    <br/>
    |${notNullString!}|<br/>
    |${notNullString!""}|<br/>
    |${notNullString!"notNullString 是 null"}|<br/>
    |${notNullString???string("notNullString 不是 null","notNullString 是 null")}|<br/>
    <br/>
    |${userinfo.password!}|<br/>
    |${userinfo.password!""}|<br/>
    |${userinfo.password!"userinfo.password 是 null"}|<br/>
    |${userinfo.password???string("userinfo.password 不是 null","userinfo.password 是 null")}
|<br/>
    <br/>
    |<#if nullString??>不是空<#else>是空</#if>|<br/>
    |<#if userinfo.username??>username 不是空<#else>username 是空</#if>|<br/>
    |<#if notNullString??>不是空<#else>是空</#if>|<br/>
    |<#if userinfo.password??>password 不是空<#else>password 是空</#if>|
</body>
```

叹号表示指定缺失变量或值为 null 的变量的默认值，双问号表示判断某个变量值是否为非 null。写法

```
${userinfo.username???string("userinfo.username 不是 null","userinfo.username 是 null")}
```

适用于只输出字符串，而写法

```
<#if userinfo.username??>username 不是空<#else>username 是空</#if>
```

适用于输出批量 HTML 代码。

程序运行结果如图 6-10 所示。

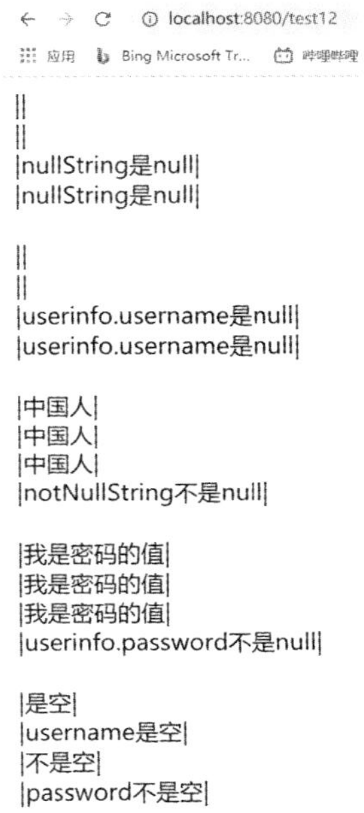

图 6-10　运行结果

下面的代码是经典的列表功能的模板标签内容。
控制层的代码如下：

```
@RequestMapping("test13")
public String test13(Model model, HttpServletRequest request, HttpServletResponse response) {
    Userinfo userinfo1 = new Userinfo();
    userinfo1.setId(100);

    Userinfo userinfo2 = new Userinfo();
    userinfo2.setUsername("usernameValue");

    Userinfo userinfo3 = new Userinfo();
    userinfo3.setPassword("passwordValue");

    Userinfo userinfo4 = new Userinfo();
    userinfo4.setAge(200);

    Userinfo userinfo5 = new Userinfo();
    userinfo5.setInsertDate(new Date());
```

```java
        List listUserinfo = new ArrayList();
        listUserinfo.add(userinfo1);
        listUserinfo.add(userinfo2);
        listUserinfo.add(userinfo3);
        listUserinfo.add(userinfo4);
        listUserinfo.add(userinfo5);
        model.addAttribute("listUserinfo", listUserinfo);

        return "test13";
}
```

模板文件的代码如下：

```
<body>
    <#list listUserinfo>
        <table border="1">
            <#items as eachUserinfo>
                <tr>
                    <td>${eachUserinfo.id}</td>
                    <td>${eachUserinfo.username!}</td>
                    <td>${eachUserinfo.password!}</td>
                    <td>${eachUserinfo.age!}</td>
                    <td><#if eachUserinfo.insertdate??>${eachUserinfo.insertdate?string("yyyy年MM月dd日 HH时mm分ss秒")}</#if></td>
                    <td><a href='deleteUserinfoById?id=${eachUserinfo.id}'>删除</a></td>
                    <td><a href='editUserinfoById?id=${eachUserinfo.id}'>编辑</a></td>
                </tr>
            </#items>
        </table>
    </#list>
</body>
```

程序运行结果如图 6-11 所示。

FreeMarker 默认不支持对 null 值直接打印，运行时会出现异常

The following has evaluated to null or missing:

可以使用叹号将 null 值作为空字符串""输出。

双问号表示 insertdate 属性值不为 null 时，对日期进行格式转换。

图 6-11 运行结果

6.1.10 实现隔行变色

控制层的代码如下：

```java
@RequestMapping("test14")
public String test14(Model model, HttpServletRequest request, HttpServletResponse response) {
    Userinfo userinfo1 = new Userinfo(1, "中国1", "中国人1", 11, new Date());
    Userinfo userinfo2 = new Userinfo(2, "中国2", "中国人2", 22, new Date());
    Userinfo userinfo3 = new Userinfo(3, "中国3", "中国人3", 33, new Date());
    Userinfo userinfo4 = new Userinfo(4, "中国4", "中国人4", 44, new Date());
    Userinfo userinfo5 = new Userinfo(5, "中国5", "中国人5", 55, new Date());
    Userinfo userinfo6 = new Userinfo(6, "中国6", "中国人6", 66, new Date());
    Userinfo userinfo7 = new Userinfo(7, "中国7", "中国人7", 77, new Date());
```

```
        Userinfo userinfo8 = new Userinfo(8, "中国8", "中国人8", 88, new Date());
        Userinfo userinfo9 = new Userinfo(9, "中国9", "中国人9", 99, new Date());

        List listBean = new ArrayList();
        listBean.add(userinfo1);
        listBean.add(userinfo2);
        listBean.add(userinfo3);
        listBean.add(userinfo4);
        listBean.add(userinfo5);
        listBean.add(userinfo6);
        listBean.add(userinfo7);
        listBean.add(userinfo8);
        listBean.add(userinfo9);

        model.addAttribute("rowColorList", listBean);

        return "test14";
}
```

模板文件的代码如下:

```
<body>
    <table border="1">
        <tr>
            <td>index</td>
            <td>counter</td>
            <td>id</td>
            <td>username</td>
            <td>password</td>
            <td>age</td>
            <td>insertDate?datetime</td>
            <td>insertDate?string("yyyy-MM-dd")}</td>
            <td>insertDate?string("HH:mm:ss")</td>
            <td>has_next?c</td>
            <td>is_first?c</td>
            <td>is_last?c</td>
        </tr>
        <#list rowColorList as userinfo>
            <#if userinfo?is_even_item>
                <tr style="background-color:red">
            </#if>
            <#if userinfo?is_odd_item>
                <tr style="background-color:green">
            </#if>
            <td>${userinfo?index}</td>
            <td>${userinfo?counter}</td>
            <td>${userinfo.id}</td>
            <td>${userinfo.username}</td>
            <td>${userinfo.password}</td>
            <td>${userinfo.age}</td>
            <td>${userinfo.insertDate?datetime}</td>
            <td>${userinfo.insertDate?string("yyyy-MM-dd")}</td>
            <td>${userinfo.insertDate?string("HH:mm:ss")}</td>
            <td>${userinfo?has_next?c}</td>
            <td>${userinfo?is_first?c}</td>
            <td>${userinfo?is_last?c}</td>
            </tr>
        </#list>
```

```
        </table>
    </body>
```

程序运行结果如图 6-12 所示。

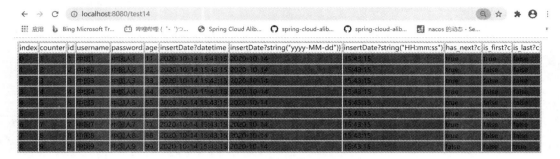

图 6-12 运行结果

6.1.11 对象嵌套有 null 值的处理

创建实体类 A 代码如下:

```
package com.ghy.www.entity;

public class A {
    private B b;

    public B getB() {
        return b;
    }

    public void setB(B b) {
        this.b = b;
    }
}
```

A 类中的 b 属性值为 null。

创建实体类 B 代码如下:

```
package com.ghy.www.entity;

public class B {
    private C c = new C();

    public C getC() {
        return c;
    }

    public void setC(C c) {
        this.c = c;
    }
}
```

创建实体类 C 代码如下:

```
package com.ghy.www.entity;

public class C {
}
```

控制层的代码如下：

```
@RequestMapping("test15")
public String test15(Model model, HttpServletRequest request, HttpServletResponse response) {
    model.addAttribute("aObject", new A());
    return "test15";
}
```

模板文件的代码如下：

```
<body>
    ${aObject.b.c}
</body>
```

程序运行结果出现异常，信息如下：

```
The following has evaluated to null or missing:
==> aObject.b  [in template "test15.ftlh" at line 8, column 11]

----
Tip: It's the step after the last dot that caused this error, not those before it.
----
Tip: If the failing expression is known to legally refer to something that's sometimes
null or missing, either specify a default value like myOptionalVar!myDefault, or use <#if
myOptionalVar??>when-present<#else>when-missing</#if>. (These only cover the last step of
the expression; to cover the whole expression, use parenthesis: (myOptionalVar.foo)!myDefault,
(myOptionalVar.foo)??
----
```

该信息提示 aObject 中的 b 属性值为 null，在多级嵌套的情况下，某一级对象值为 null 时可以使用圆括号的写法进行处理。编辑模板文件，代码如下：

```
<body>
    ${(aObject.b.c)!"某一个属性值为null"}<br/><br/>
    <#if (aObject.b.c)??>值不为null<#else>某一个属性值为null</#if>
</body>
```

程序运行结果如图 6-13 所示。

图 6-13　运行结果

6.1.12　比较运算符

模板文件的代码如下：

```
<body>
    比较运算1：<#if 1 lt 100>1<100</#if>___<#if 100 gt 1>100>1</#if><br/>
    比较运算2：<#if 1 lte 1>1<=1</#if>___<#if 100 gte 100>100>=100</#if><br/>
    比较运算3：<#if (1 <= 1)>1<=1</#if>___<#if (100 >= 100)>100>=100</#if><br/>
    比较运算4：<#if (0 < 1)>0<1</#if>___<#if (101 > 100)>101>100</#if><br/>
</body>
```

如果使用>或>=符号进行比较运算，则需要将表达式放入圆括号中，否则会认为>或>=符号是标签内容，而不是表达式，如代码

```
<#if 2 > 1>2 > 1</#if>
```

会解析成

```
<#if 2 >
```

6.1.13 遗拾增补

控制层的代码如下：

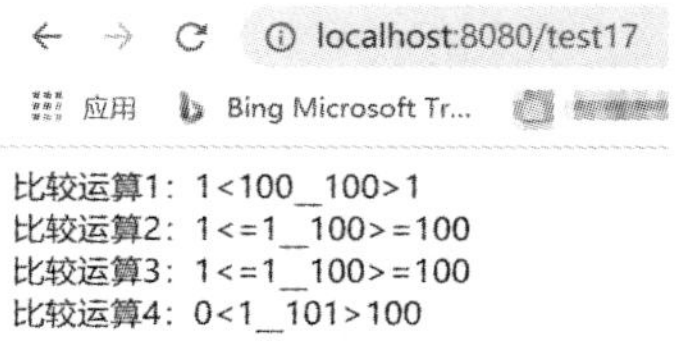

```
@RequestMapping("test18")
public String test18(Model model, HttpServletRequest request, HttpServletResponse response) {
    model.addAttribute("num1", 1);
    model.addAttribute("num2", 99);
    model.addAttribute("username", "高洪岩");
    model.addAttribute("ptpressA", "<a href='http://www.ptpress.com.cn'>我要去人民邮电出版社</a>");

    return "test18";
}
```

模板文件的代码如下：

```
<body>
    num1+num2=${num1+num2}<br/>
    1+9=${1+9}<br/>
    String 常量=${"我\\\"是常量"}<br/>
    String 拼接=${"hello:"+username}<br/>
    原生字符串：${"${username}"}__${r"${username}"}<br/><#--r 代表 raw 原始的意思-->
    不取整：${(5/2)}<br/>
    取整：${(5/2)?int}<br/>
    ptpressA1：${ptpressA}<br/>
    ptpressA2：${ptpressA?no_esc}
</body>
```

程序运行结果如图 6-15 所示。

6.1.14 填充 select 中的 option

控制层的代码如下：

```
@RequestMapping("test19")
public String test19(Model model, HttpServletRequest request, HttpServletResponse response) {
    int randomInt = (int) (Math.random() * 10);
    boolean isEven = true;// 不放数据
    if (randomInt % 2 == 0) {
        isEven = true;
    } else {
        isEven = false;
    }
    List listString = null;
    if (isEven) {

    } else {
        listString = new ArrayList();
        listString.add("a");
        listString.add("b");
        listString.add("c");
        listString.add("d");
        listString.add("e");
```

程序运行结果如图 6-14 所示。

图 6-14 运行结果

图 6-15 运行结果

```
        }
        model.addAttribute("listString", listString);

        return "test19";
    }
```

模板文件的代码如下：

```
<body>
    <select>
        <#if listString?? && listString?size!=0>
            <#list listString as eachString>
                <option value="${eachString}">${eachString}</option>
            </#list>
        <#else>
            <option value="-1">无值</option>
        </#if>
    </select>
</body>
```

程序运行结果如图 6-16 所示。

图 6-16　运行结果

6.1.15　实现自动选中 select 中的 option

控制层的代码如下：

```
public class listSelected extends HttpServlet {
    protected void doGet(HttpServletRequest request, HttpServletResponse response)
            throws ServletException, IOException {
        List listString = new ArrayList();
        listString.add("a");
        listString.add("b");
        listString.add("c");
        listString.add("d");
        listString.add("e");

        request.setAttribute("listString", listString);
        request.getRequestDispatcher("listSelected.ftl").forward(request, response);
    }
}
```

模板文件的代码如下：

```
<body>
    <select>
        <#if listString?? && listString?size!=0>
            <#list listString as eachString>
```

```
                <#if RequestParameters.selectedValue==eachString>
                    <option value="${eachString}" selected="selected">${eachString}</option>
                </#if>
                <#if RequestParameters.selectedValue!=eachString>
                    <option value="${eachString}">${eachString}</option>
                </#if>
            </#list>
        </#if>
    </select>
</body>
```

执行如下 URL 地址：

http://localhost:8080/test20?selectedValue=d

程序运行结果如图 6-17 所示。

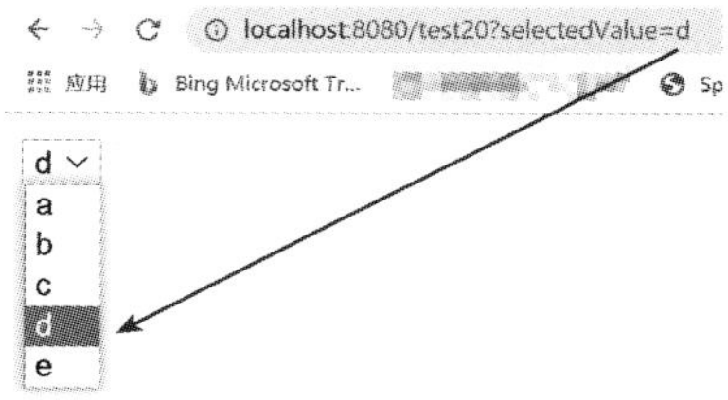

图 6-17 运行结果

6.1.16 实现页面静态化

实现页面静态化完全可以不经过 Tomcat 的处理，而让用户直接访问 *.html 文件，这样可以大大增加服务器的吞吐量，提高用户访问的速度。

控制层的代码如下：

```
@RequestMapping("test21")
public void test21(Model model, HttpServletRequest request, HttpServletResponse response) throws IOException, TemplateException {
    Configuration config = new Configuration(Configuration.VERSION_2_3_30);
    config.setClassForTemplateLoading(TestController.class, "/");
    config.setEncoding(Locale.getDefault(), "utf-8");
    config.setTemplateExceptionHandler(TemplateExceptionHandler.HTML_DEBUG_HANDLER);
    config.setNumberFormat("0.##########");
    Template template = config.getTemplate("test21.ftl");//ftl 文件在 src 路径下

    Map map = new HashMap();
    map.put("username", "张三");
    map.put("age", 100);
    map.put("address", "北京");

    String htmlFile = "d:\\staticHTML.html";
    File newFile = new File(htmlFile);
    FileOutputStream fileOutputStream = new FileOutputStream(htmlFile);
    OutputStreamWriter outputStreamWriter = new OutputStreamWriter(fileOutputStream, "UTF-8");
    Writer out = new BufferedWriter(outputStreamWriter);
    template.process(map, out);
    out.close();
    outputStreamWriter.close();
    fileOutputStream.close();
}
```

模板文件的代码如下：

```
<body>
    我是一个 html 页面<br/>
    姓名：${username}<br>
    年龄：${age}<br>
    地址：${address}<br>
</body>
```

程序运行后生成的 HTML 静态文件的代码如下：

```
<!DOCTYPE html>
<html>
    <head>
        <meta charset="UTF-8">
        <title>Insert title here</title>
    </head>
    <body>
        我是一个 html 页面<br/>
        姓名：张三<br>
        年龄：100<br>
        地址：北京<br>
    </body>
</html>
```

6.1.17　将 ftlh 文件中的内容输出到内存中

前面介绍的都是将 ftlh 文件中的内容输出到 Web 界面或*.html 文件中，本节将实现输出到内存中，并可以下一步把内存中的数据通过 Socket 传输给其他服务器或保存到数据库中。

控制层的代码如下：

```
@RequestMapping("test22")
public void test22(Model model, HttpServletRequest request, HttpServletResponse response)
        throws IOException, TemplateException {
    Configuration config = new Configuration(Configuration.VERSION_2_3_30);
    config.setClassForTemplateLoading(TestController.class, "/");
    config.setEncoding(Locale.getDefault(), "utf-8");
    config.setTemplateExceptionHandler(TemplateExceptionHandler.HTML_DEBUG_HANDLER);
    config.setNumberFormat("0.##########");
    Template template = config.getTemplate("/templates/test22.ftlh");

    Map map = new HashMap();
    map.put("username", "张三");
    map.put("age", 100);
    map.put("address", "北京");

    StringWriter stringWriter = new StringWriter();
    Writer out = new BufferedWriter(stringWriter);
    template.process(map, out);
    out.close();
    stringWriter.close();

    System.out.println(stringWriter.toString());
}
```

控制台输出结果如下：

```
<!DOCTYPE html>
<html>
    <head>
        <meta charset="UTF-8">
        <title>Insert title here</title>
    </head>
    <body>
        我是一个 html 页面<br/>
        姓名：张三<br>
        年龄：100<br>
        地址：北京<br>
```

```
        </body>
</html>
```

6.2 使用 Thymeleaf 模板引擎

Thymeleaf 是一款用于渲染 XML/XHTML/HTML5 内容的模板引擎，它类似于 JSP、Velocity、FreeMarker 等，可以作为 Web 应用的模板引擎与 Spring MVC 等 Web 框架集成。与其他模板引擎相比，Thymeleaf 的最大特点是能够直接在浏览器中打开并正确显示模板页面，而不需要启动整个 Web 应用，便于原型设计。

6.2.1 整合 Spring Boot 与常见的使用方式

创建测试项目 thymeleafTest。最终的项目结构如图 6-18 所示。

图 6-18　项目结构

6.2 使用 Thymeleaf 模板引擎

控制层的代码如下：

```
@RequestMapping("test1")
public String test1(Model model, HttpServletRequest request, HttpServletResponse response) {
    model.addAttribute("firstShowMessage", "第\\\\\"一次显示消息，成功了！");
    model.addAttribute("nowDate", new Date());
    return "test1";
}
```

创建运行类，代码如下：

```
package com.ghy.www;

import org.springframework.boot.SpringApplication;
import org.springframework.boot.autoconfigure.SpringBootApplication;

@SpringBootApplication
public class Application {
    public static void main(String[] args) {
        SpringApplication.run(Application.class, args);
    }
}
```

标签文件 test1.html 的代码如下：

```
<!DOCTYPE html>
<html xmlns:th="http://www.thymeleaf.org">
    <head>
        <meta http-equiv="Content-Type" content="text/html; charset=UTF-8"/>
        <title>Insert title here</title>
        <script src="js/myjs.js"></script>
        <link href="css/mycss.css" rel="stylesheet"/>
    </head>
    <body>
        <span class="myColorSize">测试样式的文字</span>
        <br/>
        [[${firstShowMessage}]]
        <br/>
        <span th:text="${firstShowMessage}"></span>
        <br/>
        <br/>
        <span th:text="我是普通的文本"></span>
        <br/>
        <span th:text="${'我是普通的文本'}"></span>
        <br/>
        <br/>
        <span th:text="'我\\是&quot;普通的文本'"></span><!--想显示特殊符号就要加单引号-->
        <br/>
        <span th:text="${'我\\是&quot;普通的文本'}"></span>
        <br/>
        <br/>
        <input type="text" th:value="${firstShowMessage}">
        <br/>
        <br/>
        [[${nowDate}]]
        <br/>
        <span th:text="${nowDate}"></span>
        <br/>
        <br/>
```

```
[[${0+1}]]
<br/>
[[${3-1}]]
<br/>
[[${1*3}]]
<br/>
[[${20/5}]]
<br/>
[[${105%100}]]
<br/>
<br/>
<span th:text="${0+1}"></span>
<br/>
<span th:text="${3-1}"></span>
<br/>
<span th:text="${1*3}"></span>
<br/>
<span th:text="${20/5}"></span>
<br/>
<span th:text="${105%100}"></span>
<br/>
<br/>
[['a'+'b']]
<br/>
[[${'a'+'b'}]]
<br/>
[[${"a"+"b"}]]
<br/>
[['ab'+'cd']]
<br/>
[[${'ab'+'cd'}]]
<br/>
[[${"ab"+"cd"}]]
<br/>
<span th:text="'a'+'b'"></span>
<br/>
<span th:text="${'a'+'b'}"></span>
<br/>
<span th:text='${"a"+"b"}'></span>
<br/>
<span th:text="'ab'+'cd'"></span>
<br/>
<span th:text="${'ab'+'cd'}"></span>
<br/>
<span th:text='${"ab"+"cd"}'></span>
<br/>
<br/>
[[${#dates.format(nowDate, 'yyyy年MM月dd日 hh时mm分ss秒')}]]
<br/>
<span th:text="${#dates.format(nowDate, 'yyyy年MM月dd日 hh时mm分ss秒')}"></span>
<br/>
<br/>
<th:block th:if="100 < 200">
    <span>100 &lt; 200</span>
</th:block>
<br/>
<th:block th:if="100 <= 100">
    <span>100 &le; 100</span>
```

```
        </th:block>
        <br/>
        <th:block th:if="201 > 100">
            <span>201 &gt; 100</span>
        </th:block>
        <br/>
        <th:block th:if="201 >= 201">
            <span>201 &ge; 201</span>
        </th:block>
        <br/>
        <th:block th:if="201 == 201">
            <span>201 == 201</span>
        </th:block>
        <br/>
        <th:block th:if="201 != 202">
            <span>201 != 202</span>
        </th:block>
        <br/>
        <th:block th:if="!false">
            <span>true</span>
        </th:block>
        <br/>
        <!-- (A)我是注释,我在 html 代码中-->
        <br/>
        <!--/*(B)我是注释,我不在 html 代码中*/-->
        <br/>
        <!--/*-->(C) 开始-我不在 html 代码中<span th:text="${'testText'}"></span>结束-我不在 html 代码中<!--*/-->
        <br/>
        <!--/*/
        (D) <br />原型注释开始-我在 html 代码中
        <br />
        <span th:text="${'testText'}"></span>
        <br />原型注释结束-我在 html 代码中
        <br />本注释块的作用是在浏览器中直接打开*.html 文件时,并不显示<span>标签
        <br />但是在结合 thyemleaf 框架时,解析此标签,并显示出<span>
        /*/-->
    </body>
</html>
```

配置文件 application.yml 的代码如下:

```
spring:
  thymeleaf:
    enabled: true    #开启 thymeleaf 视图解析
    encoding: utf-8  #编码
    prefix: classpath:/templates/  #前缀
    cache: false    #是否使用缓存
    mode: HTML   #严格的 HTML 语法模式
    suffix: .html   #后缀名
  messages:
    basename: i18n
```

文件 mycss.css 的代码如下:

```
.myColorSize {
    color: red;
    font-size: 30px;
}
```

文件 myjs.js 的代码如下：

```javascript
setTimeout(function () {
    alert("自动弹出");
}, 3000)
```

配置文件 pom.xml 的核心代码如下：

```xml
<parent>
    <groupId>org.springframework.boot</groupId>
    <artifactId>spring-boot-starter-parent</artifactId>
    <version>2.3.4.RELEASE</version>
</parent>

<properties>
    <project.build.sourceEncoding>UTF-8</project.build.sourceEncoding>
    <maven.compiler.source>1.8</maven.compiler.source>
    <maven.compiler.target>1.8</maven.compiler.target>
</properties>

<dependencies>
    <dependency>
        <groupId>junit</groupId>
        <artifactId>junit</artifactId>
        <version>4.11</version>
        <scope>test</scope>
    </dependency>

    <dependency>
        <groupId>org.springframework.boot</groupId>
        <artifactId>spring-boot-starter-web</artifactId>
    </dependency>

    <dependency>
        <groupId>org.springframework.boot</groupId>
        <artifactId>spring-boot-starter-thymeleaf</artifactId>
    </dependency>
</dependencies>
```

创建实体类 Userinfo，代码如下：

```java
package com.ghy.www.entity;

public class Userinfo {
    private String username;
    private long age;

    public Userinfo() {
    }

    public Userinfo(String username, long age) {
        this.username = username;
        this.age = age;
    }

    public String getUsername() {
        return username;
    }

    public void setUsername(String username) {
        this.username = username;
    }
```

```
    public long getAge() {
        return age;
    }

    public void setAge(long age) {
        this.age = age;
    }
}
```

创建实体类 A，代码如下：

```
package com.ghy.www.entity;

public class A {
    private B b = new B();

    public B getB() {
        return b;
    }

    public void setB(B b) {
        this.b = b;
    }
}
```

创建实体类 B，代码如下：

```
package com.ghy.www.entity;

public class B {
    private C c = new C();

    public C getC() {
        return c;
    }

    public void setC(C c) {
        this.c = c;
    }
}
```

创建实体类 C，代码如下：

```
package com.ghy.www.entity;

public class C {
    private String username = "username 属性值";

    public String getUsername() {
        return username;
    }

    public void setUsername(String username) {
        this.username = username;
    }
}
```

6.2.2 处理复杂数据类型

控制层的代码如下：

```
@RequestMapping("test2")
public String test2(Model model, HttpServletRequest request, HttpServletResponse response) {
    model.addAttribute("userinfo", new Userinfo("中国", 123L));
    return "test2";
}
```

模板文件的代码如下：

```
<!DOCTYPE html>
<html xmlns:th="http://www.thymeleaf.org">
    <head>
        <meta http-equiv="Content-Type" content="text/html; charset=UTF-8"/>
        <title>Insert title here</title>
    </head>
    <body>
        A:
        <br/>
        [[${userinfo.username}]]
        <br/>
        [[${userinfo.age}]]
        <br/>
        <br/>
        B:
        <br/>
        [[${userinfo['username']}]]
        <br/>
        [[${userinfo['age']}]]
        <br/>
        <br/>
        C:
        <br/>
        <th:block th:object="${userinfo}">
            [[*{username}]]___[[*{age}]]
        </th:block>
        <br/>
        <br/>
        D:
        <br/>
        <span th:text="${userinfo.username}"></span>
        <br/>
        <span th:text="${userinfo.age}"></span>
        <br/>
        <br/>
        E:
        <br/>
        <span th:text="${userinfo['username']}"></span>
        <br/>
        <span th:text="${userinfo['age']}"></span>
        <br/>
        <br/>
        F:
        <br/>
        <span th:object="${userinfo}" th:text="*{username}"></span>
        <br/>
        <span th:object="${userinfo}" th:text="*{age}"></span>
        <br/>
        <br/>
        G:
```

```
        <br/>
        <span th:object="${userinfo}" th:text="*{['username']}"></span>
        <br/>
        <span th:object="${userinfo}" th:text="*{['age']}"></span>
        <br/>
        <br/>
        H:
        <br/>
        [[${userinfo.username=='中国'}?'我是中国':'我不是中国']]
        <br/>
        <span th:text="${userinfo.username=='中国'}?'我是中国':'我不是中国'"></span>
        <br/>
        <br/>
        I:
        <br/>
        [[${'[[${userinfo.age}]]'}]]
        <br/>
        <span th:text="${'[[${userinfo.age}]]'}"></span>
        <br/>
        <span th:inline="none">[[${userinfo.age}]]</span><!--原样输出-->
        <br/>
        <span th:inline="text">[[${userinfo.age}]]</span><!--解析表达式并输出-->
        <br/>
        <br/>
        J:
        <br/>
        [[|我的姓名是：${userinfo.username}|]]
        <br/>
        <span th:text="|我的姓名是：${userinfo.username}|"></span>
        <br/>
        [['我的年龄是 40 岁'+' '+|我的姓名是：${userinfo.username}|]]
        <br/>
        <span th:text="'我的年龄是 40 岁'+' '+|我的姓名是：${userinfo.username}|"></span>
        <br/>
        <br/>
        K:
        <br/>
        <th:block th:switch="${userinfo.username}">
            <span th:case="'中国 1'">1 等于中国 1</span>
            <span th:case="${userinfo.age}">2 等于${userinfo.age}</span>
            <span th:case="*">3 其他值</span>
        </th:block>
    </body>
</html>
```

6.2.3 处理嵌套数据类型

控制层的代码如下：

```
@RequestMapping("test3")
public String test3(Model model, HttpServletRequest request, HttpServletResponse response) {
    model.addAttribute("a", new A());
    return "test3";
}
```

模板文件的代码如下：

```html
<!DOCTYPE html>
<html xmlns:th="http://www.thymeleaf.org">
    <head>
        <meta http-equiv="Content-Type" content="text/html; charset=UTF-8"/>
        <title>Insert title here</title>
    </head>
    <body>
        A:
        <br/>
        [[${a.b.c.username}]]
        <br/>
        <br/>
        B:
        <br/>
        [[${a['b']['c']['username']}]]
        <br/>
        <br/>
        C:
        <br/>
        <span th:text="${a.b.c.username}"></span>
        <br/>
        <br/>
        D:
        <br/>
        <span th:text="${a['b']['c']['username']}"></span>
    </body>
</html>
```

6.2.4 访问 Array

控制层的代码如下：

```java
@RequestMapping("test4")
public String test4(Model model, HttpServletRequest request, HttpServletResponse response) {
    String[] stringArray = new String[]{"a", "b", "c"};
    Userinfo[] userinfoArray = new Userinfo[]{new Userinfo("账号1", 1), new Userinfo("账号2", 2),
            new Userinfo("账号3", 3)};
    model.addAttribute("stringArray", stringArray);
    model.addAttribute("userinfoArray", userinfoArray);
    return "test4";
}
```

模板文件的代码如下：

```html
<!DOCTYPE html>
<html xmlns:th="http://www.thymeleaf.org">
    <head>
        <meta http-equiv="Content-Type" content="text/html; charset=UTF-8"/>
        <title>Insert title here</title>
    </head>
    <body>
        A:
        <br/>
        [[${stringArray[0]}]]
```

```html
            <br/>
            [[${stringArray[1]}]]
            <br/>
            [[${stringArray[2]}]]
            <br/>
            <br/>
            B:
            <br/>
            <span th:text="${stringArray[0]}"></span>
            <br/>
            <span th:text="${stringArray[1]}"></span>
            <br/>
            <span th:text="${stringArray[2]}"></span>
            <br/>
            <br/>
            C:
            <br/>
            [[${userinfoArray[0].username}]]__[[${userinfoArray[0].age}]]
            <br/>
            [[${userinfoArray[1].username}]]__[[${userinfoArray[1].age}]]
            <br/>
            [[${userinfoArray[2].username}]]__[[${userinfoArray[2].age}]]
            <br/>
            <br/>
            D:
            <br/>
            <span th:text="${userinfoArray[0].username}"></span>__<span th:text="${userinfoArray[0].age}"></span>
            <br/>
            <span th:text="${userinfoArray[1].username}"></span>__<span th:text="${userinfoArray[1].age}"></span>
            <br/>
            <span th:text="${userinfoArray[2].username}"></span>__<span th:text="${userinfoArray[2].age}"></span>
        </body>
    </html>
```

6.2.5　访问 List

控制层的代码如下：

```java
@RequestMapping("test5")
public String test5(Model model, HttpServletRequest request, HttpServletResponse response) {
    ArrayList listString = new ArrayList();
    listString.add("listString1");
    listString.add("listString2");
    listString.add("listString3");
    ArrayList listUserinfo = new ArrayList();
    listUserinfo.add(new Userinfo("listUserinfo1", 1));
    listUserinfo.add(new Userinfo("listUserinfo2", 2));
    listUserinfo.add(new Userinfo("listUserinfo3", 3));
    model.addAttribute("listString", listString);
    model.addAttribute("listUserinfo", listUserinfo);
    return "test5";
}
```

模板文件的代码如下：

```html
<!DOCTYPE html>
<html xmlns:th="http://www.thymeleaf.org">
    <head>
        <meta http-equiv="Content-Type" content="text/html; charset=UTF-8"/>
        <title>Insert title here</title>
    </head>
    <body>
        A:
        <br/>
        [[${listString[0]}]]
        <br/>
        [[${listString[1]}]]
        <br/>
        [[${listString[2]}]]
        <br/>
        <br/>
        B:
        <br/>
        <span th:text="${listString[0]}"></span>
        <br/>
        <span th:text="${listString[1]}"></span>
        <br/>
        <span th:text="${listString[2]}"></span>
        <br/>
        <br/>
        C:
        <br/>
        [[${listUserinfo[0].username}]]__[[${listUserinfo[0].age}]]
        <br/>
        [[${listUserinfo[1].username}]]__[[${listUserinfo[1].age}]]
        <br/>
        [[${listUserinfo[2].username}]]__[[${listUserinfo[2].age}]]
        <br/>
        <br/>
        D:
        <br/>
        <span th:text="${listUserinfo[0].username}"></span>__ <span th:text="${listUserinfo[0].age}"></span>
        <br/>
        <span th:text="${listUserinfo[1].username}"></span>__ <span th:text="${listUserinfo[1].age}"></span>
        <br/>
        <span th:text="${listUserinfo[2].username}"></span>__ <span th:text="${listUserinfo[2].age}"></span>
    </body>
</html>
```

6.2.6 访问 Map

控制层的代码如下：

```java
@RequestMapping("test6")
public String test6(Model model, HttpServletRequest request, HttpServletResponse response) {
    Map mapString = new LinkedHashMap();
    mapString.put("key1", "mapString1");
```

6.2 使用 Thymeleaf 模板引擎

```
        mapString.put("key2", "mapString2");
        mapString.put("key3", "mapString3");
        Map mapUserinfo = new LinkedHashMap();
        mapUserinfo.put("key1", new Userinfo("mapUserinfo1", 1));
        mapUserinfo.put("key2", new Userinfo("mapUserinfo2", 2));
        mapUserinfo.put("key3", new Userinfo("mapUserinfo3", 3));
        model.addAttribute("mapString", mapString);
        model.addAttribute("mapUserinfo", mapUserinfo);
        return "test6";
    }
```

模板文件的代码如下:

```
<!DOCTYPE html>
<html xmlns:th="http://www.thymeleaf.org">
    <head>
        <meta http-equiv="Content-Type" content="text/html; charset=UTF-8"/>
        <title>Insert title here</title>
    </head>
    <body>
        A:
        <br/>
        [[${mapString['key1']}]]
        <br/>
        [[${mapString['key2']}]]
        <br/>
        [[${mapString['key3']}]]
        <br/>
        <br/>
        Bs:
        <br/>
        <span th:text="${mapString['key1']}"></span>
        <br/>
        <span th:text="${mapString['key2']}"></span>
        <br/>
        <span th:text="${mapString['key3']}"></span>
        <br/>
        <br/>
        C:
        <br/>
        [[${mapUserinfo['key1'].username}]]__[[${mapUserinfo['key1'].age}]]
        <br/>
        [[${mapUserinfo['key2'].username}]]__[[${mapUserinfo['key2'].age}]]
        <br/>
        [[${mapUserinfo['key3'].username}]]__[[${mapUserinfo['key3'].age}]]
        <br/>
        <br/>
        D:
        <br/>
        <span th:text="${mapUserinfo['key1'].username}"></span>__<span th:text="${mapUserinfo['key1'].age}"></span>
        <br/>
        <span th:text="${mapUserinfo['key2'].username}"></span>__<span th:text="${mapUserinfo['key2'].age}"></span>
        <br/>
        <span th:text="${mapUserinfo['key3'].username}"></span>__<span th:text="${mapUserinfo['key3'].age}"></span>
    </body>
</html>
```

6.2.7 访问 request-session-application 作用域

控制层的代码如下：

```
@RequestMapping("test7")
public String test7(Model model, HttpServletRequest request, HttpServletResponse response) {
    request.setAttribute("requestKey", "requestValue 值");
    request.getSession().setAttribute("sessionKey", "sessionValue 值");
    request.getServletContext().setAttribute("applicationKey", "applicationValue 值");
    return "test7";
}
```

模板文件的代码如下：

```
<!DOCTYPE html>
<html xmlns:th="http://www.thymeleaf.org">
    <head>
        <meta http-equiv="Content-Type" content="text/html; charset=UTF-8"/>
        <title>Insert title here</title>
    </head>
    <body>
        requestValue:
        <br/>
        [[${requestKey}]]
        <br/>
        <span th:text="${requestKey}"></span>
        <br/>
        <br/>
        sessionValue:
        <br/>
        [[${session.sessionKey}]]
        <br/>
        <span th:text="${session.sessionKey}"></span>
        <br/>
        <br/>
        applicationValue:
        <br/>
        [[${application.applicationKey}]]
        <br/>
        <span th:text="${application.applicationKey}"></span>
    </body>
</html>
```

6.2.8 访问 URL 参数值

控制层的代码如下：

```
@RequestMapping("test8")
public String test8(Model model, HttpServletRequest request, HttpServletResponse response) {
    return "test8";
}
```

模板文件的代码如下：

```
<!DOCTYPE html>
<html xmlns:th="http://www.thymeleaf.org">
```

```
<head>
    <meta http-equiv="Content-Type" content="text/html; charset=UTF-8"/>
    <title>Insert title here</title>
</head>
<body>
    [[${param.username}]]__<span th:text="${param.username}"></span>
</body>
</html>
```

6.2.9 循环 Array

控制层的代码如下:

```
@RequestMapping("test9")
public String test9(Model model, HttpServletRequest request, HttpServletResponse response) {
    String[] stringArray = new String[]{"a", "b", "c"};
    Userinfo[] userinfoArray = new Userinfo[]{new Userinfo("账号1", 1), new Userinfo("账号2", 2),
            new Userinfo("账号3", 3)};
    model.addAttribute("stringArray", stringArray);
    model.addAttribute("userinfoArray", userinfoArray);
    return "test9";
}
```

模板文件的代码如下:

```
<!DOCTYPE html>
<html xmlns:th="http://www.thymeleaf.org">
    <head>
        <meta http-equiv="Content-Type" content="text/html; charset=UTF-8"/>
        <title>Insert title here</title>
    </head>
    <body>
        <th:block th:each="each:${stringArray}">
            [[${each}]]
            <br/>
        </th:block>
        <br/>
        <th:block th:each="each:${userinfoArray}">
            [[${each.username}]]__[[${each.age}]]
            <br/>
        </th:block>
    </body>
</html>
```

6.2.10 循环 List

控制层的代码如下:

```
@RequestMapping("test10")
public String test10(Model model, HttpServletRequest request, HttpServletResponse response) {
    ArrayList listString = new ArrayList();
    listString.add("listString1");
    listString.add("listString2");
    listString.add("listString3");
    ArrayList listUserinfo = new ArrayList();
```

```
        listUserinfo.add(new Userinfo("listUserinfo1", 1));
        listUserinfo.add(new Userinfo("listUserinfo2", 2));
        listUserinfo.add(new Userinfo("listUserinfo3", 3));
        model.addAttribute("listString", listString);
        model.addAttribute("listUserinfo", listUserinfo);
        return "test10";
}
```

模板文件的代码如下:

```
<!DOCTYPE html>
<html xmlns:th="http://www.thymeleaf.org">
    <head>
        <meta http-equiv="Content-Type" content="text/html; charset=UTF-8"/>
        <title>Insert title here</title>
    </head>
    <body>
        <th:block th:each="each:${listString}">
            [[${each}]]
            <br/>
        </th:block>
        <br/>
        <th:block th:each="each:${listUserinfo}">
            [[${each.username}]]__[[${each.age}]]
            <br/>
        </th:block>
    </body>
</html>
```

6.2.11 循环 Set

控制层的代码如下:

```
@RequestMapping("test11")
public String test11(Model model, HttpServletRequest request, HttpServletResponse response) {
        Set setString = new LinkedHashSet();
        setString.add("setString1");
        setString.add("setString2");
        setString.add("setString3");
        Set setUserinfo = new LinkedHashSet();
        setUserinfo.add(new Userinfo("setUserinfo1", 1));
        setUserinfo.add(new Userinfo("setUserinfo2", 2));
        setUserinfo.add(new Userinfo("setUserinfo3", 3));
        model.addAttribute("setString", setString);
        model.addAttribute("setUserinfo", setUserinfo);
        return "test11";
}
```

模板文件的代码如下:

```
<!DOCTYPE html>
<html xmlns:th="http://www.thymeleaf.org">
    <head>
        <meta http-equiv="Content-Type" content="text/html; charset=UTF-8"/>
        <title>Insert title here</title>
    </head>
    <body>
        <th:block th:each="each:${setString}">
```

```
            [[${each}]]
            <br/>
        </th:block>
        <br/>
        <th:block th:each="each:${setUserinfo}">
            [[${each.username}]]__[[${each.age}]]
            <br/>
        </th:block>
    </body>
</html>
```

6.2.12 循环 Map

控制层的代码如下:

```
@RequestMapping("test12")
public String test12(Model model, HttpServletRequest request, HttpServletResponse response) {
    Map mapString = new LinkedHashMap();
    mapString.put("key1", "mapString1");
    mapString.put("key2", "mapString2");
    mapString.put("key3", "mapString3");
    Map mapUserinfo = new LinkedHashMap();
    mapUserinfo.put("key1", new Userinfo("mapUserinfo1", 1));
    mapUserinfo.put("key2", new Userinfo("mapUserinfo2", 2));
    mapUserinfo.put("key3", new Userinfo("mapUserinfo3", 3));
    model.addAttribute("mapString", mapString);
    model.addAttribute("mapUserinfo", mapUserinfo);
    return "test12";
}
```

模板文件的代码如下:

```
<!DOCTYPE html>
<html xmlns:th="http://www.thymeleaf.org">
    <head>
        <meta http-equiv="Content-Type" content="text/html; charset=UTF-8"/>
        <title>Insert title here</title>
    </head>
    <body>
        <th:block th:each="each:${mapString}">
            [[${each.key}]]__[[${each.value}]]
            <br/>
        </th:block>
        <br/>
        <th:block th:each="each:${mapUserinfo}">
            [[${each.key}]]__[[${each.value.username}]]__[[${each.value.age}]]
            <br/>
        </th:block>
    </body>
</html>
```

6.2.13 生成 Table

控制层的代码如下:

```
@RequestMapping("test13")
public String test13(Model model, HttpServletRequest request, HttpServletResponse response) {
```

```
        ArrayList colorListUserinfo = new ArrayList();
        colorListUserinfo.add(new Userinfo("listUserinfo1", 1));
        colorListUserinfo.add(new Userinfo("listUserinfo2", 2));
        colorListUserinfo.add(new Userinfo("listUserinfo3", 3));
        colorListUserinfo.add(new Userinfo("listUserinfo4", 4));
        colorListUserinfo.add(new Userinfo("listUserinfo5", 5));
        colorListUserinfo.add(new Userinfo("listUserinfo6", 6));
        colorListUserinfo.add(new Userinfo("listUserinfo7", 7));
        colorListUserinfo.add(new Userinfo("listUserinfo8", 8));
        model.addAttribute("listUserinfo", colorListUserinfo);
        return "test13";
}
```

模板文件的代码如下：

```
<!DOCTYPE html>
<html xmlns:th="http://www.thymeleaf.org">
    <head>
        <meta http-equiv="Content-Type" content="text/html; charset=UTF-8"/>
        <title>Insert title here</title>
    </head>
    <body>
        <table border="1">
            <tr th:each="eachUserinfo : ${listUserinfo}">
                <td th:text="${eachUserinfo.username}">
                </td>
                <td th:text="${eachUserinfo.age}">
                </td>
            </tr>
        </table>
    </body>
</html>
```

6.2.14 循环生成<input type=text>

控制层的代码如下：

```
@RequestMapping("test14")
public String test14(Model model, HttpServletRequest request, HttpServletResponse response) {
        ArrayList colorListUserinfo = new ArrayList();
        colorListUserinfo.add(new Userinfo("listUserinfo1", 1));
        colorListUserinfo.add(new Userinfo("listUserinfo2", 2));
        colorListUserinfo.add(new Userinfo("listUserinfo3", 3));
        colorListUserinfo.add(new Userinfo("listUserinfo4", 4));
        colorListUserinfo.add(new Userinfo("listUserinfo5", 5));
        colorListUserinfo.add(new Userinfo("listUserinfo6", 6));
        colorListUserinfo.add(new Userinfo("listUserinfo7", 7));
        colorListUserinfo.add(new Userinfo("listUserinfo8", 8));
        model.addAttribute("listUserinfo", colorListUserinfo);
        return "test14";
}
```

模板文件的代码如下：

```
<!DOCTYPE html>
<html xmlns:th="http://www.thymeleaf.org">
    <head>
        <meta http-equiv="Content-Type" content="text/html; charset=UTF-8"/>
```

```html
        <title>Insert title here</title>
    </head>
    <body>
        <th:block th:each="eachUserinfo : ${listUserinfo}">
            <input type="text" th:value="${eachUserinfo.username}">
            <br/>
        </th:block>
    </body>
</html>
```

6.2.15　获得状态变量

控制层的代码如下：

```java
@RequestMapping("test15")
public String test15(Model model, HttpServletRequest request, HttpServletResponse response) {
    ArrayList colorListUserinfo = new ArrayList();
    colorListUserinfo.add(new Userinfo("listUserinfo1", 1));
    colorListUserinfo.add(new Userinfo("listUserinfo2", 2));
    colorListUserinfo.add(new Userinfo("listUserinfo3", 3));
    colorListUserinfo.add(new Userinfo("listUserinfo4", 4));
    colorListUserinfo.add(new Userinfo("listUserinfo5", 5));
    colorListUserinfo.add(new Userinfo("listUserinfo6", 6));
    colorListUserinfo.add(new Userinfo("listUserinfo7", 7));
    colorListUserinfo.add(new Userinfo("listUserinfo8", 8));
    model.addAttribute("listUserinfo", colorListUserinfo);
    return "test15";
}
```

模板文件的代码如下：

```html
<!DOCTYPE html>
<html xmlns:th="http://www.thymeleaf.org">
    <head>
        <meta http-equiv="Content-Type" content="text/html; charset=UTF-8"/>
        <title>Insert title here</title>
    </head>
    <body>
        <table border="1">
            <tr th:each="eachUserinfo,myStatus: ${listUserinfo}"
                th:styleappend="${myStatus.odd==true}?'color:red':'color:blue'">
                <td th:text="${eachUserinfo.username}">
                </td>
                <td th:text="${eachUserinfo.age}">
                </td>
                <td th:text="'odd='+${myStatus.odd}"><!-- 是否为奇数 -->
                </td>
                <td th:text="'even='+${myStatus.even}"><!-- 是否为偶数 -->
                </td>
                <td th:text="'size='+${myStatus.size}"><!-- 一共循环次数 -->
                </td>
                <td th:text="'count='+${myStatus.count}"><!-- 现在循环的次数 -->
                </td>
                <td th:text="'index='+${myStatus.index}"><!-- 当前循环的索引 -->
                </td>
                <td th:text="'current='+${myStatus.current}"><!-- 当前循环的对象 -->
                </td>
```

```
                <td th:text="'first='+${myStatus.first}"><!--是否为第一个-->
                </td>
                <td th:text="'last='+${myStatus.last}"><!--是否为最后一个-->
                </td>
            </tr>
        </table>
    </body>
</html>
```

6.2.16 获得状态变量的简化版

控制层的代码如下:

```
@RequestMapping("test16")
public String test16(Model model, HttpServletRequest request, HttpServletResponse response) {
    ArrayList colorListUserinfo = new ArrayList();
    colorListUserinfo.add(new Userinfo("listUserinfo1", 1));
    colorListUserinfo.add(new Userinfo("listUserinfo2", 2));
    colorListUserinfo.add(new Userinfo("listUserinfo3", 3));
    colorListUserinfo.add(new Userinfo("listUserinfo4", 4));
    colorListUserinfo.add(new Userinfo("listUserinfo5", 5));
    colorListUserinfo.add(new Userinfo("listUserinfo6", 6));
    colorListUserinfo.add(new Userinfo("listUserinfo7", 7));
    colorListUserinfo.add(new Userinfo("listUserinfo8", 8));
    model.addAttribute("listUserinfo", colorListUserinfo);
    return "test16";
}
```

模板文件的代码如下:

```
<!DOCTYPE html>
<html xmlns:th="http://www.thymeleaf.org">
    <head>
        <meta http-equiv="Content-Type" content="text/html; charset=UTF-8"/>
        <title>Insert title here</title>
    </head>
    <body>
        <table border="1">
            <tr th:each="eachUserinfo: ${listUserinfo}"
                th:styleappend="${eachUserinfoStat.odd==true}?'color:red':'color:blue'">
                <td th:text="${eachUserinfo.username}">
                </td>
                <td th:text="${eachUserinfo.age}">
                </td>
                <td th:text="'odd='+${eachUserinfoStat.odd}">
                </td>
                <td th:text="'even='+${eachUserinfoStat.even}">
                </td>
                <td th:text="'size='+${eachUserinfoStat.size}">
                </td>
                <td th:text="'count='+${eachUserinfoStat.count}">
                </td>
                <td th:text="'index='+${eachUserinfoStat.index}">
                </td>
                <td th:text="'current='+${eachUserinfoStat.current}">
                </td>
                <td th:text="'first='+${eachUserinfoStat.first}">
```

```
                </td>
                <td th:text="'last='+${eachUserinfoStat.last}">
                </td>
            </tr>
        </table>
    </body>
</html>
```

6.2.17 实现国际化

控制层的代码如下:

```
@RequestMapping("test17")
public String test17(Model model, HttpServletRequest request, HttpServletResponse response) {
    model.addAttribute("param1", "参数1");
    model.addAttribute("param2", "参数2");
    model.addAttribute("param3", "参数3");
    model.addAttribute("showPropName", "propTextHasParam");
    return "test17";
}
```

模板文件的代码如下:

```
<!DOCTYPE html>
<html xmlns:th="http://www.thymeleaf.org">
    <head>
        <meta http-equiv="Content-Type" content="text/html; charset=UTF-8"/>
        <title>Insert title here</title>
    </head>
    <body>
        根据key取得国际化文本需要使用#符号,而不是$。
        <br/>
        (1)显示国际化文本:
        <br/>
        [[#{welcome.text}]]
        <br/>
        <span th:text="#{welcome.text}"></span>
        <br/>
        <br/>
        (2)向property传递字面参数值:
        <br/>
        [[#{propTextHasParam(${'参数1'},${'参数2'},${'参数3'})}]]
        <br/>
        <span th:text="#{propTextHasParam(${'参数1'},${'参数2'},${'参数3'})}"></span>
        <br/>
        <br/>
        (3)向property传递变量参数值:
        <br/>
        [[#{propTextHasParam(${param1},${param2},${param3})}]]
        <br/>
        <span th:text="#{propTextHasParam(${param1},${param2},${param3})}"></span>
        <br/>
        <br/>
        (4)显示属性文件中的key值来自变量:
        <br/>
        [[#{${showPropName}(${param1},${param2},${param3})}]]
        <br/>
```

```html
            <span th:text="#{${showPropName}(${param1},${param2},${param3})}"></span>
            <br/>
            <br/>
            （5）使用[[${'[[]]'}]]和th:text显示信息：
            <br/>
            [[#{showHTMLCode}]]
            <br/>
            <span th:text="#{showHTMLCode}"></span>
            <br/>
            <br/>
            （6）使用[[${'[()]'}]]和th:utext显示信息：
            <br/>
            [(#{showHTMLCode})]
            <br/>
            <span th:utext="#{showHTMLCode}"></span>
    </body>
</html>
```

6.2.18　处理 URL

控制层的代码如下：

```
@RequestMapping("test18")
public String test18(Model model, HttpServletRequest request, HttpServletResponse response) {
    model.addAttribute("userinfo", new Userinfo("中国", 123L));
    model.addAttribute("url", "findUserinfo3");
    return "test18";
}
```

模板文件的代码如下：

```
<!DOCTYPE html>
<html xmlns:th="http://www.thymeleaf.org">
    <head>
        <meta http-equiv="Content-Type" content="text/html; charset=UTF-8"/>
        <title>Insert title here</title>
    </head>
    <body>
        <a th:href="@{http://a:8080/aaa/findUserinfo(userId=123)}">aaa</a>
        <br/>
        <a th:href="@{http://b:8080/bbb/findUserinfo(userId=123,age=456)}">bbb</a>
        <br/>
        <a th:href="@{http://c:8080/ccc/findUserinfo(username=${userinfo.username})}">ccc</a>
        <br/>
        <a th:href="@{http://d:8080/ddd/findUserinfo(username=${userinfo.username},age=${userinfo.age})}">ddd</a>
        <br/>
        <a th:href="@{/findUserinfo1(username=${userinfo.username})}">findUserinfo1</a>
        <br/>
        <a th:href="@{/findUserinfo2/{username}(username=${userinfo.username})}">findUserinfo2</a><!--生成REST风格的URL-->
        <br/>
        <a th:href="@{${url}(username=${userinfo.username})}">findUserinfo3</a>
        <br/>
        <a th:href="@{${url}+'/'+${userinfo.username}}">newURL</a>
        <br/>
```

```html
            <a th:href="@{'/findUserinfo4/'+${userinfo.username}(age=${userinfo.age})}">findUserinfo4</a>
            <br/>
            <a th:href="@{~/otherPorject/findUserinfo5}">findUserinfo5</a>
            <br/>
            <a th:href="@{~/otherProject/{username}(username=${userinfo.username},age=${userinfo.age})}">findUserinfo6</a>
        </body>
    </html>
```

6.2.19 处理布尔值

控制层的代码如下：

```java
@RequestMapping("test19")
public String test19(Model model, HttpServletRequest request, HttpServletResponse response) {
    model.addAttribute("booleanValue", false);
    return "test19";
}
```

模板文件的代码如下：

```html
<!DOCTYPE html>
<html xmlns:th="http://www.thymeleaf.org">
    <head>
        <meta http-equiv="Content-Type" content="text/html; charset=UTF-8"/>
        <title>Insert title here</title>
    </head>
    <body>
        <span th:if="${booleanValue}==false">${booleanValue}==false==1</span>
        <br/>
        <span th:if="${booleanValue}==true">${booleanValue}==true==2</span>
        <br/>
        <span th:if="${booleanValue==false}">${booleanValue==false}==3</span>
        <br/>
        <span th:if="${booleanValue==true}">${booleanValue==true}==4</span>
        <br/>
        <span th:unless="${booleanValue==true}">${booleanValue==true}==5</span>
        <br/>
        <span th:unless="${booleanValue==false}">${booleanValue==false}==6</span>
        <br/>
        可以使用如下代码实现 if 判断，不需要 span 标记
        <br/>
        <th:block th:if="${booleanValue==false}">
            booleanKey==false end!
        </th:block>
        <th:block th:unless="${booleanValue!=true}">
            booleanKey==true end!
        </th:block>
    </body>
</html>
```

6.2.20 操作属性

控制层的代码如下：

```
@RequestMapping("test20")
public String test20(Model model, HttpServletRequest request, HttpServletResponse response) {
    model.addAttribute("userinfo", new Userinfo("中国", 123L));
    model.addAttribute("checkboxStatus1", false);
    model.addAttribute("checkboxStatus2", true);
    return "test20";
}
```

模板文件的代码如下:

```
<!DOCTYPE html>
<html xmlns:th="http://www.thymeleaf.org">
    <head>
        <meta http-equiv="Content-Type" content="text/html; charset=UTF-8"/>
        <title>Insert title here</title>
    </head>
    <body>
        (1) <span th:class="'myClass'">myspan1</span>
        <br/>
        (2) <span th:class="${userinfo.username}">myspan2</span>
        <br/>
        (3) <input type="text" th:value="${userinfo.username}">
    </input>
        <br/>
        (4) <input type="text" th:attr="value=${userinfo.username}">
    </input>
        <br/>
        (5) <a th:attr="href=@{findUsername5(username=${userinfo.username})}">5</a>
        <br/>
        (6) <a th:attr="href=@{findUsername6/{username}(username=${userinfo.username})}">6</a>
        <br/>
        (7) <input type="button" th:attr="value=#{submitText}"/>
        <br/>
        (8) <input type="text" th:attr="a=${userinfo.username},b=${userinfo.age}"/>
        <br/>
        (9) <span class="a b" th:attrappend="class=${' '}+'c'">span</span>
        <br/>
        (10) <span class="a b" th:attrprepend="class=${'c'}+' '">span</span>
        <br/>
        (11) <input type="checkbox" th:checked="${checkboxStatus1}"/>
        <br/>
        (12) <input type="checkbox" th:checked="${checkboxStatus2}"/>
        <br/>
        (13) <span th:myProperty="'myvalue'">span</span>
    </body>
</html>
```